ENGINEERING PHYSICS II
(Theory & Experiments)

ENGINEERING PHYSICS II
(Theory & Experiments)

(According to Latest Syllabus of Rajasthan Technical University, Kota)

G D Ladiwala

Senior Professor of Physics
Arya College of Engg. & I.T., Jaipur (Raj.)
Former Principal
&
Head, P. G. Department of Physics
Agrawal P.G. College, Jaipur (Raj.)

S S Sharma

Head, Department of Physics
Govt. Women Engineering College
Ajmer (Raj.)

NEW AGE
RTU

An Imprint of

NEW AGE INTERNATIONAL (P) LIMITED, PUBLISHERS

New Delhi • Bangalore • Chennai • Cochin • Guwahati
Hyderabad • Kolkata • Lucknow • Mumbai
Visit us at **www.newagepublishers.com**

Published by New Age International (P) Ltd., Publishers
First Edition: 2014
Reprint: 2014

BRANCHES

- **Bangalore** 37/10, 8th Cross (Near Hanuman Temple), Azad Nagar, Chamarajpet, Bangalore-560 018
  Tel.: (080) 26756823, Telefax: 26756820, **E-mail: bangalore@newagepublishers.com**
- **Chennai** 26, Damodaran Street, T. Nagar, Chennai-600 017
  Tel.: (044) 24353401, Telefax: 24351463, **E-mail: chennai@newagepublishers.com**
- **Cochin** CC-39/1016, Carrier Station Road, Ernakulam South, Cochin-682 016
  Tel.: (0484) 2377004, Telefax: 4051303, **E-mail: cochin@newagepublishers.com**
- **Guwahati** Hemsen Complex, Mohd. Shah Road, Paltan Bazar, Near Starline Hotel, Guwahati-781 008
  Tel.: (0361) 2513881, Telefax: 2543669, **E-mail: guwahati@newagepublishers.com**
- **Hyderabad** 105, 1st Floor, Madhiray Kaveri Tower, 3-2-19, Azam Jahi Road, Nimboliadda, Hyderabad-500 027
  Tel.: (040) 24652456, Telefax: 24652457, **E-mail: hyderabad@newagepublishers.com**
- **Kolkata** RDB Chambers (Formerly Lotus Cinema) 106A, 1st Floor, S.N. Banerjee Road, Kolkata-700 014
  Tel.: (033) 22273773, Telefax: 22275247, **E-mail: kolkata@newagepublishers.com**
- **Lucknow** 16-A, Jopling Road, Lucknow-226 001
  Tel.: (0522) 2209578, 4045297, Telefax: 2204098, **E-mail: lucknow@newagepublishers.com**
- **Mumbai** 142C, Victor House, Ground Floor, N.M. Joshi Marg, Lower Parel, Mumbai-400 013
  Tel.: (022) 24927869, Telefax: 24915415, **E-mail: mumbai@newagepublishers.com**
- **New Delhi** 22, Golden House, Daryaganj, New Delhi-110 002
  Tel.: (011) 23262368, 23262370, Telefax: 43551305, **E-mail: sales@newagepublishers.com**

ISBN : 978-81-224-3561-0

₹ 195.00

C-14-02-7413

Printed in India at Kanishik Printing Press, Delhi.
Typeset at In-house, Delhi.

PUBLISHING FOR ONE WORLD

NEW AGE INTERNATIONAL (P) LIMITED, PUBLISHERS
7/30 A, Darya Ganj, New Delhi-110002
Visit us at **www.newagepublishers.com**

PREFACE

Today the Technical Education in India is expanding rigorously throughout the country. A large number of Engineering and Technical Institutions already exist, and many more are in the pipeline to be opened in near future. Physics is a mandatory subject for all branches of Engineering upon which the huge course of technical education is based. The book entitled **Engineering PHYSICS-II** for B. Tech. II Sem. has been written keeping in mind the need of undergraduate students of Rajasthan Technical University, Kota. It caters to the complete syllabus of second semester of R.T.U.

The aim of writing this book has been to present the material in a concise and simple way so that even the weak students can easily grasp the fundamentals of the course. Every chapter starts with a simple introduction and then related topics are covered with a detailed description alongwith the help of diagrams. A large number of solved problems, compiled from University Question Papers, have been presented at the end of each topic.

The Practical Experiments prescribed by R.T.U. for second semester of B. Tech. have been presented at the end of the book. Each experiment carries sufficient number of short questions with answers to be asked in viva-voce at the time of University Practical Examination.

We take this opportunity to express our gratitude to **Prof. M.P. Saksena, Prof. B.K. Sharma, Prof. B.K. Srivastava, Prof. Y.K. Vijay, Prof. A. Krishnamurthi, Prof. Sardar Singh, Prof. N.S. Saxena, Prof. Kanan Bala, Prof. Deepak Bhatnagar and Prof. G. D. Sharma** who have influenced our knowledge of Physics in one way or the other.

We are very much grateful to **Prof. M.C. Govil**. Principal, Government Women Engineering College, Ajmer, who has inspired us to write this book.

We are thankful to all relatives and friends who have directly or indirectly helped us in the preparation of the book in its present form.

We would like to specially thank **New Age International (P) Ltd. Publishers, New Delhi** for accepting our manuscript for publication. We really appreciate the dynamism shown by the entire team of **New Age International (P) Ltd. Publishers, New Delhi** in bringing out the book in the present form in such a short period of time.

Finally, we would like to thank the members of our family for their support and encouragement. Suggestions for improvement of the book from learned colleagues and students will be thankfully acknowledged.

—Dr. S. S. Sharma & Dr. G. D. Ladiwala

SYLLABUS

203 ENGINEERING PHYSICS-II

Unit 1: Quantum Mechanics: Compton effect & quantum nature of light, Derivation of time dependent and time independent Schrödinger's Wave Equation, Physical interpretation of wave function and its properties, boundary conditions, Particle in one-dimensional box.

Unit 2: Applications of Schrödinger's Equation, Particle in three-dimensional box and Degeneracy, Barrier penetration and tunnel effect, Tunneling probability, Alpha Decay, Sommerfeld's Free electron gas model Postulates, Density of energy states, Fermi energy level.

Unit 3: Coherence and Optical Fibres, Spatial and temporal coherence, Coherence length, Coherence time and 'Q' factor for light, Visibility as a measure of coherence, Spatial Coherence and size of the source, Temporal coherence and spectral purity, Optical fiber as optical wave-guide, Numerical aperture , maximum angle of acceptance and applications of Optical Fiber.

Unit 4: Lasers and Holography: Theory of laser action, Einstein's coefficients, Components of a laser, Threshold conditions for laser action; Theory, Design and applications of He-Ne and semiconductor lasers; Holography versus photography, Basic theory of holography, Basic requirement of a holographic laboratory; Applications of holography in microscopy and interferometry.

Unit 5: Nuclear Radiation Detectors, Characteristics of gas filled detectors: general considerations, Constructions, Working and properties of: Ionization chamber, proportional counter, G. M. Counter and Scintillation Counter.

207 ENGINEERING PHYSICS LAB-II

1. To determine the height of water tank with the help of a Sextant.
2. To determine the dispersive power of material of a Prism for Violet Red and yellow colours of Mercury light with the help of a spectrometer.
3. To measure the Numerical Aperture of an Optical Fibre.
4. To determine the ferromagnetic constants retentivity, permeability and susceptibility by tracing B-H curve using C.R.O.

5. To study the Charge & Discharge of a condenser and hence determine time constant (Both current and voltage graphs are to be plotted).
6. To determine the high resistance by method of leakage, using a Ballistic galvanometer.
7. To verify the expression for the resolving power of a Telescope.
8. To determine the specific resistance of the material of a wire by Carey Fosters bridge.

CONTENTS

Preface *v*
Syllabus *vii*

CHAPTER-1
QUANTUM MECHANICS

1.1 INTRODUCTION 1.1
1.2 BLACK BODY RADIATION 1.1
1.2.1 Planck's Quantum Theory 1.2
1.3 PHOTOELECTRIC EFFECT 1.3
1.3.1 Explanation of Photoelectric Effect 1.3
1.4 MATTER WAVE 1.9
1.4.1 The de Broglie Wavelength 1.10
1.4.2 Important Points to Note 1.10
1.4.3 de Broglie Waves Associated with Macroscopic and Microscopic Particles 1.11
1.5 WAVE PACKET AND UNCERTAINTY PRINCIPLE 1.14
1.5.1 The Uncertainty Principle 1.15
1.5.2 Mathematical Relations 1.15
1.6 COMPTON EFFECT 1.18
1.6.1 Theory of Compton Scattering 1.19
1.6.2 Angle of Scattering of the Electron (Direction of Recoiled Electron) 1.21
1.6.3 Kinetic Energy of Recoiled Electron 1.22
1.6.4 Frequency of the Scattered Photon 1.23
1.6.5 Experimental Demonstration of Compton Effect 1.24
1.6.6 Comparison of Compton Effect and Photoelectric Effect 1.25
1.7 CONCEPT OF WAVE FUNCTION 1.33
1.8 WAVE MECHANICS 1.35
1.8.1 Postulates of Wave Mechanics 1.35
1.9 SCHRÖDINGER'S EQUATION 1.36
1.9.1 Time Dependent Schrödinger's Equation 1.37
1.9.2 Time-Independent Schrödinger's Equation 1.38
1.9.3 Physical Interpretation of Wave Function $\psi(\vec{r}, t)$ 1.39

1.9.4 Normalization of Wave Function $\psi(\vec{r},t)$ 1.39
1.9.5 Orthogonal Wave Functions 1.40
1.9.6 Properties of Wave function ψ 1.40
1.10 APPLICATION OF SCHRÖDINGER'S EQUATION 1.40
1.10.1 The Particle in One Dimensional Potential Box 1.40
QUESTIONS AND PROBLEMS 1.48
QUESTIONS 1.48
PROBLEMS 1.49
REFERENCES 1.51

CHAPTER-2
APPLICATIONS OF SCHRÖDINGER'S EQUATION

2.1 INTRODUCTION 2.1
2.2 THE PARTICLE IN A THREE DIMENSIONAL BOX 2.2
2.2.1 Degeneracy of Energy Levels 2.5
2.3 THE POTENTIAL BARRIER (TUNNEL EFFECT) 2.7
2.4 THEORY OF α- DECAY 2.18
2.5 SOMMERFELD'S FREE ELECTRON GAS MODEL OF SOLIDS (METALS) 2.22
2.5.1 Postulates of Sommerfeld Model 2.22
2.5.2 Density of Energy States and Fermi Energy 2.23
2.5.3 The Fermi Distribution Function and Effect of Temperature on it 2.26
2.5.4 Mean Energy of Electron Gas at Absolute Zero 2.27
QUESTIONS AND PROBLEMS 2.35
QUESTIONS 2.35
PROBLEMS 2.37
REFERENCES 2.38

CHAPTER-3
COHERENCE AND OPTICAL FIBRES

3.1 COHERENCE 3.1
3.1.1 Spatial Coherence 3.1
3.1.2 Temporal Coherence 3.1
3.2 DETAILED CONCEPT OF TEMPORAL AND SPATIAL COHERENCE 3.2
3.2.1 Temporal Coherence and Monochromaticity of the Source 3.2
3.2.2 Spatial Coherence and Size of the Source 3.4
3.3 VISIBILITY AS A MEASURE OF COHERENCE 3.5
3.4 OPTICAL FIBRE 3.12
3.4.1 Importance of Optical Fibres 3.13
3.4.2 Structure of Optical Fibre 3.13

3.5	PROPAGATION OF LIGHT WAVES THROUGH FIBRE	3.14
	3.5.1 Mechanism	3.14
	3.5.2 Conditions	3.14
3.6	TYPES OF OPTICAL FIBRE	3.15
	3.6.1 Step Index (SI) Optical Fibre	3.15
	3.6.2 Graded Index (GRIN) Optical Fibre	3.16
3.7	ACCEPTANCE ANGLE AND ACCEPTANCE CONE OF A STEP INDEX FIBRE	3.17
	3.7.1 Numerical Aperture	3.18
	3.7.2 V-Number	3.19
3.8	APPLICATIONS OF OPTICAL FIBRE	3.25
	3.8.1 Optical Fibre Communication	3.26
	3.8.2 Fibre Optics in Medical Science	3.26
	QUESTIONS AND PROBLEMS	3.27
	QUESTIONS	3.27
	PROBLEMS	3.28
	REFERENCES	3.28

CHAPTER-4
LASER AND HOLOGRAPHY

4.1	LASER	4.1
	4.1.1 Comparision of Laser Source with Ordinary Source	4.1
4.2	ABSORPTION, SPONTANEOUS EMISSION AND INDUCED EMISSION OF RADIATION	4.2
4.3	PRINCIPLE OF LASER	4.5
	4.3.1 Pumping and Population Inversion	4.6
	4.3.2 Cavity Resonators and Shaping of a Beam (Operation of Laser)	4.7
4.4	THE HELIUM-NEON LASER	4.8
4.5	SEMICONDUCTOR LASER	4.10
	4.5.1 Introduction	4.10
	4.5.2 Theory of Semiconductor	4.10
	4.5.3 Homojunction Laser	4.11
	4.5.4 Construction	4.11
	4.5.5 Principle of Operation	4.12
	4.5.6 Heterojunction Laser	4.13
	4.5.7 Double Heterostructure Lasers	4.13
4.6	APPLICATIONS OF LASERS	4.14
4.7	CHARACTERISTICS OF THE LASER LIGHT	4.14
4.8	Q-SWITCHING	4.16
	4.8.1 Principle of Q-Switching	4.17
	4.8.2 Evolution of a Q-Switched Laser Pulse	4.17
	4.8.3 Types of Q-Switching	4.18
	4.8.4 Applications	4.19

4.9 MODE LOCKING 4.20
4.9.1 Longitudinal Modes of the Laser Cavity 4.20
4.9.2 Mode Locking Theory 4.21
4.9.3 Mode-Locking Methods 4.23
4.9.4 Application of Mode Locked Laser Output 4.23

HOLOGRAPHY

4.10 INTRODUCTION 4.25
4.11 THE BASIC PRINCIPLE OF HOLOGRAPHY 4.25
4.11.1 Obtaining a Hologram (Construction of Hologram) 4.26
4.11.2 Viewing the Object (Reconstruction of an Image from Hologram) 4.28
4.12 TYPES OF HOLOGRAMS 4.29
4.13 HOLOGRAPHY VERSUS PHOTOGRAPHY 4.30
4.14 BASIC REQUIREMENTS OF A HOLOGRAPHIC LABORATORY 4.30
4.15 SOME SPECIAL FEATURES OF A HOLOGRAM 4.32
4.16 APPLICATIONS OF HOLOGRAPHY 4.33
4.16.1 Holographic Interferometry 4.33
4.16.2 Holographic Microscopy 4.33
4.16.3 Acoustic Holography 4.34
QUESTIONS AND PROBLEMS 4.35
QUESTIONS 4.35
PROBLEMS 4.37
REFERENCES 4.38

CHAPTER-5
NUCLEAR RADIATION DETECTORS

5.1 INTRODUCTION 5.1
5.2 GAS FILLED DETECTORS 5.1
5.3 IONIZATION CHAMBER 5.3
5.3.1 Advantages 5.4
5.3.2 Disadvantages 5.5
5.4 PROPORTIONAL COUNTER 5.7
5.4.1 Construction 5.7
5.4.2 Operation 5.8
5.4.3 Gas Multiplication Factor 5.8
5.4.4 Advantages of Proportional Counter 5.9
5.4.5 Disadvantage of Proportional Counter 5.9
5.5 GEIGER-MÜLLER COUNTER 5.11
5.5.1 Construction 5.11
5.5.2 Operation 5.12
5.5.3 Quenching 5.13

5.5.4 Dead Time and Recovery Time 5.13
5.5.5 Main Features of a G.M. Counter 5.14
5.5.6 Precautions in using G.M. Tube 5.14
5.5.7 Comparison between Proportional Counter and G.M. Counter 5.15
5.6 SCINTILLATION DETECTOR 5.18
5.6.1 Introduction 5.18
5.6.2 Construction 5.18
5.6.3. Advantages and Disadvantages of Scintillation Detectors 5.22
QUESTIONS AND PROBLEMS 5.25
QUESTIONS 5.25
PROBLEMS 5.26
REFERENCES 5.27

LIST OF EXPERIMENTS

Experiment 0. General Equipments 0.1 – 0.8
Experiment 1. Sextant 1.1 – 1.8
Experiment 2. Dispersive Power 2.1 – 2.13
Experiment 3. Numerical Aperture 3.1 – 3.5
Experiment 4. Magnetic Hysteresis 4.1 – 4.8
Experiment 5. Charging and Discharging of a Condenser 5.1 – 5.5
Experiment 6. High Resistance by Method of Leakage 6.1 – 6.7
Experiment 7. Resolving Power 7.1 – 7.7
Experiment 8. Carey-Foster's Bridge 8.1 – 8.8

APPENDICES

Appendix A A1-A2
Appendix B A2-A3
Appendix C A4-A9
INDEX

1

QUANTUM MECHANICS

1.1 INTRODUCTION

The failureness of classical mechanics to explain about observed spectrum of black body radiation was a main cause to develop a new mechanics, called **Quantum Mechanics**. Although the stability of atoms, the observed variation of the specific heat of solids or gases, the origin of discrete spectra of atoms, the detailed description of the motion of atomic particles and photoelectric effect, compton effect, raman effect *etc*. were some other facts which could not be explained by the classical mechanics.

The basic idea of quantum theory was introduced by Max Planck in 1900 to explain the black body radiations, but the subsequent developments and interpretations were made by a number of physicists including Einstein, Bohr, deBroglie, Heisenberg, Schrödinger, Born and Dirac.

1.2 BLACK BODY RADIATION

A black body is an object that absorbs all the radiations that fall on it, *i.e.* no radiation passes through it and none is reflected. This body looks black under outside illumination. A hollow object with a small hole, acts like a black body. If the black body is heated to a high temperature, it emits exactly same wavelengths and intensities which would be obsorbed by the body. The radiation emitted by a black body is called black body radiation.

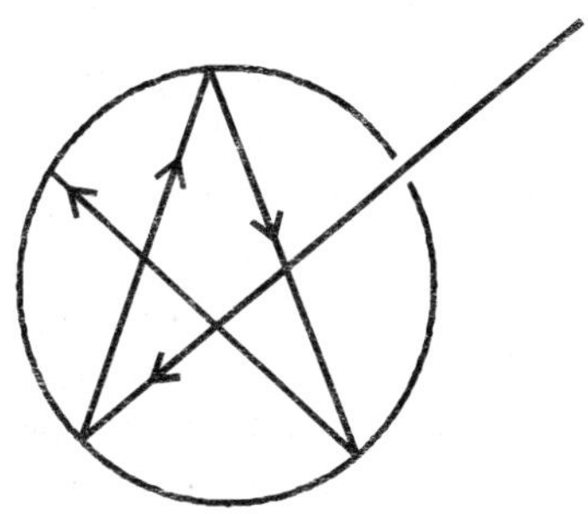

Fig. 1.1 Black body model

In 1899, Lummer and Pringsheim studied the distribution of energy amongst the different wavelengths or frequencies of a thermal spectrum of a black body radiation. The experimental results of this study are illustrated in Figs. 1.2 (*a*) and (*b*).

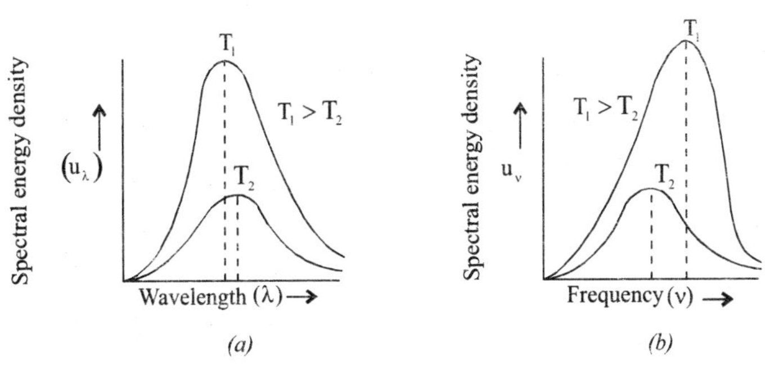

Fig. 1.2 (*a*) and (*b*) Black body radiation spectrum

Wien and Rayleigh-Jeans studied the black body radiations and derived a formula to explain these results. But the Wien's formula* holds good only in the region of shorter wavelengths (higher frequencies) and the Rayleigh-Jean's fromula** holds good in the region of longer wavelengths (lower frequencies). Hence Wien's formula and Rayleigh-Jean's formula do not precisely agree with the experimental results given by Lummer and Pringsheim.

1.2.1 Planck's Quantum Theory

In order to explain black body radiation, Max Planck, in the year 1900, proposed the quantum theory. According to this theory, matter is composed of a large number of oscillating particles which vibrate with different frequencies. In quantum theory, the energy of oscillating particle is quantized. The oscillating particle cannot have any energy but only those energies which are given by

$$E = nh\nu$$

where h is Planck's constant whose value is 6.625×10^{-34} Joule-second, ν is the oscillator frequency and n is an integer (n = 1, 2, 3. . . .). In this theory, it is also assumed that the vibrating particle does not radiate energy continuosuly but discretely, *i.e.*, in the form of some minimum quantity of energy known as quanta or photon.

Karl Ernst Ludwig Max Planck (April 23, 1858 – October 4, 1947) was a German physicist. He was considered the founder of quantum theory. According to planck the energy could only be a multiple of an elementary unit $E = h\nu$. where h is Planck's constant and ν is the frequency of the radiation. It was in recognition of his monumental accomplishment that Planck was awarded the Nobel Prize in Physics in 1918.

While studing black body radiations, Max Planck in 1901 concluded that the emission and absorption of thermal energy is not a continuous process but it takes place in discrete amount *i.e.* an integral multiple of a certain energy unit '$h\nu$'. According to this postulate, the exchange of energy between light and matter is not continuous, but it is in the form of small *bundles* or *packets* of definite energy proportional to frequency of light. These small packets of energy are called photons or quanta.

Max Planck derived a theoretical expression for the energy distribution on the basis of quantum theory of heat radiation and Maxwell-Boltzmann distribution Law. He deduced following formula

$$u_\nu d\nu = \frac{8\pi h\nu^3}{c^3} \frac{d\nu}{\left[e^{h\nu/kT} - 1\right]} \qquad \textbf{(1.1)}$$

This is known as Planck's radiation formula. It solved the problem of ultraviolet catastrophe. Planck's radiation formula explained the spectral energy density of black body completely. Hence, the concept of quantum of energy (quantum theroy) was the foundation of modern physics. Einstein built on this idea and proposed the quantization of electromagnetic radiation in 1905 to explain the photoelectric effect.

---

* Wien's formula: $u(\lambda)d\lambda = \frac{8\pi hc}{\lambda^5} e^{-hc/\lambda kT} d\lambda$, (for small values of λT).

** Rayleigh-Jean's formula: $u(\lambda)d\lambda = \frac{8\pi kT}{\lambda^4} d\lambda$, (for large values of λT).

1.3 PHOTOELECTRIC EFFECT

The emission of electrons from a metal plate when illuminated by electromagnetic radiation of suitable frequency or wavelength is called photoelectric effect. The emitted electrons are called photoelectrons.

The phenomenon was discovered by Hertz in 1887 when he allowed ultraviolet light to fall on zinc plate. The phenomenon was experimentally verified by Hallwachs and Lenard. Afterwards, it was discovered that alkali metals like lithium, sodium, potassium, rubidium and cesium eject electrons when visible light falls on them. Millikan investigated the effect with a number of alkali metals over a wide range of light frequencies and was given Nobel Prize in 1923.

1.3.1 Explanation of Photoelectric Effect

Failure of Classical Theory: The classical theory, which assumes that light travels in the form of electromagnetic waves, fails to explain the experimental observations on photoelectric effect. The main drawbacks of the classical theory are:

(*i*) According to the classical theory, if the intensity of incident radiation were increased, *i.e.* more energy is given to the electorns, the energy of emitted electrons must increase. This is contrary to the experimental observation. On increasing the intensity of incident radiation the photoelectrons are ejected with the same kinetic energy.

(*ii*) The existence of a threshold frequency for a given material cannot be explained on the basis of wave theory of radiation.

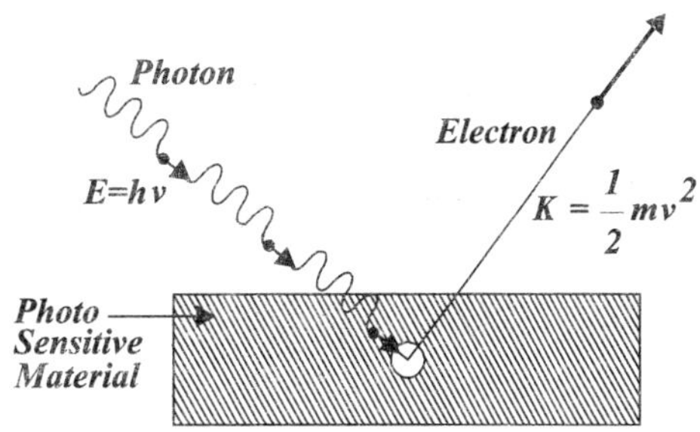

Fig. 1.3

(*iii*) The instantaneous ejection of photoelectrons from the metal surface cannot be explained by classical theory. If the illumination is faint, the classical theory predicts that time of several minutes would have lapsed before a single electron is released. This is contrary to the experimental observation.

Einstein's Explanation of Photoelectric Effect: Einstein applied Planck's quantum theory to explain the phenomenon of photoelectric effect in 1905. According to this theory, light travels in the form of energy packats called photons (or quanta). The velocity of these photons is equal to that of light ($c = 3 \times 10^8$ m/s), and their energy is $h\nu$, where 'h' is Planck's constant and ν the frequency of light. Thus, when light of frequency ν is incident on a metallic surface, the energy of the photon is $E = h\nu$. This energy is supplied to an electron present in the surface of photosensitive material, where this energy is utilised in two parts (Fig. 1.3).

(*i*) A part of its energy is used to release electron from the atom and away from the metal surface. This energy is known as photoelectric work function of the metal and is denoted by W.

(*ii*) The rest of the energy is utilised in imparting kinetic energy to the ejected electron. If the ejected electron moves with a velocity ν, its kinetic energy is $K = \frac{1}{2}m\nu^2$.

$\therefore$ Energy of incident photon = Energy spent in removing an electron (work-function) + Kinetic energy of electron.

i.e., $E = W + K$

$$\Rightarrow \quad hv = W + \frac{1}{2}mv^2 \quad (1.2)$$

Further, if the frequency of incident light is equal to the threshold frequency v_0, then the electron will be just-ejected from the metal surface but its velocity (or kinetic energy) will be zero.

Hence put $\nu = v_0$ and $v = 0$ in Eq. (1.2), we get

$$W = hv_0 \quad (1.3)$$

Putting this value of W in Eq. (1.2), we get

$$hv = hv_0 + \frac{1}{2}mv^2$$

$$\Rightarrow \quad h(v - v_0) = \frac{1}{2}mv^2 \quad (1.4)$$

This equation is called Einstein's photoelectric equation. In terms of stopping potential V_0, we have

$$eV_0 = \frac{1}{2}mv^2 \quad (1.5)$$

and hence $h(v - v_0) = eV_0$ **(1.6)**

$$\Rightarrow \quad V_0 = \frac{h}{e}v - \frac{hv_0}{e} \quad (1.7)$$

This is another form of Einstein's photoelectric equation. Eq. (1.7) represents a straight line shown in Fig. 1.4. The slope of this straight is *h/e*. As *h* and *e* are both constant, the slope of all such straight lines is constant.

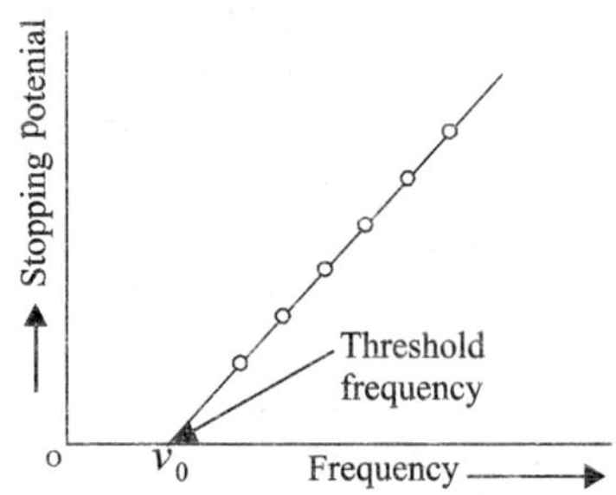

Fig. 1.4

Corresponding to the threshold frequency, we define long wavelength limit (λ_0). It represents the upper limit of the wavelength for photoelectric effect. Its physical significance is that radiation having wavelength longer than λ_0 would not be able to eject electorns from a given metal surface whereas those having $\lambda < \lambda_0$ will eject. The value of λ_0 is given by

$$c = \nu_0\lambda_0$$

$$\Rightarrow \quad \lambda_0 = \frac{c}{v_0} = \frac{ch}{W}$$

$$\Rightarrow \quad \lambda_0 = \frac{3 \times 10^8 \times 6.625 \times 10^{-34}}{W}$$

$$= \frac{19.875 \times 10^{-26}}{W} \text{metre}$$

$$= \frac{19.875 \times 10^{-16}}{W} \text{ Å} \tag{1.8}$$

Here W is expressed in Joules. If W expressed in eV, then

$$\lambda_0 = \frac{19.875 \times 10^{-16}}{W \times 1.602 \times 10^{-19}}$$

$$\approx \frac{12400}{W(\text{eV})} \text{Å} \tag{1.9}$$

From Eq. (1.9), the value of photoelectric work function is given by

$$W \approx \frac{12400}{\lambda_0(\text{Å})} \text{eV} \tag{1.10}$$

The Einstein's photoelectric equation (Eq. 1.2) can explain all experimental results. For a particular material, work function W is constant, hence

$$K.E = \frac{1}{2} mv^2 \propto h\nu$$

$$v^2 \propto \nu \tag{1.11}$$

Thus the increase in frequency of incident light causes increase in velocity of photoelectrons provided its intensity remains cosntant.

An increase in the intensity of incident radiation is equivalent to an increase in the number of photons falling on the emitting surface. If the frequency of incident radiation is above the threshold frequency $(\nu > \nu_0)$, *the number of emitted photoelectrons will increase*. In this way intensity of emitted electrons (photoelectric current) is directly proportional to the intensity of incident radiation.

Example 1.1: A certain spectral line has wavelength 4000 Å. Calculate the frequency and energy in eV of the photon associated with it. (Given $c = 3 \times 10^8$ m/s and $h = 6.63 \times 10^{-34}$ J-s)

Solution: Frequency of the photon is given by $\nu = \frac{c}{\lambda}$

Here $c = 3 \times 10^8$ m/s and $\lambda = 4 \times 10^{-7}$ m

$$\therefore \quad \nu = \frac{3 \times 10^8}{4 \times 10^{-7}} = 7.5 \times 10^{14} \text{Hz}$$ **Ans.**

Energy of photon is given by

$$E = h\nu$$

$$= 6.63 \times 10^{-34} \times 7.5 \times 10^{14} \text{ Joule}$$

$$= \frac{6.63 \times 10^{-34} \times 7.5 \times 10^{14}}{1.6 \times 10^{-19}} \text{eV}$$

$$= 3.108 \text{ eV}$$ **Ans.**

Example 1.2: Radiations of wavelength 5400 Å fall on a metal plate whose work function is 1.9 eV. Find the kinetic energy of the emitted photoelectrons and their stopping potential.

Solution: Kinetic energy of emitted photoelectrons is given by

$$K = \frac{1}{2}mv^2 = E - W$$

But $E = h\nu = \frac{hc}{\lambda}$

Here $h = 6.62 \times 10^{-34}$, $c = 3 \times 10^8$ m/s, $\lambda = 5400$ Å $= 5.4 \times 10^{-7}$ m and $W = 1.9$ eV

$$\therefore \quad E = \frac{6.63 \times 10^{-34} \times 3 \times 10^8}{5.4 \times 10^{-7} \times 1.6 \times 10^{-19}} = 2.3 \text{ eV}$$

$$\therefore \quad K = E - W = 2.3 - 1.9 = 0.4 \text{ eV}$$ **Ans.**

Stopping potential is given by

$$V_0 = \frac{K}{e} = \frac{0.4 \times 1.6 \times 10^{-19}}{1.6 \times 10^{-19}} = 0.4 \text{ Volt}$$ **Ans.**

Example 1.3: If the speed of photoelectrons is 10^6 m/s, what should be the frequency of the incident radiation on a potassium metal having work function 2.3 eV?

Solution: According to Einstein's photoelectric equation

$$hv = W + \frac{1}{2}mv^2$$

Here $W = 2.3$ eV $= 2.3 \times 1.610^{-19}$ J, $m = 9.1 \times 10^{-31}$ kg, $v = 10^6$ m/s

$$hv = 2.3 \times 1.6 \times 10^{-19} + \frac{1}{2} \times 9.1 \times 10^{-31} \times 10^{12}$$

$$= 8.3 \times 10^{-19} \text{ J}$$

$$\therefore \text{ Frequency } \nu = \frac{8.3 \times 10^{-19}}{h}$$

$$= \frac{8.3 \times 10^{-19}}{6.62 \times 10^{-34}}$$

$$= 1.24 \times 10^{15} \text{ Hz}$$ **Ans.**

Example 1.4: The energy of photoelectrons emitted from a sensitive plate is 1.46 eV. If its threshold is 4500 Å, calculate the wavelength of incident radiation.

Solution: Kinetic energy of emitted electron is given by

$$K = h(\nu - \nu_0) = hc\left(\frac{1}{\lambda} - \frac{1}{\lambda_0}\right)$$

or
$$\frac{1}{\lambda} = \frac{K}{hc} + \frac{1}{\lambda_0}$$

Here K= 1.46 eV $= 1.46 \times 1.6 \times 10^{-19}$ J, $h = 6.63 \times 10^{-34}$ J-s, $c = 3 \times 10^8$ m/s, $\lambda_0 = 4.5 \times 10^{-7}$ m

$$\therefore \quad \frac{1}{\lambda} = \frac{1.46 \times 1.6 \times 10^{-19}}{6.62 \times 10^{-34} \times 3 \times 10^8} + \frac{1}{4.5 \times 10^{-7}}$$

$$= 2.222 \times 10^6 + 1.176 \times 10^6$$

$$= 3.398 \times 10^6 \text{ m}^{-1}$$

$\therefore$ Wavelength of incident radiation

$$\lambda = \frac{1}{3.398 \times 10^6}$$

$$= 2.943 \times 10^{-7} \text{ m} = 2943 \text{ Å}$$ **Ans.**

Example 1.5: In an experiment tungsten cathode which has a threshold 2300 Å is irradiated by ultraviolet light of wavelength 1800 Å. Calculate (*i*) the maximum energy of emitted , photoelectrons and (*ii*) the work function for tungsten. Give your results in electron-volts. Given 1 eV = 1.6×10^{-19} J and velocity of light $c = 3 \times 10^8$ m/s.

Solution: From Einstein's photoelectric equaiton

$$\frac{1}{2} m v^2_{\max} = h\nu - h\nu_0 = h\left(\frac{c}{\lambda} - \frac{c}{\lambda_0}\right) = hc\left(\frac{\lambda_0 - \lambda}{\lambda \lambda_0}\right)$$

Here $h = 6.62 \times 10^{-34}$ J-s, $\lambda = 1800$ Å $= 1.8 \times 10^{-7}$ m, $\lambda_0 = 2.3 \times 10^{-7}$ m

$$\Rightarrow \quad \frac{1}{2} m v^2_{\max} = \frac{6.62 \times 10^{-34} \times 3 \times 10^8}{1.6 \times 10^{-19}} \left[\frac{2.3 - 1.8}{2.3 \times 1.8 \times 10^{-7}}\right]$$

$$= \frac{6.62 \times 3 \times 0.5}{1.6 \times 2.3 \times 1.8} = 1.5 \text{ eV}$$ **Ans.**

Work function $W = h\nu_0 = \dfrac{hc}{\lambda_0}$

$$\Rightarrow \quad W = \frac{6.62 \times 10^{-34} \times 3 \times 10^8}{2.3 \times 10^{-7} \times 1.6 \times 10^{-19}} = 5.39 \text{ eV}$$ **Ans.**

Example 1.6: A metallic surface, when illumintated with light of wavelength λ_1, emits electrons with energies up to a maximum value E_1 and when illuminated with light of wavelength $\lambda_2 (\lambda_2 < \lambda_1)$, it emits electrons with energies up to a maximum value E_2. Prove that Planck's constant h and the work function W are given by,

$$h = \frac{(E_2 - E_1)\lambda_1\lambda_2}{c(\lambda_1 - \lambda_2)} \text{ and } W = \frac{E_2\lambda_2 - E_1\lambda_1}{\lambda_1 - \lambda_2}$$

Solution: Maximum kinetic energy E_1 is given by

$$E_1 = \frac{hc}{\lambda_1} - W \quad \textbf{(1)}$$

and E_2 is given by

$$E_2 = \frac{hc}{\lambda_2} - W \quad \textbf{(2)}$$

Subtracting (1) from (2), we get

$$(E_2 - E_1) = hc\left(\frac{1}{\lambda_2} - \frac{1}{\lambda_1}\right) = hc\left(\frac{\lambda_1 - \lambda_2}{\lambda_1\lambda_2}\right)$$

$$\therefore \quad h = \frac{(E_2 - E_1)\lambda_1\lambda_2}{c(\lambda_1 - \lambda_2)}$$

Hence proved.

From Eqs. (1) and (2), we have

$$E_1\lambda_1 = hc - W\lambda_1 \quad \textbf{(3)}$$

$$E_2\lambda_2 = hc - W\lambda_2 \quad \textbf{(4)}$$

Subtracting (3) from (4), we get

$$E_2\lambda_2 - E_1\lambda_1 = W(\lambda_1 - \lambda_2)$$

$$\Rightarrow \quad W = \frac{E_2\lambda_2 - E_1\lambda_1}{\lambda_1 - \lambda_2}$$

Hence proved.

Example 1.7: A metallic surface, when illuminated with light of wavelength 3333 Å, emits electrons with energies upto 0.6 eV, and when illuminated with light of wavelength 2400 Å, it emits electrons with energies up to 2.04 eV. Calculate Planck's constant and the work function of the metal.

Solution: Utilising the results of the previous example, Planck's constant is given by

$$h = \frac{(E_2 - E_1)\lambda_1\lambda_2}{c(\lambda_1 - \lambda_2)}$$

Here $E_1 = 0.6 \text{ eV} = 0.6 \times 1.6 \times 10^{-19} \text{ J}$

$$E_2 = 2.04 \text{ eV} = 2.04 \times 1.6 \times 10^{-19} \text{ J}$$

$$\lambda_1 = 3.333 \times 10^{-7} \text{m}$$

$$\lambda_2 = 2.4 \times 10^{-7} \text{m}$$

$$\therefore \quad h = \frac{(2.04 - 0.6) \times 1.6 \times 10^{-19} \times 3.333 \times 2.4 \times 10^{-14}}{3 \times 10^8 \times (3.333 - 2.4) \times 10^{-7}}$$

$$= \frac{1.44 \times 1.6 \times 3.333 \times 2.4 \times 10^{-34}}{3 \times 0.933} = 6.58 \times 10^{-34} \text{ J-s}$$

Work function is given by

$$W = \frac{E_2\lambda_2 - E_1\lambda_1}{c(\lambda_1 - \lambda_2)} = \frac{(2.04 \times 2.4 - 0.6 \times 3.333) \times 10^{-7}}{(3.33 - 2.4) \times 10^{-7}} = 3.1 \text{ eV}$$ **Ans.**

Example 1.8: The work function of a metal surface is 1.2 eV. Calculate the kinetic energy of the fastest and slowest photoelectrons and the retarding potential when light of frequency $(5.5 \times 10^{14}\,\text{Hz})$ falls on the surface.

Solution: The kinetic energy of fastest photoelectrons is given by

$$K_{max} = h\nu - W$$

Here $h = 6.62 \times 10^{-34}$ J-s, $\nu = 5.5 \times 10^{14}$ Hz, $W = 1.2$ eV

$$\therefore \quad K_{max} = \frac{6.62 \times 10^{-34} \times 5.5 \times 10^{14}}{1.6 \times 10^{-19}} - 1.2$$

$$= 2.27 - 1.2 = 1.07 \text{ eV}$$ **Ans.**

The kinetic energy of sowest electrons is zero.

The stopping potential is given by

$$V_0 = \frac{K_{max}}{e} = \frac{1.07 \times 1.6 \times 10^{-19}}{1.6 \times 10^{-19}} = 1.07 \text{ Volt}$$ **Ans.**

1.4 MATTER WAVE

After the discovery of Planck's quantum theory of radiation, it became an established fact that radiation has dual character behaving either as waves or as particles under suitable circumstances. Interference and Diffraction phenomena demonstrate their wave behaviour while Photoelectric Effect, Compton Effect etc. (*i.e.,* emission and absorption of photons) demonstrate their particle behaviour.

Towards the middle of the nineteenth century, it became clear to the physicists that the *physical universe is made up of two great entities viz. the energy and the matter.* The dual nature of radiation stimulated the scientists, as to why the other physical entity, *i.e., matter* must also not possess the *dual nature* like radiant energy. *Louis de Broglie*, a French physicist put forward a daring suggestion in 1924 that just like radiation, *matter also shows dual nature*. In other words, particles of matter, e.g., electrons also show wave like properties. His duality hypothesis was based on the following observations.

(*i*) The whole energy in this universe is in the form of electromagnetic radiation and matter.

(*ii*) As nature loves symmetry, hence matter and energy must be mutually symmetrical. If radiant energy has dual character, matter must also possess dual character.

Louis-Victor-Pierre-Raymond, 7th duc de Broglie (August 15, 1892 – March 19, 1987) was a French physicist and a Nobel laureate. In physics the **de Broglie hypothesis** is the statement that all matter (any object) has a wave-like nature (wave-particle duality). The **de Broglie relations** show that the wavelength is inversely proportional to the momentum of a particle and that the frequency is directly proportional to the particle's kinetic energy. The hypothesis was advanced by Louis de Broglie in 1924 in his PhD thesis. He was awarded the Nobel Prize for Physics in 1929 for this work, which made him the first person to receive a Nobel Prize on a PhD thesis.

The waves which were assumed to be associated with matter on the basis of de Broglie's hypothesis are called as *Matter Waves* or *de Broglie Waves.* The principle of *Complimentarity* is also applicable here. This implies that in no experiment, matter is found to exist both as a particle and as a wave simultaneously. The two aspects of *wave and particle are always complimentary to each other.*

1.4.1 The de Broglie Wavelength

According to Planck's quantum theory of radiation, the energy of a photon of radiation of frequency ν is given by

$$E = h\nu \qquad (h = \text{Planck's constant}) \qquad \textbf{(1.12)}$$

If photon is considered as a particle whose effective mass is m, then according to *Einstein's mass-energy equivalence principle*, the energy E of photon is

$$E = mc^2 \qquad (c = \text{velocity of light}) \qquad \textbf{(1.13)}$$

Since the energy of the photon in the two cases is same, hence

$$h\nu = mc^2$$

$$\Rightarrow \qquad mc = \frac{h\nu}{c}$$

The quantity mc is the momentum p (say) of the photon of effective mass m and travelling with the speed of light, *i.e.*, c.

$$\therefore \qquad p = mc = \frac{h}{\lambda} \qquad (\because c = \nu\lambda)$$

$$\Rightarrow \qquad \lambda = \frac{h}{p} \qquad \textbf{(1.14)}$$

This equation provides a connection between a *wave like property of radiation, the wavelength* (λ) *and a particle like property, the momentum* (p).

de Broglie suggested that this same relationship connects the particle like and wave like properties of matter. That is, there is a *sinusoidal wave having a wavelength* associated with a *free particle moving with linear momentum 'p'*. Thus the wavelength of the wave associated with a particle having mass 'm' and moving with velocity 'v' is given by

$$\lambda = \frac{h}{p} = \frac{h}{m\nu} \qquad \textbf{(1.15)}$$

The wavelength of a particle computed according to Eq. (1.15) is called its *de Broglie wavelength*. Note that Planck's constant provides the connecting link between the wave and particle natures of both matter and energy.

1.4.2 Important Points to Note

(1) In the expression for de Broglie wavelength $\lambda = \dfrac{h}{m\nu}$, m represents the relativistic mass of the particle. If the velocity of the particle is comparable to that of light, then mass of the particle is given by:

$$m = \frac{m_0}{\sqrt{1 - \frac{v^2}{c^2}}}$$

where m_0 is the rest mass of the particle.

(2) In the expression $\lambda = \frac{h}{mv}$

If $v = 0$, $\lambda \to \infty$ and if $v \to \infty$, $\lambda \to 0$

This shows that matter waves are associated with moving material particles only.

(3) The de Broglie wavelength of a moving particle is independent of the charge or nature of the particle.

(4) The matter waves are not electromagnetic waves. Following observations confirm this fact:

(*i*) The matter waves are associated with the moving particles irrespective of whether the particles are charged or not. Electromagnetic waves are produced only by accelerated charged particles.

(*ii*) The velocity of matter waves depends upon the velocity of material particles. However, velocity of electromagnetic waves is constant in a given medium.

(*iii*) Matter waves travel faster as compared to electromagnetic waves.

(5) The velocity of de Broglie waves associated with a moving particle is greater than the velocity of the particle itself. Therefore it was difficult to grasp as to how the de Broglie waves are associated with a particle. If a large number of waves of nearly same frequencies are superimposed, then a wave packet is formed. The velocity of the wave packet is the same as that of the particle. Hence a moving particle is associated with a wave packet, rather than a wave.

1.4.3 de Broglie Waves Associated with Macroscopic and Microscopic Particles

The De Broglie waves are associated, in general, with every moving particle irrespective of whether it is big or small, charged or uncharged. But, even then, ordinary objects around us are found not to show any wave like properties like diffraction and interference. For example, a bullet from a gun is associated with a wave and therefore it should bend or deviate, *i.e.*, show diffraction as it passes through an aperture or a slit. However, the observation shows that the bullet passes through the aperture without bending. The reason behind this observation is that the de Broglie wavelength 'λ' associated with moving macroscopic bodies is extremely small. For example, a bullet of mass 100 gm and travelling with a velocity of 400 m/s will have de Broglie wavelength

$$\lambda = \frac{h}{mv} = \frac{6.62 \times 10^{-34}}{0.1 \times 400} = 1.66 \times 10^{-35}\,\text{m} = 1.66 \times 10^{-25}\ \text{Å}$$

which is 10^{-19} times smaller than the radius of proton, quite beyond measurement. *As it is practically impossible to prepare an aperture of the size of this wavelength* $(= 1.66 \times 10^{-35}\,\text{m})$ *associated with macroscopic bodies, hence the wave like properties like diffraction cannot be observed in case of such big bodies.*

On the other hand microscopic bodies like molecules, atoms, neutrons, electrons *etc.* have much larger wavelength ($\approx 10^{-10}$ m) as compared to that of macroscopic bodies. Hence diffraction effects can easily be observed by a beam of fast moving neutrons or electrons showing that *matter has wave nature.*

Example 1.9: Calculate the order of wavelength associated with thermal neutrons at 27°C. Their energy is of the order of kT, where k is Boltzmann constant.

Solution: de Broglie wavelength associated with particles of energy E is given by

$$\lambda = \frac{h}{\sqrt{2mE}} = \frac{h}{\sqrt{2mkT}}$$

Here $m = 1.67 \times 10^{-27}$ kg, $h = 6.62 \times 10^{-34}$ Joule-sec, $k = 1.38 \times 10^{-23}$ J/K, T = 273 + 27 = 300 K

$$\therefore \quad \lambda = \frac{6.62 \times 10^{-34}}{\sqrt{2 \times 1.67 \times 10^{-27} \times 1.38 \times 10^{-23} \times 300}}$$

$$\lambda = 1.78 \times 10^{-10} \text{m} = 1.78 \text{Å}$$ **Ans.**

Example 1.10: A neutron beam of energy 0.04 eV is diffracted by a crystal with $d = 3.14$ Å. Calculate the de Broglie wavelength associated with neutrons. Hence determine the glancing angle at which a first order maximum will be observed.

Solution: de Broglie wavelength of neutrons with energy E eV is given by

$$\lambda = \frac{0.286}{\sqrt{E}} \text{Å}$$

Here $E = 0.04$ eV

$$\therefore \quad \lambda = \frac{12.27}{\sqrt{0.04}} \text{Å} = \frac{0.286}{0.2} = 1.43 \text{Å}$$ **Ans.**

From Bragg's law for first order

$$2d \sin\theta = \lambda$$

$$\Rightarrow \quad \sin\theta = \frac{\lambda}{2d}$$

Here $d = 3.14$ Å and $\lambda = 1.43$ Å

$$\therefore \quad \sin\theta = \frac{1.43}{2 \times 3.14} = 0.2277$$

$$\therefore \quad \theta = 13.16°$$ **Ans.**

Example 1.11: Using Bohr's model find the wavelength of the electron in the ground state of hydrogen atom.

Solution: In the ground state the principal quantum number $n = 1$. By the quantum condition

$$mvr = \frac{nh}{2\pi}$$

$$\Rightarrow \quad mv = \frac{nh}{2\pi r_1} = \frac{h}{2\pi r_1}$$

But $r_1 = \frac{1}{K}\frac{h^2}{4\pi^2 e^2 m}$

$$\therefore \qquad \lambda = \frac{h}{mv} = \frac{1}{K}\frac{h^2}{2\pi m e^2}$$

Here $m = 9.1 \times 10^{-31}$ kg, $h = 6.62 \times 10^{-34}$ Joule -sec, $K = 9 \times 10^9$ S.I. units, $e = 1.6 \times 10^{-19}$ C

$$\therefore \qquad \lambda = \frac{\left(6.62 \times 10^{-34}\right)^2}{9 \times 10^9 \times 2 \times 3.14 \times 9.1 \times 10^{-31} \times \left(1.6 \times 10^{-19}\right)^2}$$

$$= 3.37 \times 10^{-10}\,\text{m} = 3.37\,\text{Å}$$

Ans.

Example 1.12: Calculate the de Broglie wavelength of a virus particle of mass 1.0×10^{-15} kg at a speed of 2 mm/s. Can this wavelength be detected practically?

Solution: The de Broglie wavelength is given by

$$\lambda = \frac{h}{mv}$$

Here $m = 1.0 \times 10^{-15}$ kg, $h = 6.62 \times 10^{-34}$ Joule -sec, $v = 2 \times 10^{-3}$ m/s

$$\therefore \qquad \lambda = \frac{6.62 \times 10^{-34}}{1.0 \times 10^{-15} \times 2 \times 10^{-3}}$$

$$= 3.31 \times 10^{-16}\,\text{m}$$

Ans.

As no slit or aperture of size $\approx 10^{-16}$ m can be practically constructed, hence it will not be possible to detect such wavelengths practically. **Ans.**

Example 1.13: An electron and a photon each have a wavelength of 2.0 Å. Compute their (*a*) momenta (*b*) total energies.

Solution: (*a*) According to de Broglie formula

$$\lambda = \frac{h}{mv} = \frac{h}{p}$$

$$\Rightarrow \qquad p = \frac{h}{\lambda}$$

Since the wavelength of electron and the photon is the same, hence their momentum will also be same.

here $\lambda = 2\,\text{Å} = 2 \times 10^{-10}$ m

$$\therefore \qquad p = \frac{6.62 \times 10^{-34}}{2 \times 10^{-10}} = 3.31 \times 10^{-24}\,\text{kg -m/s}$$

(*b*) Energy of photon is given by

$$E = h\nu = \frac{hc}{\lambda} = pc$$

here $p = 3.31 \times 10^{-24}$ kg -m/s and $c = 3 \times 10^8$ m/s

$$\therefore \qquad E = 3.31 \times 10^{-24} \times 3 \times 10^8\,\text{Joule}$$

$$\Rightarrow \qquad E = \frac{3.31 \times 10^{-24} \times 3 \times 10^{8}}{1.6 \times 10^{-19}} \text{eV} = 6.2 \times 10^{3} \text{eV} = 6.2 \text{ KeV}$$ **Ans.**

Kinetic energy of electron is given by

$$K_e = \frac{p^2}{2m} = \frac{\left(3.31 \times 10^{-24}\right)^2}{2 \times 9.1 \times 10^{-31}} \text{Joule}$$

$$= \frac{3.31^2 \times 10^{-48}}{2 \times 9.1 \times 10^{-31} \times 1.6 \times 10^{-19}} \text{eV}$$

$$= 37.6 \text{ eV}$$

Rest mass energy of electron is

$$E_r = m_0 c^2 = \frac{9.1 \times 10^{-31} \times \left(3 \times 10^{8}\right)^2}{1.6 \times 10^{-19}} = 0.511 \text{ MeV}$$

$$= 5.11 \times 10^{5} \text{eV}$$

As $K_e << E_r$, therefore the total energy is almost the same as the rest energy 0.511 MeV. **Ans.**

1.5 WAVE PACKET AND UNCERTAINTY PRINCIPLE

According to classical mechanics, the position and momentum of a moving particles can be determined with great accuracy. But according to wave mechanics (de Broglie's Theory) a moving material particle has waves associated with it. Actually a material particle in motion is equivalent to a wave packet rather than a single wave.

A wave packet is a group of waves, each with slightly different frequency and propagation constant. The phase and amplitudes of these waves are so chosen that they interfere constructively over a shell region of space where the particle is likely to be. Outside this region they produce destructive interference, so that the amplitude reduces to zero rapidly.

Actually the square of wave amplitude at a point is related to the probability of finding the particles at that point. So in the region of wave packet there is a high probability of finding the particle, the probability outside it is negligible.

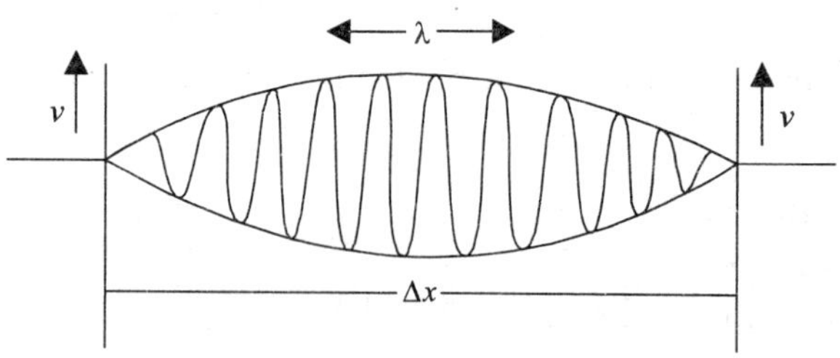

Fig. 1.5

A single particle moving with a uniform velocity is associated with a wave group moving with a group velocity v as shown in Fig. 1.5.

Now, the question is "where is that particle located." The answer is "within the wave packet." The particle has a definite position but the wave packet has a finite spread. The more the number of waves we superimpose, lesser will be the spread of the wave packet. The more precise we try to locate the particle, more shall be the variation in wavelength or momentum. Thus, the momentum and position of a particle cannot simultaneously be measured with any desired accuracy. This is Heisenberg's uncertainty principle.

1.5.1 The Uncertainty Principle

According to the classical mechanics the position and momentum of a moving particle can be determined with great accuracy. When an electron is considered to be a wave (according to wave mechanics), it is, however, not possible to know the exact location of the electron on the wave as it is extending throughout a region of space. A question, then, arises.

"If an electron is exhibiting dual nature (wave and particle), is it possible to know the exact position of the electron in space at same given point."

The answer to this question was given by Heisenberg's as follows:

"It is impossible to determine simultaneously the position and momentum of the electron with any desired accuracy." The above statement is known as *Heisenberg's uncertainty principle.* The principle of uncertainty explains a large number of facts which could not be explained by classical mechanics like non-existance of free electrons in the nucleus*, value of ground state energy of hydrogen atom, zero point energy of harmonic oscillator, finite width of spectral lines *etc.*

1.5.2 Mathematical Relations

(*i*) Momentum and Position Uncertainty

Heisenberg's uncertainty principle can be stated mathematically as

$$\Delta p \Delta x \geq \frac{\hbar}{2} \tag{1.16}$$

where Δp is the uncertainty in the determination of the momentum and Δx, the uncertainty in the determination of position. Eq. (1.16) is known as Heisenberg's equation which can be stated in words as *"the product of uncertainty in the simultaneous measurement of the position and momentum of a particle is equal to or greater than the $\hbar/2$"*. Here

$$\hbar = \frac{h}{2\pi}$$

where h is Planck's constant.

The momentum in Eq. (1.16) should actually be the component of the momentum in the x-direction, and it is more properly written as

$$\Delta p_x \, \Delta x \geq \frac{\hbar}{2} \tag{1.17}$$

Pairs of variables such as p_x and x, in which the uncertainty in one has placed a limit on the accuracy of measurement of the other, are called conjugate variables. There occurs no limitation in the precision of measurement of the components of the momentum along y and z directions (Δp_y and Δp_z) and the position along the x-direction.

* If the electrons are present inside the nucleus, their energy must be of the order of 10 MeV. But the experimental data reveal that no β-particle (electron) in the atom has energy greater than 4MeV. This clearly reveals that electrons do not exist inside the nucleus.

Due to small value of h, this uncertainty is not concerned with natural uncertainty that appears in the macroscopic world, namely, the interaction that appears between the measuring instrument and the magnitude which we are trying to measure.

The problem of uncertainty is a logical consequence of the dual behaviour of matter.

(*ii*) Time-Energy Uncertainty

Just as momentum and displacement could not be determined simultaneously with infinite precision, so the energy E and the time t (which are another set of conjugate variables) could not be determined simultaneously with infinite precision .The more precise we are in our measurement of time, the less precise we are in the determination of energy. A more sophisticated derivation would give, *i.e.,* if the time for which the system remains in a particular energy state is short, the energy will not be well defined.

$$\Delta E \cdot \Delta t \geq \frac{\hbar}{2} \qquad \textbf{(1.18)}$$

Eq. (1.18) applies to complex molecules and to internal molecular energy levels as well. For example, according to Bohr's theory, an electron in an atom goes to an excited state from where it jumps back to the ground state to emit a photon of radiation. If Δt is the time it remains in this excited state, that will give the error in the measurement of time. Then the error in the measurement of energy cannot be less than $\Delta E = \hbar / 2\,\Delta t$. Thus, there will be always an error in the measurement of the energy of the photon, and hence in the values of frequency and wavelength.

Even inside the nucleus, when it gets excited, it remains in the excited state for a short duration of time Δt. Then it returns to the more stable state emitting γ -ray photon. In estimating the energy of the photon obtained during the transition, there will always be a fundamental error in the measurement of energy of the order of

$$\Delta E = \frac{\hbar}{2\Delta t}$$

(*iii*) Angular Momentum and Angular Displacement Uncertainty

One other way this principle can be applied is

$$\Delta J \cdot \Delta\theta \geq \frac{\hbar}{2} \qquad \textbf{(1.19)}$$

where ΔJ is the error in the measurement of the angular momentum and $\Delta\theta$ is the error in the measurement of angular displacement.

Example 1.14: What is the uncertainty in the location of a photon if its wavelength of 5000 Å is known to an accuracy of one part in 10^7 ?

Solution: The momentum of a photon is given by

$$p = \frac{h}{\lambda}$$

Then $|\Delta p| = \dfrac{h}{\lambda^2}\Delta\lambda$

If this momentum is in the x-direction, then the uncertainty in the x-position is

$$\Delta x \geq \frac{h}{|\Delta p|} = \frac{\lambda^2}{\Delta\lambda}$$

Here $\lambda = 5000\text{Å} = 5 \times 10^{-7}\text{m}$

$$\Delta\lambda = \frac{5000}{10^7} = 5 \times 10^{-4}\,\text{Å} = 5 \times 10^{-14}\text{m}$$

$$\therefore \quad \Delta x \approx \frac{\left(5 \times 10^{-7}\right)^2}{5 \times 10^{-14}} = 5\text{m}$$ **Ans.**

Example 1.15: If the life-time of an electronic excited state is 1×10^{-9} s, what is the uncertainty in the energy of the state?

Solution: According to Energy-time uncertainty

$$\Delta E \ \Delta t \approx \frac{\hbar}{2}$$

$$\Rightarrow \quad \Delta E \approx \frac{h}{4\pi}$$

Here $\Delta t = 10^{-9}$ sec and $h = 6.63\times10^{-34}$ J-s

$$\Delta E = \frac{6.63 \times 10^{-34}}{4\pi \times 10^{-9}}\text{ J}$$

$$= \frac{6.63\times10^{-34}}{4\times3.14\times10^{-9}\times1.6\times10^{-19}}$$

$$= 3.3 \times 10^{-7}\text{eV}$$ **Ans.**

Example 1.16: The velocity of (*i*) an electron (*ii*) a proton is 10^6 m/s and the fundamental error in its measurement is 0.01%. Calculate the fundamental error in the measurement of its position. Given, mass of electron = 9.1×10^{-31} kg and mass of proton $= 1.67 \times 10^{-27}$ kg.

Solution: (*i*) Error in the measurement of momentum of electron

$$\Delta p = \Delta(mv) = m\Delta v$$

$$= 9.1\times10^{-31}\times10^{-4}\times10^{6}$$

$$= 9.1 \times 10^{-29}\text{ kg m/s}$$

$\therefore$ Error in the measurement of position

$$\Delta x = \frac{\hbar}{2\times\Delta p} = \frac{1.054\times10^{-34}}{2\times9.1\times10^{-29}}$$

$$= 0.58 \times 10^{-6}\text{ m}$$

(*ii*) Error in the measurement of momentum of proton

$$\Delta p = \Delta(mv) = m\Delta v$$

$$= 1.67\times10^{-27}\times10^{-4}\times10^{6}$$

$$= 1.67 \times 10^{-25}\text{ kg m/s}$$

$\therefore$ Error in the measurement of position

$$\Delta x = \frac{\hbar}{2\times\Delta p} = \frac{1.054\times10^{-34}}{2\times1.67\times10^{-25}}$$

$$= 3.15\times10^{-10}\text{ m}$$ **Ans.**

Example 1.17: Suppose the uncertainty in the momentum of a particle is equal to the particle's momentum. How is the minimum uncertainty in the particle's location is related to its de Broglie wavelength ?

Solution: According to uncertainty principle

$$\Delta x \; \Delta p \geq \frac{\hbar}{2}$$

Given that $\Delta p = p$

∴ Uncertainty in particle's location

$$\Delta x \geq \frac{\hbar}{2\Delta p} = \frac{\hbar}{2p} = \frac{h}{4\pi p}.$$

de Broglie wavelength $\lambda = \dfrac{h}{p}$

$$\therefore \quad \Delta x \geq \frac{\lambda}{4\pi}$$

Thus minimum uncertainty in the particle's location is $\dfrac{\lambda}{4\pi}$. **Ans.**

1.6 COMPTON EFFECT

According to classical theory of X-ray scattering the following conclusions were drawn.

(*i*) The scattered X-ray should possess the same wavelength as the incident one.

(*ii*) The scattering constant (denoted by σ) should have a constant value 0.2 for all types of X-ray scatterings. It means that the scattering constant is independent of the wavelength of incident radiation.

(*iii*) The intensity distribution of X-rays should be symmetrical in the scattered X-rays.

Arthur Holly Compton (September 10, 1892 – March 15, 1962) was an American physicist and Nobel laureate in physics for his discovery of the Compton effect.

When the experimental results of scattering of hard X-rays and γ-rays were analysed, the following points of discrepancy from classical expectation were brought to notice.

(*i*) The scattered radiation was found to have a greater wave length than that of the incident radiation.

(*ii*) The scattering coefficient σ was found to vary with the wavelength of incident radiation. Its value is diminished as the value of incident radiation is decreased.

(*iii*) The distribution of scattered radiation was not symmetrical. Scattering was taking place in the same direction as that of the incident radiation.

In order to explain the above experimental facts about the scattering of X-rays, Compton in 1923 proposed that the phenomenon of scattering might be regarded as an elastic collision of two particles, photon of X-rays and the scatterer. When a photon of energy $h\nu$ collides with an electron, it will lose a part of its energy in the form of kinetic energy to the electron. Thus the scattered photon will have a smaller energy $h\nu'$ and in consequence, a lower frequency or greater wavelength than that of incident photon. *This observed change in frequency or the wavelength of the scattered radiation is known as*

Compton effect. In such a case an electron is also ejected from the scatterer with an energy depending upon its direction. This ejected electron is known as Compton recoil electron.

1.6.1 Theory of Compton Scattering

Suppose the incident *X*-rays consist of photons each of energy ($h\nu$) is much greater than that required to eject a free electron from the target or the scatterer. Let the photon collides with an electron at rest in the target. A part of its energy is imparted to the electron which is ejected with a velocity v in a direction making an angle θ with that of the incident *X*-ray photon (Fig. 1.6).

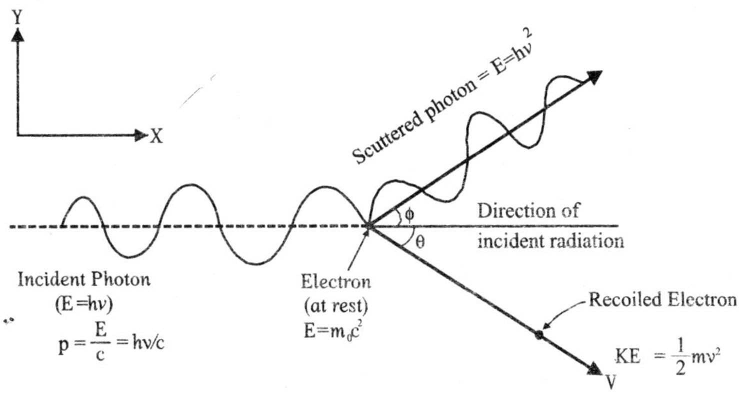

Fig. 1.6 Compton Scattering

The remaining energy is associated with the scattered photon, having lower frequency ν' and moving in a direction making an angle ϕ with that of the incident photon. If m_0 is the rest mass of the electron, its mass, when moving with a velocity v, will be given by the theory of relativity as:

$$m = \frac{m_0}{\left(1 - \frac{v^2}{c^2}\right)^{1/2}} \tag{1.20}$$

where c is the velocity of light.

On applying the principle of conservation of energy, we get

$$h\nu + m_0c^2 = h\nu' + mc^2$$

$$mc^2 = h(\nu - \nu') + m_0c^2 \tag{1.21}$$

Squaring the above equation, we get

$$m^2c^4 = h^2(\nu^2 + \nu'^2 - 2\nu\nu') + 2h(\nu - \nu')m_0c^2 + m^2{}_0c^4 \tag{1.22}$$

On applying the principle of conservation of momentum in the direction of incident photon, we get

$$\frac{h\nu}{c} = \frac{h\nu'}{c}\cos\phi + mv\cos\theta \tag{1.23}$$

and in a direction perpendicular to the direction of the incident photon

$$0 = \frac{h\nu'}{c}\sin\phi - mv\sin\theta \tag{1.24}$$

Eqs. (1.23) and (1.24) on simplification yield

$$mvc\ \cos\theta = h(\nu - \nu'\cos\phi) \tag{1.25}$$

$$mvc\ \sin\theta = h\nu'\sin\phi \tag{1.26}$$

Squaring and adding Eq. (1.25) and (1.26) we get

$$m^2v^2c^2\left(\sin^2\theta + \cos^2\theta\right) = h^2\left(\nu - \nu'\cos\phi\right)^2 + h^2\nu'^2\sin^2\phi$$

$$\Rightarrow \quad m^2v^2c^2 = h^2\left(\nu^2 + \nu'^2\cos^2\phi - 2\nu\nu'\cos\phi + \nu'^2\sin^2\phi\right)$$

$$= h^2\left(\nu^2 + \nu'^2 - 2\nu\nu'\cos\phi\right) \tag{1.27}$$

Subtracting Eq. (1.27) from Eq. (1.28), we get

$$m^2c^2\left(c^2 - v^2\right) = -2\nu\nu'h^2\ (1-\cos\phi) + 2h(\nu-\nu')\ m_0c^2 + m^2{}_0c^4$$

$$\Rightarrow \quad \frac{m^2{}_0c^2}{\left(1-\dfrac{v^2}{c^2}\right)}\left(c^2 - v^2\right) = m^2{}_0c^4 + 2h(\nu-\nu')\ m_0c^2 - 2\nu\nu'h^2(1-\cos\phi)$$

$$\Rightarrow \quad 2h(\nu-\nu')\ m_0c^2 = 2\nu\nu'h^2(1-\cos\phi)$$

$$\Rightarrow \quad \frac{\nu-\nu'}{\nu\nu'} = \frac{h(1-\cos\phi)}{m_0c^2}$$

$$\Rightarrow \quad \frac{1}{\nu'} - \frac{1}{\nu} = \frac{h}{m_0c^2}(1-\cos\phi)$$

$$\Rightarrow \quad \frac{c}{\nu'} - \frac{c}{\nu} = \frac{h}{m_0c}(1-\cos\phi) \tag{1.28}$$

If λ and λ' are the wavelengths corresponding to the frequencies ν and ν' respectively, then Eq. (1.28) may be written as

$$(\lambda' - \lambda) = \Delta\lambda = \frac{h}{m_0c}(1-\cos\phi) = 0.0242\,(1-\cos\phi)\ \text{Å} \tag{1.29}$$

From Eq. (1.29), it follows that the increase in wavelength $\Delta\lambda$ is independent of the wavelength of incident radiation and the nature of the scattering substance, but depends upon the angle of scattering.

The quantity $\dfrac{h}{m_0c} = 0.0242$Å is the maximum wavelength shift and is called the Compton wavelength. The ratio $\dfrac{\Delta\lambda_{max}}{\lambda}$ is too small to be detected for longer wavelength of incident photons in the visible region. Hence, Compton effect is observable only for photons whose wavelength is very small of the order of 1Å. This is the reason why *x*-rays are used for observing Compton effect and not the visible or UV light.

$\because \qquad \cos\phi = 1 - 2\sin^2\dfrac{\phi}{2}$

$\therefore \qquad$ Eq. (1.29) becomes

$$(\lambda' - \lambda) = \Delta\lambda = \frac{2h}{m_0 c}\sin^2\frac{\phi}{2} \tag{1.30}$$

The following three cases may arise :

(*i*) Where the scattering angle is zero, *i.e.*, $\phi = 0$, then

$$\sin\frac{\phi}{2} = \sin 0° = 0$$

$$\therefore \qquad \lambda' = \lambda \tag{1.31}$$

Thus there is no scattering.

(*ii*) When $\phi = 90°$, $\sin^2\dfrac{\phi}{2} = \dfrac{1}{2}$

$$\therefore \qquad \lambda' - \lambda = \frac{h}{m_0 c} = \text{Compton wavelength} \tag{1.32}$$

As h, m_0 and c are constants, Eq. (1.32) becomes

$$(\lambda' - \lambda) = \Delta\lambda = \frac{h}{m_0 c} = \text{constant} \tag{1.33}$$

(*iii*) When $\phi = 180°$, $\sin^2\dfrac{\phi}{2} = 1$, then

$$\lambda' - \lambda = \frac{2h}{m_0 c} \tag{1.34}$$

1.6.2 Angle of Scattering of the Electron (Direction of Recoiled Electron)

Dividing Eq. (1.26) by Eq. (1.25), we get

$$\tan\theta = \frac{v'\sin\phi}{v - v'\cos\phi} = \frac{\sin\phi}{\dfrac{v}{v'} - \cos\phi}$$

$$= \frac{\sin\phi}{\dfrac{\lambda'}{\lambda} - \cos\phi} = \frac{\sin\phi}{1 + \dfrac{h(1-\cos\phi)}{m_0 c\lambda} - \cos\phi}$$

$$= \frac{\sin\phi}{(1-\cos\phi)(1+\alpha)} \tag{1.35}$$

$$\text{where} \quad \alpha = \frac{h}{m_0 c\lambda} = \frac{hv}{m_0 c^2} \tag{1.36}$$

Simplifying further we get

$$\tan\theta = \frac{2\sin\frac{\phi}{2}\cos\frac{\phi}{2}}{2\sin^2\frac{\phi}{2}(1+\alpha)} = \frac{\cot\frac{\phi}{2}}{(1+\alpha)} \tag{1.37}$$

Thus, when $\phi = 0$, $\tan\theta = \infty$

$$\therefore \quad \theta = \frac{\pi}{2} \tag{1.38}$$

When $\phi = \frac{\pi}{2}$, we get

$$\tan\theta = \frac{1}{1+\alpha} \tag{1.39}$$

When $\phi = \pi$, we get $\tan\theta = 0$

$$\therefore \quad \theta = 0 \tag{1.40}$$

1.6.3 Kinetic Energy of Recoiled Electron

The kinetic energy imparted to the electron by incident photon is given by

$$K = (m - m_0)c^2 = h(\nu - \nu')$$

$$= h\nu\left(1 - \frac{\nu'}{\nu}\right) = h\nu\left(1 - \frac{\lambda}{\lambda'}\right)$$

$$= h\nu\frac{\lambda' - \lambda}{\lambda'} = h\nu\frac{d\lambda}{\lambda + d\lambda} = \frac{h\nu\left(\frac{d\lambda}{\lambda}\right)}{1+\left(\frac{d\lambda}{\lambda}\right)}$$

Putting the value of $\left(\frac{d\lambda}{\lambda}\right)$ from Eq. (1.29) we get

$$K = \frac{(h\nu)\frac{h}{m_0c\lambda}(1-\cos\phi)}{1+\frac{h}{m_0c\lambda}(1-\cos\phi)}$$

$$\Rightarrow \quad K = \frac{h\nu\alpha(1-\cos\phi)}{1+\alpha(1-\cos\phi)} \tag{1.41}$$

where $\alpha = \frac{h\nu}{m_0c^2}$

The fraction of incident energy imparted to electron in the form of kinetic energy is

$$\frac{K}{h\nu} = \frac{\alpha(1-\cos\phi)}{1+\alpha(1-\cos\phi)} \tag{1.42}$$

In case of backward scattering, the angle of scattering is $\phi = 180°$ i.e. maximum energy is imparted to the electron.

$$K_{max} = \frac{2h\nu\alpha}{1+2\alpha} \tag{1.43}$$

The kinetic energy of the electrons can also be expressed in terms of the angle of scattering of the electron. From Eq. (1.42)

$$K = h\nu \cdot \frac{\alpha.2\sin^2(\phi/2)}{1+\alpha.2\sin^2(\phi/2)}$$

$$= \frac{h\nu.2\alpha}{\text{cosec}^2(\phi/2)+2\alpha}$$

$$= \frac{h\nu.2\alpha}{1+\cot^2(\phi/2)+2\alpha} \qquad \left(\because 1+\cot^2(\phi/2) = \text{cosec}^2\left(\frac{\phi}{2}\right)\right)$$

But from Eq. (1.37), $\cot\frac{\phi}{2} = (1+\alpha)\tan\theta$

$$\therefore \quad K = \frac{h\nu.2\alpha}{(1+\alpha)^2\tan^2\theta + (1+2\alpha)}$$

$$= h\nu\,\frac{2\alpha.\cos^2\theta}{(1+\alpha)^2\sin^2\theta + (1+2\alpha)\cos^2\theta}$$

$$= h\nu\frac{2\alpha.\cos^2\theta}{(1+\alpha)^2 - \alpha^2\cos^2\theta} \tag{1.44}$$

From Eq. (1.41), we can easily see that th e energy of the recoiled electron will be $K = 0$, when $\phi = 0$, $E = \frac{h\nu\alpha}{1+\alpha}$, when $\theta = \frac{\pi}{2}$; $E = h\nu\frac{2\alpha}{1+2\alpha}$, when $\theta = \pi$.

1.6.4 Frequency of the Scattered Photon

The frequency of the scattered photon is given by

$$\nu = \frac{c}{\lambda'} = \frac{c}{\lambda + \frac{h}{m_0 c}(1-\cos\phi)} = \frac{c/\lambda}{1+\frac{h}{m_0 c\lambda}(1-\cos\phi)}$$

$$= \frac{\nu}{1+\frac{h\nu}{m_0 c^2}(1-\cos\phi)}$$

$$= \frac{\nu}{1+\alpha(1-\cos\phi)} \tag{1.45}$$

where $\alpha = \frac{h\nu}{m_0 c^2} = \frac{h}{m_0 c\lambda}$

Thus when $\phi = 0$, we get $\nu' = \nu$ **(1.46)**

When $\phi = \frac{\pi}{2}$, we get $\nu' = \frac{\nu}{1+\alpha}$ **(1.47)**

and when $\phi = \pi$, we get $\nu' = \frac{\nu}{1+2\alpha}$ **(1.48)**

From these equations it is inferred that the frequency of scattered photon is always less than that of the incident photon except when $\phi = 0$.

The energy of scattered photon is give by

$$h\nu' = \frac{h\nu}{1+\alpha(1-\cos\phi)} \tag{1.49}$$

1.6.5 Experimental Demonstration of Compton Effect

The schematic representation of the experimental set up used by Compton to study Compton effect is given in Fig. 1.7. A beam of monochromatic *X*-rays from Molybdenum target of known wavelength (λ) is made incident on a graphite scatterer.

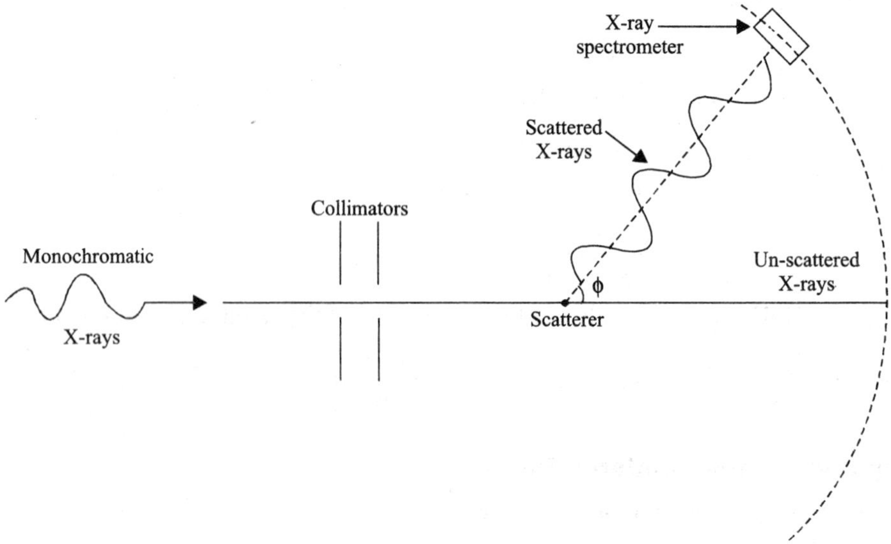

Fig. 1.7 Experimental verification of compton effect

The distribution of intensity with wavelength of the monochromatic *X*-rays scattered at various angles is measured by *X*-ray spectrometer. The intensity distribution versus wavelength curves for various scattering angles are shown in Fig. 1.7. Obviously, the curves have two peaks, one corresponding to modified X-rays and the other corresponding to unmodified radiations. The difference between two peaks on wavelength axis gives the Compton effect or shift.

From the various curves it is obvious that greater is the scattering angle, greater is the Compton shift in accordance with Eq. (1.29). At $\phi = 90°$, $\Delta\lambda = 0.024\,\text{Å}$ is in good agreement with the value obtained from Compton formula given by Eq. (1.29).

Compton's work provided rather convincing evidence that a photon can undergo particle like collision with both atoms and unbound electrons. Later, studies of the recoil electrons and their energies added further confirmation to the prediction of the theory.

1.6.6 Comparison of Compton Effect and Photoelectric Effect

(*i*) In Compton effect, a photon cannot transfer all of its energy to a free electron, whereas in photoelectric effect, an electron absorbs the incident photon completely before it leaves the photometal surface. Let us consider a Compton collision between a photon and free electron (initially at rest). The incident photon has an energy $h\nu$ and a momentum $\frac{h\nu}{c}$. Let us suppose that this photon transfers whole of its energy to the electron which acquires a velocity *v*. From the law of conservation of energy and momentum, we have

$$h\nu = \frac{1}{2}mv^2$$

and $$\frac{h\nu}{c} = mv$$

Eliminating $h\nu$ from these equations, one obtains.

$$v = 2c$$

But *v* cannot exeed *c* (velocity of light) and hence a free electron cannot absorb all the energy of a photon.

(*ii*) Photoelectric effect can be explained by mere quantum considerations, whereas one has to use relativistic considerations in applying laws of conservation of momentum and energy in case of Compton sccatterig, with quantum theory. Obviously, we can say that Compton effect is relativistic photoelectric effect.

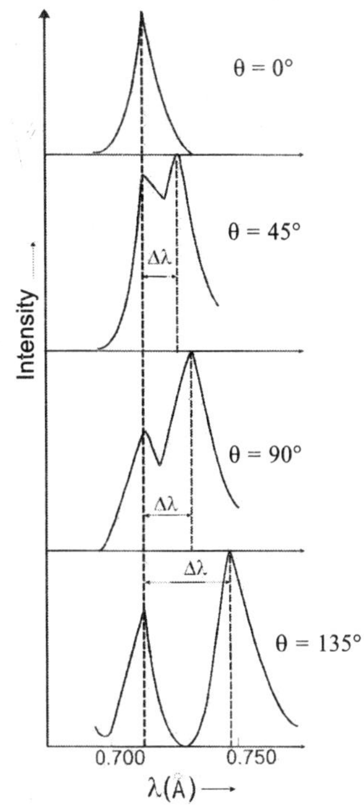

Fig. 1.8 Compton Shift *

Example 1.18: A photon of wavelength 3Å suffers Compton scattering by a free electron originally at rest. If the angle of scattering is 90°, what is the Compton wavelength and kinetic energy of the recoil electron ?

Solution: Compton shift is given by

$$\Delta\lambda = \frac{h}{m_0 c}(1 - \cos\phi)$$

Here $h = 6.62 \times 10^{-34}$ J-s, $m_0 = 9.1 \times 10^{-31}$ kg,

$c = 3 \times 10^8$ m/s and $\phi = \frac{\pi}{2}$

$$\therefore \quad \Delta\lambda = \frac{6.62 \times 10^{-34}}{9.1 \times 10^{-31} \times 3 \times 10^8}(1 - 0)$$

$$= 0.0242 \text{ Å}$$

K.E. of recoiled electron

$$K = hc\left(\frac{\lambda' - \lambda}{\lambda\lambda'}\right) = \frac{hc\Delta\lambda}{\lambda(\lambda + \Delta\lambda)}$$

Here $\lambda = 3\text{Å}, \ \Delta\lambda = 0.0242\text{Å}$

$$\therefore \quad K = \frac{6.62 \times 10^{-34} \times 3 \times 10^8 \times 0.0242 \times 10^{-10}}{3 \times 10^{-10} \times 3.0242 \times 10^{-10}} \text{ Joule}$$

* The figure has been adapted from A.H.Compton's paper on "The spectrum of scattered X-ray" Physical Review , 1923, 22, 409.

$$= \frac{6.62 \times 3 \times 0.0242 \times 10^{-16}}{3 \times 3.0242 \times 1.6 \times 10^{-19}} \text{eV}$$

$$= 33.1 \text{ eV}$$ **Ans.**

Example 1.19: *X*-rays of wavelength 0.7080 Å are scattered from a carbon block through an angle of 90° and are analysed with a calcite crystal, the interplanar distance of whose reflecting planes in 3.13Å. Determine the angular separation, in the first order, between the modified and unmodified rays.

Solution: Wavelength of the modified rays is

$$\lambda' = \lambda + \frac{h}{m_0 c}(1 - \cos\phi)$$

Here $\lambda = 0.708\text{Å}$, $h = 6.62 \times 10^{-34}$ J-s, $m_0 = 9.1 \times 10^{-31}$ kg.

$$c = 3 \times 10^8 \text{ m/s}, \ \phi = \frac{\pi}{2}$$

$$\therefore \quad \lambda' = 0.708 + \frac{6.62 \times 10^{-34} \times 10^{10}}{9.1 \times 10^{-31} \times 3 \times 10^8} \times (1 - 0)$$

$$= 0.7323\text{Å}$$

Let θ and θ′ be the angles of Bragg reflections corresponding to the wavelengths λ and λ′. Then, for n = 1 (First order)

$$\sin\theta = \frac{\lambda}{2d} \text{ and } \sin\theta' = \frac{\lambda'}{2d}$$

Here d = 3.313 Å

$$\therefore \quad \sin\theta = \frac{0.708}{2 \times 3.13} = 0.1131$$

$$\therefore \quad \theta = \sin^{-1}(0.1131) = 6.5°$$

$$\therefore \quad \sin\theta' = \frac{0.7323}{2 \times 3.13} = 0.1170$$

$$\therefore \quad \theta' = \sin^{-1}(0.1170) = 6.72°$$

$\therefore$ Angular separation = $\theta' - \theta = 0.22°$ **Ans.**

Example 1.20: Monochromatic *X*-rays of wavelength $\lambda = 0.124$ Å are scattered from a carbon block. Determine the wavelength of the *X*-rays scattered through an angle such that the recoil electron has the maximum kinetic energy.

Solution: The electron has maximum kinetic energy when radiation is scattered through 180°.

$$\therefore \quad d\lambda = \frac{h}{m_0 c}(1 - \cos\theta)$$

Here $h = 6.62 \times 10^{-34}$ J-s, $m_0 = 9.1 \times 10^{-31}$ kg.

$c = 3 \times 10^8$ m/s and θ = 180°

$$\therefore \quad d\lambda = \frac{6.62 \times 10^{-34} \times 10^{10}}{9.1 \times 10^{-31} \times 3 \times 10^8} \times (1 - \cos 180°) = 0.0485 \text{ Å}$$

$$\therefore \quad \lambda' = \lambda + d\lambda = 0.124 + 0.0485 = 0.173 \text{ Å}$$ **Ans.**

Example 1.21: An *X*-ray of wavelength 0.3 Å undergoes a 60° Compton scattering. Find the wavelength of the scattered photon and the energy of the electron after scattering.

Solution: The wavelength of scattered photon is given by

$$\lambda' = \lambda + \frac{h}{m_0 c}(1-\cos\phi)$$

$$= 0.3 + \frac{6.62\times10^{-34}\times10^{10}}{9.1\times10^{-31}\times3\times10^{8}}\times(1-\cos 60°)$$

$$= 0.3121 \text{ Å}$$ **Ans.**

The energy of the electron after scattering

$$E = h\nu - h\nu' = hc\left[\frac{1}{\lambda}-\frac{1}{\lambda'}\right]$$

$$= \frac{6.62\times10^{-34}\times3\times10^{8}}{1.6\times10^{-19}}\times\left[\frac{1}{3\times10^{-11}}-\frac{1}{3.121\times10^{-11}}\right]$$

$$= \frac{6.62\times3\times0.121\times10^{4}}{1.6\times3\times3.121}\text{eV}$$

$$= 1.6\times10^{3}\text{eV} = 1.6 \text{ KeV}$$ **Ans.**

Example 1.22: *X*-rays with $\lambda = 1$ Å are scattered from a carbon block. The scattered radiation is viewed at 90° to the indcident beam.

(*i*) What is Compton wavelength?

(*ii*) What kinetic energy is imparted to the recoiling electron? **(RTU B. Tech. I Sem. 2009)**

Solution: (*i*) The Compton wavelength is given by

$$\lambda_c = \frac{h}{m_0 c}$$

Wavelength shift is given by

$$d\lambda = \frac{h}{m_0 c}(1-\cos\phi)$$

Here $\phi = 90°$

$$d\lambda = \frac{h}{m_0 c} = \lambda_c$$

Here $h = 6.62\times10^{-34}$ J-s, $m_0 = 9.1\times10^{-31}$kg.

and $c = 3\times10^{8}$ m/s

$$\therefore \quad \lambda_c = \frac{6.62\times10^{-34}}{9.1\times10^{-31}\times3\times10^{8}}$$

$$= 0.02426 \text{ Å}$$

(*ii*) The kinetic energy of the recoil electron is given by

$$K = hc\left[\frac{1}{\lambda}-\frac{1}{\lambda'}\right] = \frac{hc(d\lambda)}{\lambda(\lambda+d\lambda)}$$

Here $d\lambda = 0.02426$ Å, $\lambda = 1$ Å

$$\therefore \quad K = \frac{6.62 \times 10^{-34} \times 3 \times 10^{8} \times 0.02426 \times 10^{-10}}{10^{-10} \times 1.02426 \times 10^{-10} \times 1.6 \times 10^{-19}} \text{eV}$$

$$= \frac{6.62 \times 3 \times 0.02426 \times 10^{2}}{0.02426 \times 1.6}$$

$$= 294.4 \text{ eV}$$ **Ans.**

Example 1.23: A photon recoils back after striking an electron at rest. What is Compton shift for the photon?

Solution: The Compton shift is given by

$$d\lambda = \frac{h}{m_0 c}(1 - \cos\phi)$$

Here $h = 6.62 \times 10^{-34}$ J-s, $m_0 = 9.1 \times 10^{-31}$ kg.

$c = 3 \times 10^{8}$ m/s, $\phi = 180°$

$$\therefore \quad d\lambda = \frac{6.62 \times 10^{-34} \times 10^{10}}{9.1 \times 10^{-31} \times 3 \times 10^{9}}(1 - \cos 180°)\text{Å}$$

$$= 4.85 \times 10^{-2} \text{Å}$$ **Ans.**

Example 1.24: An *X*-ray photon is found to have its wavelength doubled on being scattered through 180°. Find the wavelength and energy of the incident photon. What is the energy of the recoiled electron? (Given Compton wavelength $\lambda_c = 0.02426$ Å)

Solution: The wavelength shift is given by

$$d\lambda = \lambda' - \lambda = \frac{h}{m_0 c}(1 - \cos\phi) = 2\lambda - \lambda = \lambda$$

$\therefore$ Wavelength of incident photon

$$\lambda = \frac{h}{m_0 c}(1 - \cos\phi)$$

$$\Rightarrow \quad \lambda = \lambda_c (1 - \cos\phi)$$

Here $\lambda_c = 0.02426$ Å and $\phi = 180°$

$$\therefore \quad \lambda = 0.02426 \times (1 - \cos 180°)$$

$$\Rightarrow \quad \lambda = 0.04852 \text{ Å}$$ **Ans.**

Energy of incident photon

$$E_{ph} = h\nu = \frac{hc}{\lambda} = \frac{6.62 \times 10^{-34} \times 3 \times 10^{8}}{4.852 \times 10^{-12} \times 1.9 \times 10^{-19}}$$

$$= 2.15 \times 10^{5} \text{eV}$$

$$= 0.215 \text{MeV}$$ **Ans.**

In this case maximum kinetic energy is imparted to the recoiling electron

$$K_{max} = \frac{hc(d\lambda)}{\lambda(\lambda + d\lambda)} = \frac{hc\lambda}{2\lambda^2} = \frac{hc}{2\lambda} = 0.1075 \text{ MeV}$$ **Ans.**

Example 1.25: In Compton experiment, the wavelength of the *X*-ray radiation scattered at an angle of 45° is 0.022 Å. Calculate the wavelength of indident *X*-rays.

Solution: The Compton shift is given by

$$d\lambda = \frac{h}{m_0 c}(1-\cos\phi)$$

Here $\phi = 45°$

$$\therefore \quad d\lambda = \frac{6.62\times10^{-34}}{9.1\times10^{-31}\times3\times10^{8}}\times(1-\cos 45°)$$

Wavelength of incident *X*-rays

$$\lambda = \lambda' - d\lambda = 0.022 - 0.0071$$

$$= 0.0149 \text{ Å}$$ **Ans.**

Example 1.26: (*a*) A beam of *X*-rays of wavelength 0.1 nm is incident on a target in a Compton scattering experiment. The *X*-rays scattered by the target through a right angle are detected. Calculate the Compton shift, the wavelength of scattered radiation and the fractional change in wavelength in the scattering.

(*b*) Repeat the calculations for the right angle scattering for electrons with γ-ray photon of energies 0.492 MeV.

Solution: (*a*) Compton shift is given by

$$d\lambda = \frac{h}{m_0 c}(1-\cos\phi)$$

Here $\phi = 90°$

$$\therefore \quad d\lambda = \frac{6.62\times10^{-34}\times10^{9}}{9.1\times10^{-31}\times3\times10^{8}}\times(1-\cos 90°)$$

$$= 0.00243 \text{ nm}$$ **Ans.**

The wavelength of scattered radiation

$$\lambda' = \lambda + d\lambda = 0.1 + 0.00243$$

$$= 0.10243 \text{ nm}$$ **Ans.**

Fractional change is wavelength

$$\frac{\lambda'-\lambda}{\lambda} = \frac{0.00243}{0.1} = 0.0243$$ **Ans.**

(*b*) The Compton shift does not depend on the wavelength of the incident radiation. Hence

$$\lambda' - \lambda = d\lambda = 0.00243 \text{ nm}$$ **Ans.**

The wavelength of incident photon is given by

$$\lambda = \frac{hc}{E}$$

Here $E = 0.492$ MeV

$$= 0.492\times1.61\times10^{-13}\text{ J}$$

$$\therefore \quad \lambda = \frac{6.62\times10^{-34}\times3\times10^{8}}{0.492\times1.6\times10^{-13}}$$

$$= 0.002524 \text{ nm}$$

Wavelength of scattered photon

$$\lambda' = \lambda + d\lambda = 0.002524 + 0.00243$$

$$= 0.00495 \text{ nm}$$ **Ans.**

Fractional change in wavelength

$$\frac{\lambda' - \lambda}{\lambda} = \frac{0.00243}{0.002524} = 0.9628$$ **Ans.**

Example 1.27: Show that the minimum value of energy which a photon must have so that it may transfer half of its energy to an electron at rest is about 256 KeV. **(Raj. B.E. I 2000)**

Solution: Energy of the scattered photon $\frac{hc}{\lambda'} = \frac{hc}{2\lambda}$

$\therefore \quad \lambda' = 2\lambda$

Compton shift $d\lambda = \lambda' - \lambda = 2\lambda - \lambda = \lambda = \frac{h}{m_0 c}(1 - \cos 180°)$

For minimum energy of incident photon, λ should be maximum i.e. $\phi = 180°$

$$\therefore \quad \lambda_{max} = \frac{h}{m_0 c}(1 - \cos 180°) = \frac{2h}{m_0 c}$$

$$= \frac{2 \times 6.62 \times 10^{-34}}{9.1 \times 10^{-31} \times 3 \times 10^{8}}$$

$$= 0.0484 \times 10^{-10} \text{ m}$$

$\therefore$ Minimum energy of incident photon

$$E_{min} = \frac{hc}{\lambda_{max}} = \frac{6.62 \times 10^{-34} \times 3 \times 10^{8}}{0.0484 \times 10^{-10} \times 1.6 \times 10^{-19}}$$

$$= 256 \times 10^{3} \text{ eV}$$

$$= 256 \text{ KeV}$$ **Hence proved**

Example 1.28: Calculate the maximum percentage change in wavelength due to compton scattering for incident photons of wavelength 1 Å and 10 Å. What inference do you draw from the results? **[RTU B. Tech. II Sem. 2010]**

Solution: Wavelength shift is given by

$$\Delta\lambda = \frac{h}{m_0 c}(1 - \cos\phi)$$

Maximum change in λ will occur when $\phi = 180°$

$$\therefore \quad \Delta\lambda = \frac{h}{m_0 c}(1 - \cos 180°) = \frac{2h}{m_0 c}$$

Percentage change in λ is given by

$$\frac{\Delta\lambda}{\lambda} \times 100 = \frac{2h}{m_0 \lambda c} \times 100$$

For $\lambda = 1$ Å

$$\frac{\Delta\lambda}{\lambda} \times 100 = \frac{2 \times 6.62 \times 10^{-34} \times 100}{9.1 \times 10^{-31} \times 10^{-10} \times 3 \times 10^{8}}$$

$$= 4.85\%$$

When $\lambda = 10\text{ Å} = 10^{-9}\text{m}$

$$\frac{\Delta\lambda}{\lambda} \times 100 = \frac{2 \times 6.62 \times 10^{-34} \times 100}{9.1 \times 10^{-31} \times 10^{-9} \times 3 \times 10^{8}}$$

$$= 0.485\%$$ **Ans.**

The percentage change in wavelength is negligibly small for larger wavelengths *i.e.,* compton effect can be observed only for photon whose wavelength is small of the order of 1 Å.

Example 1.29: *X*-rays of wavelength 0.1 Å are scattered by a block of a material. Find the wavelength of the *X*-ray scattered along a direction inclined at 90° with the incident direction. Also calculate the angle of scattering of the electron and its kinetic energy.

Solution: Wavelength of the scattered *X*-ray

$$\lambda' = \lambda + \frac{h}{m_0 c}(1 - \cos\phi)$$

Here $\lambda = 0.1\text{ Å}$, $h = 6.62 \times 10^{-34}$ J-s

$$m_0 = 9.1 \times 10^{-31}\text{kg},\ c = 3 \times 10^{8}\ \text{m/s},\ \phi = \frac{\pi}{2}$$

$$\therefore \quad \lambda' = 0.1 + \frac{6.62 \times 10^{-34}}{9.1 \times 10^{-31} \times 3 \times 10^{8}}(1 - \cos 90°)$$

$$= 0.1 + 0.0242 = 0.1242\text{ Å}$$

Kinetic energy of scattered electron

$$E = hc\left[\frac{1}{\lambda} - \frac{1}{\lambda'}\right]$$

$$= \frac{6.62 \times 10^{-34} \times 3 \times 10^{8}}{1.6 \times 10^{-19}} \times \left[\frac{1}{10^{-11}} - \frac{1}{1.242 \times 10^{-11}}\right]$$

$$= 2.41 \times 10^{4}\text{eV}$$ **Ans.**

Angle of scattering of electron is given by

$$\tan\theta = \frac{\cot(\phi/2)}{1 + \dfrac{h}{m_0 \lambda c}} = \frac{\cot 45°}{1 + \dfrac{6.62 \times 10^{-34}}{9.1 \times 10^{-31} \times 10^{-11} \times 3 \times 10^{8}}}$$

$$= 0.8052$$

$$\theta = 38.84°$$ **Ans.**

Example 1.30: Calculate the value of the least and maximum values of wavelength obtained when γ-rays of energy 0.5 MeV are scattered by matter.

Solution: Wavelength of the incident photon

$$\lambda = \frac{hc}{E} = \frac{6.62 \times 10^{-34} \times 3 \times 10^{8}}{0.5 \times 1.6 \times 10^{-13}}$$

$$= 0.02475\text{ Å}$$

Wavelength of scattered photon

$$\lambda' = \lambda + \Delta\lambda = \lambda + \frac{h}{m_0 c}(1-\cos\phi)$$

Minimum wavelength will be when $\phi = 0$

$\therefore \quad \lambda' = \lambda == 0.02475$ Å **Ans.**

Maximum wavelegth will be when $\phi = 180°$

$$\lambda' = \lambda + \frac{h}{m_0 c}(1-\cos 180°) = \lambda + \frac{2h}{m_0 c}$$

$$= 0.02475 + \frac{2\times 6.62\times 10^{-34}\times 10^{10}}{9.1\times 10^{-31}\times 3\times 10^{8}}$$

$= 0.02475 + 0.0484$

$= 0.07315$ Å **Ans.**

Example 1.31: In Compton scattering, the energy of an incident X-ray photon is 150 KeV and that of scattered photon is 130 KeV. Determine the angle of scattering.

[Raj. B.E. I 2002, RTU B. Tech. Back, 2011]

Solution: If E and E' are energies of incident and scattered photons, then

$$\frac{1}{E'} - \frac{1}{E} = \frac{1}{m_0 c^2}(1-\cos\phi)$$

or
$$(1-\cos\phi) = \frac{m_0 c^2 (E-E')}{EE'}$$

$$= \frac{9.1\times 10^{-31}\times (3\times 10^8)^2 \times (150-130)\times 1.6\times 10^{-16}}{150\times 130\times (1.6\times 10^{-16})^2}$$

$= 0.525$

or $\quad \cos\phi = 0.475$

$\phi = 61.64°$ **Ans.**

Example 1.32: X-ray photon of wavelength 0.3 Å is scattered through an angle of 60° by a free electron. Find wavelength of scattered photon and recoil energy of electron.

[Raj. B.Tech. I & II sem. 2008]

Solution: Given $\lambda = 0.3$ Å and $\phi = 60°$

The compton shift is

$$\lambda' - \lambda = 0.0242(1-\cos\phi)$$

$$\lambda' = \lambda + 0.0242(1-\cos\phi)$$

$$= 0.3 + 0.0242\left(\frac{1}{2}\right)$$

$$\lambda' = 0.3 + 0.0121$$

$\lambda' = 0.3121$ Å **Ans.**

The kinetic energy of recoil electron is

$$K = h\nu - h\nu^1 = h(\nu - \nu^1)$$

$$= h\left(\frac{c}{\lambda} - \frac{c}{\lambda^1}\right)$$

$$= hc\left(\frac{\lambda^1 - \lambda}{\lambda\,\lambda^1}\right)$$

$$= \frac{6.625\times10^{-34}\times3\times10^{8}(0.3121-0.3)}{0.3\times10^{-10}\times0.3121\times10^{-10}}$$

$$= 2.5610^{-16}\text{ J}$$

$$= \frac{2.56\times10^{-16}}{1.6\times10^{-19}}\text{eV}$$

$$K = 1.6\times10^{3}\text{eV} = 1.6\text{ KeV}$$ **Ans.**

1.7 CONCEPT OF WAVE FUNCTION

We know that every moving material particle has a wave associated with it and its wavelength is given by

$$\lambda = \frac{h}{mv} = \frac{h}{p}$$

where m is mass of particle and v is velocity of moving particle.

The velocity of associated wave is known as the phase velocity (v_p) and is given by

$$v_p = \frac{\omega}{k} = \frac{2\pi\nu}{2\pi/\lambda} = \nu\lambda \qquad (1.50)$$

$\because$ $$E = h\nu = mc^2$$

$\therefore$ $$v_p = \frac{mc^2}{h}\cdot\frac{h}{mv}$$

$$v_p = \frac{c^2}{v} \qquad (1.51)$$

Since $v < c$ this implies that the phase velocity of matter wave is greater than the velocity of light as well as the velocity of moving particle. It is impossible. Therefore a wave is not associated with moving particle but the group of waves or wave packet is associated with moving particle. The group velocity is given by

$$v_g = \frac{d\omega}{dk} \qquad (1.52)$$

Now $E = \hbar\omega$ and $p = \hbar k$

$$v_g = \left(\frac{d\omega}{dE}\right)\left(\frac{dE}{dp}\right)\left(\frac{dp}{dk}\right) = \frac{1}{\hbar}\left(\frac{dE}{dp}\right)\hbar$$

$$v_g = \frac{dE}{dp} = \frac{d}{dp}\left(\frac{p^2}{2m}\right) = \frac{p}{m} = v \qquad (1.53)$$

For a relativistic particle

$$k = \frac{2\pi}{\lambda} = \frac{2\pi}{h} p = \frac{2\pi m_0 v}{h\sqrt{1 - v^2/c^2}}$$

and $$\omega = 2\pi \nu = \frac{2\pi mc^2}{h} = \frac{2\pi m_0 c^2}{h\sqrt{1 - v^2/c^2}} \quad (1.54)$$

$\therefore$ $$v_g = \frac{d\omega}{dk} = \frac{d\omega/dv}{dk/dv}$$

$\Rightarrow$ $$v_g = \frac{2\pi m_0 v}{h\left(1 - \frac{v^2}{c^2}\right)^{3/2}} \frac{h\left(1 - \frac{v^2}{c^2}\right)^{3/2}}{2\pi m_0}$$

$\Rightarrow$ $$v_g = v \quad (1.55)$$

Hence, the velocity of particle is equal to the group velocity of the wave packet associated that moving particle. Hence in wave mechanics, a particle is not completely localized due to wave nature. It must be represented by a wave that is confined to a small region of space around the particle. Such a wave is called a wave packet as shown in Fig. 1.9. Its amplitude is negligible outside the region of dimension Δx.

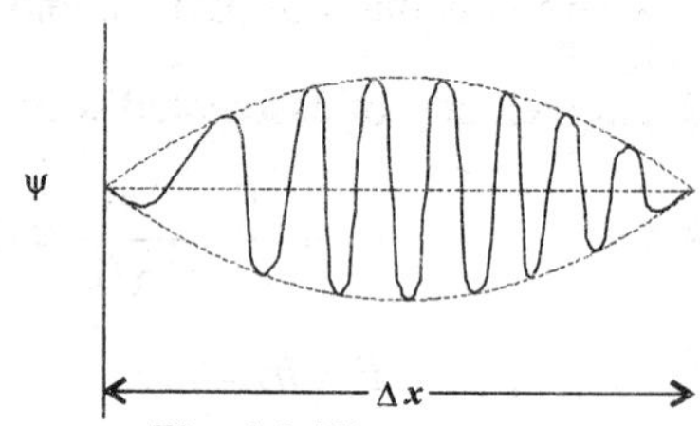

Fig. 1.9 Wave packet

The localized particle will now be represented by a wave function which will be formed by the superposition of simple harmonic functions. These continuous simple harmonic waves will have different wavelengths, *i.e., p* values and different amplitudes. If the amplitude of a particular wave having wavelength λ corresponding to a momentum p is $A(p)$ then the wave function $\psi(x, t)$ will be given by

$$\psi(\vec{x},t) = \int A(p) \exp\left[\frac{i}{\hbar}(px - Et)\right] dp \quad (1.56)$$

According to uncertainty principle, Δx is infinity for a particle with well defined momentum.

Since $\Delta p = \hbar \Delta k = 0$

then $$\Delta x = \frac{\hbar}{\Delta p} \approx \infty$$

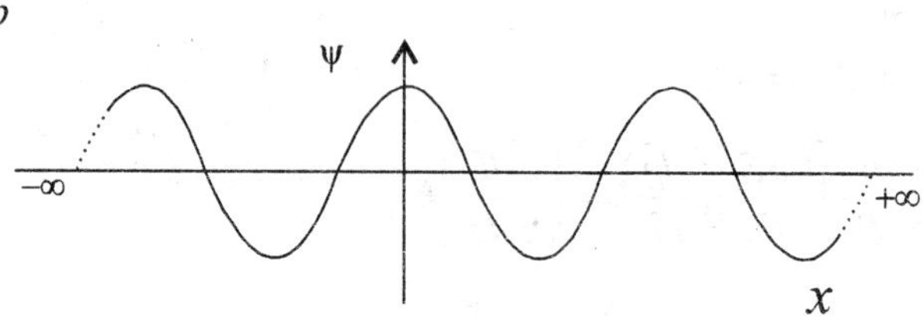

Fig. 1.10 A wave associated with particle of well defined momentum

Hence a particle moving along x-axis with a definite momentum is described by an infinite plane wave as shown in Fig. 1.10, which is given by

$$\psi\left(\vec{x},t\right) = Ae^{i(kx-\omega t)}$$

$$\psi\ (\vec{x},t) = A\ e^{\frac{i}{\hbar}(px-Et)} \qquad (1.57)$$

where k is the wave vector and ω is the angular frequency.

In three dimensions $\psi\ (\vec{r},t) = A\ e^{\frac{i}{\hbar}(\vec{p}.\vec{r}-Et)}$ (1.58)

1.8 WAVE MECHANICS

According to de Broglie's hypothesis, every moving material particle has a wave packet associated with it. These waves are known as matter waves. As we know that in water waves, the quantity that varies periodically is the height of the water surface. In sound waves, it is pressure and in light waves, variable quantities are electric and magnetic fields but what is the quantity which varies periodically in case of matter waves. The quantity whose variations make up the matter waves is called "*Wave function*". Thus the amplitude of matter waves is described by wave function and represented by $\psi(\vec{r},t)$. In 1926 Schrödinger, generalizing the concept of matter waves, discovered the equation of propagation of the wave function representing a given system (particle). The equation is known as the Schrödinger equation and it constitutes the basic element of wave mechanics.

Hence the mechanics related to wave function $\psi(\vec{r},t)$ is known as the wave mechanics. In the development of wave mechanics, there are some basic postulates, which are called the fundamental postulates of quantum mechanics.

1.8.1 Postulates of Wave Mechanics

(*i*) The state of system is completely described by a wave function $\psi(\vec{r},t)$. It is also known as the quantum mechanical state of the system.

(*ii*) Each dynamical variable (observable quantity) relating to the motion of a particle can be represented by an operator. The quantum operators for several physical quantities are shown in following table.

Quantity	Classical definition	Quantum operator
Position	$\vec{r}$	$\hat{\vec{r}} = r$
Momentum	$\vec{p}$	$\hat{\vec{p}} = -i\dfrac{h}{2\pi}\vec{\nabla}$
Angular momentum	$\vec{r} \times \vec{p}$	$\hat{\vec{L}} = -i\dfrac{h}{2\pi}\vec{r} \times \vec{\nabla}$
Kinetic energy	$p^2/2m$	$\hat{K} = -\left(h^2/8\pi^2 m\right)\nabla^2$
Total energy	E	$\hat{E} = i\dfrac{h}{2\pi}\dfrac{\partial}{\partial t}$

Hamiltonian $H = \frac{p^2}{2m} + V(\vec{r})$ $\hat{H} = -(h^2/8\pi^2 m)\nabla^2 + V(\vec{r})$

(*iii*) A linear eigenvalue equation can be always linked with each operator.

Example: The total energy operator is $i\left(\frac{h}{2\pi}\right)\frac{\partial}{\partial t}$. Consider the eigenvalue equation $i\left(\frac{h}{2\pi}\right)\frac{\partial \psi}{\partial t} = E\psi$. Here ψ is said to be an eigenfunction of the operator $i\left(\frac{h}{2\pi}\right)\frac{\partial}{\partial t}$ and E is called the corresponding energy eigenvalue.

(*iv*) In general, when a measurement of a dynamical quantity a is made on a particle for which the wave function is ψ, we get different values of a during different trials. This is in conformity with the uncertainly principle. The most probable value of a is given by

$$< a > = \int_{-\infty}^{\infty} \psi^* \hat{A} \psi \, dV$$

where $\hat{A}$ is the operator associated with the quantity a and ψ^* is the complex conjugate of ψ. The quantity $<a>$ is called the *expectation value* of $\hat{A}$ (that is the value of a obtained in the majority of the trials). The expectation value of momentum and energy may be found by using the corresponding differential operator. Thus

$$< \vec{p} > = \int_{-\infty}^{\infty} \psi^* \left(-\frac{ih}{2\pi}\nabla\right) \psi \, dxdydz$$

$$< E > = \int_{-\infty}^{\infty} \psi^* \left(i\frac{h}{2\pi}\frac{\partial}{\partial t}\right) \psi \, dxdydz$$

Here wave function is normalized, therefore the denominator term $\left(\int_{-\infty}^{\infty} \psi^* \psi \, dv\right)$ has the value 1.

(*v*) A moving particle can be represented by the following equation.

$$\left[-\frac{\hbar^2}{2m}\nabla^2 + V(\vec{r})\right]\psi(\vec{r},t) = E\psi(\vec{r},t)$$

$$\Rightarrow \quad \hat{H}\psi(\vec{r},t) = E\psi(\vec{r},t)$$

where $\widehat{H}$ is the Hamiltonian operator.

This equation is known as the Schrödinger wave equation, which is applicable to all the cases of particle motion and the nature of potential $V(\vec{r})$ decides the solution for a particular case.

1.9 SCHRÖDINGER'S EQUATION

In 1926, Schrödinger proposed a equation to calculate wave function $\psi(\vec{r},t)$ associated with a moving particle. This equation is known as the Schrödinger's equation. It has not been proved or deduced from the fundamental properties of medium. Schrödinger's equation is fundamental equation of wave mechanics in the same sense as the newton's second law of motion in classical mechanics.

The choice of the wave equation, whose solutions give the wave functions, is governed by two main conditions:

(*i*) The wave function must possess the property of superposition, which is the characteristic of all types of waves in general. This requires the wave equation to be linear and homogeneous. With a linear equation the solution can be superposed to produce interference effects and form wavepackets.

(*ii*) To satisfy the requirement that once ψ is known the dynamical state of the system be completely predictable at all other instants, the wave equation, which has to be a differential equation, must be of the first order with respect to time.

1.9.1 Time Dependent Schrödinger's Equation

A wave function that represents a particle travelling in the $+x$ direction with momentum p and total energy E is specified by

$$\psi = A\, e^{2\pi i\left(\frac{x}{\lambda}-\frac{t}{T}\right)} = A\, e^{i(kx-wt)}$$

Since $\quad p = \hbar k,\ k = \dfrac{2\pi}{\lambda}$ and $E = h\nu = \hbar\omega$

Therefore we have

$$\psi = A\, e^{\frac{i}{\hbar}(px-Et)} \tag{1.59}$$

Erwin Rudolf Josef Alexander Schrödinger (12 August 1887 – 4 January 1961) was an Austrian physicist who achieved fame for his contributions to quantum mechanics. In January 1926, Schrödinger published in the Annalen der Physik the paper Quantisation as an Eigenvalue Problem on wave mechanics and what is now known as the Schrödinger equation, for which he received the Nobel Prize in 1933.

Equation (1.59) describes the wave equivalent of a free particle of total energy E and momentum p moving in $+x$ direction. However, we are interested in the motion of a particle subjected to various restrictions. For example, an electron bound to an atom by the electric field of its nucleus.

We have to obtain the fundamental differential equation for ψ, which can be solved for ψ in a given specific situation. Now differentiate (1.59) twice with respect to x. We get

$$\frac{\partial^2\psi}{\partial x^2} = -\frac{p^2}{\hbar^2}\psi$$

$$\Rightarrow \qquad p^2\psi = -\hbar^2\frac{\partial^2\psi}{\partial x^2} \tag{1.60}$$

Differentiating equation (1.59) with respect to t, we get

$$\frac{\partial\psi}{\partial t} = \frac{-iE}{\hbar}\psi$$

$$\Rightarrow \qquad E\psi = i\hbar\frac{\partial\psi}{\partial t} \tag{1.61}$$

When the speed of a particle is small as compared to that of light, the total energy of the particle is a sum of kinetic energy $\left(\dfrac{p^2}{2m}\right)$ and its potential energy (V), where V is a function of position (x) and time (t).

Thus $E = \frac{p^2}{2m} + V(\vec{x},t)$ **(1.62)**

The function V represents the influence of the rest of the universe on the particle.

Multiplying both side of equation (1.62) by ψ gives

$$E\psi = \frac{p^2\psi}{2m} + V(\vec{x},t)\psi$$

Put the value of $p^2\psi$ and $E\psi$ from equations (1.60) and (1.61) respectively. We have

$$i\hbar\frac{\partial\psi}{\partial t} = -\frac{\hbar^2}{2m}\frac{\partial^2\psi}{\partial x^2} + V(\vec{x},t)\psi \quad \textbf{(1.63)}$$

Equation (1.63) is the time-dependent Schrödinger's equation in one dimension.

In three dimensions

$$i\hbar\frac{\partial\psi}{\partial t} = -\frac{\hbar^2}{2m}\left(\frac{\partial^2\psi}{\partial x^2} + \frac{\partial^2\psi}{\partial y^2} + \frac{\partial^2\psi}{\partial z^2}\right) + V(\vec{r},t)\psi$$

$$\Rightarrow \quad i\hbar\frac{\partial\psi}{\partial t} = -\frac{\hbar^2}{2m}\nabla^2\psi + V(\vec{r},t)\psi \quad \textbf{(1.64)}$$

where the particle's potential energy V is a function of $\vec{x},\vec{y},\vec{z}$ and t. The restrictions on the particle's motion affects the potential energy function V. Once V is known, Schrödinger equation may be solved for the wave function $\vec{x}, \vec{y}, \vec{z}$ of the particle and its probability density $|\psi|^2$ may also be determined for a specified values of $\vec{x},\vec{y},\vec{z}$ and t.

1.9.2 Time-Independent Schrödinger's Equation

In most of the situations the potential energy (V) of the particle does not depend on time explicitly and the forces acting on it, hence V varies only with the position of the particle. In this case, Schrödinger equation may be simplified by removing all reference to t.

The one dimensional wave function ψ of a free particle can be written as

$$\psi(\vec{x},t) = A\, e^{\frac{i}{\hbar}px} e^{-\frac{i}{\hbar}Et} \quad \textbf{(1.65)}$$

$$\psi(\vec{x},t) = \psi(\vec{x})e^{-\frac{i}{\hbar}Et}$$

Here ψ is the product of a time dependent function and position dependent function. Substituting equation (1.65) in equation (1.64), we get

$$E\psi(\vec{x})e^{-\frac{i}{\hbar}Et} = \frac{-\hbar^2}{2m}e^{-\frac{i}{\hbar}Et}\frac{\partial^2\psi(\vec{x})}{\partial x^2} + e^{\frac{-i}{\hbar}Et}V(\vec{x})\psi(\vec{x})$$

$$\Rightarrow \qquad E\psi(\vec{x}) = \frac{-\hbar^2}{2m}\frac{\partial^2\psi(x)}{\partial x^2} + V(\vec{x})\psi(\vec{x})$$

$$\Rightarrow \qquad \frac{\partial^2\psi(\vec{x})}{\partial x^2} + \frac{2m}{\hbar^2}\left[E - V(\vec{x})\right]\psi(\vec{x}) = 0 \qquad (1.66)$$

This is time independent form of Schrödinger equation. In three dimensions it is

$$\nabla^2\psi(\vec{r}) + \frac{2m}{\hbar^2}\left[E - V(\vec{r})\right]\psi(\vec{r}) = 0 \qquad (1.67)$$

where $\nabla^2 = \frac{\partial^2}{\partial x^2} + \frac{\partial^2}{\partial y^2} + \frac{\partial^2}{\partial z^2}$

The time Independent (Steady state) Schrödinger equation shows the energy quantization, which appears in wave machanics as a natural element of theory and a universal phenomenon for all stable systems. The discrete values of Energy (E_n) are called eigenvalues and the corresponding wave function ψ_n are called eigen functions. If there are more than one eigen functions corresponding to given energy value E, the state is called degenerate. If there is a single wave function for given energy E, the state is called non-degenerate.

1.9.3 Physical Interpretation of Wave Function $\psi(\vec{r}, t)$

The wave function $\psi(\vec{r},t)$ which is a solution of the Schrödinger wave equation provides a complete description of the behaviour of a particle of mass m situated in a potential field $V(\vec{r},t)$. The wave function has large magnitude where the particle is likely to be and small elsewhere. In statistical terms ψ is regarded as a measure of the probability of finding the particle at a particular position and at a particular time. This statistical explanation of wave function was given by Max born in 1926. Since probability can never be negative, hence the wave function has all information about the particle but itself does not have any direct physical meaning. As we know that the matter waves are not real waves therefore the wave function ψ is in general a complex quantity, but as the probability density is obtained by getting the square of the absolute magnitude of ψ.

$$P(\vec{r},t) = \left|\psi(\vec{r},t)\right|^2$$

$$= \psi^*(\vec{r},t)\,\psi(\vec{r},t) \qquad (1.68)$$

$P(\vec{r},t)$ is the probability density and $P(\vec{r},t)$ *dxdydz* is the probability of finding a particle in the volume element *dxdydz* about the point $\vec{r}$ at a time t. The value of $P(\vec{r},t)$ is therefore given by the product of ψ and its complex conjugate ψ^*.

1.9.4 Normalization of Wave Function $\psi(\vec{r}, t)$

As it is certain that a given particle is somewhere in space, total probability of finding the particle somewhere must be unity. Hence

$$\int_{-\infty}^{\infty}\left|\psi\left(\vec{r},t\right)\right|^2 dxdydz = 1$$

$$\Rightarrow \qquad \int_{-\infty}^{\infty}\psi^*\left(\vec{r},t\right)\psi\left(\vec{r},t\right)dxdydz = 1 \qquad \textbf{(1.69)}$$

If a wave function ψ is chosen to satisfy the above condition, it is said to be *normalized*. For such normalization the coefficient of ψ is adjusted to make the integral equal to unity. The coefficient so obtained must be independent of time, otherwise the normalization condition will be satisfied only at one instant of time.

1.9.5 Orthogonal Wave Functions

If the product of a function $\psi_1(x)$ and the complex conjugate $\psi_2{}^*(x)$ of a function $\psi_2(x)$ vanishes when integrated with respect to x over the interval $a \le x \le b$ that is, if

$$\int_a^b \psi_2{}^*(x)\psi_1(x)dx = 0 \qquad \textbf{(1.70)}$$

then $\psi_1(x)$ and $\psi_2(x)$ are said to be orthogonal in the interval (a, b).

1.9.6 Properties of Wave Function ψ

(*i*) ψ and its derivatives $\left(\frac{\partial\psi}{\partial x}, \frac{\partial\psi}{\partial y}, \frac{\partial\psi}{\partial z}\right)$ must be finite, continuous and single-valued everywhere.

(*ii*) ψ must be normalized *i.e.*, $\int |\psi|^2 dV = 1$ all space.

(*iii*) ψ must obey superposition principle. If $\psi_1(x), \psi_2(x), \ldots, \psi_n(x)$ are solutions of Schrödinger equation, then the linear combination

$\psi(x) = a_1\psi_1(x) + a_2\psi_2(x) + \ldots + a_n\psi_n(x)$ must be a solution.

1.10 APPLICATION OF SCHRÖDINGER'S EQUATION

1.10.1 The Particle in One Dimensional Potential Box

Let us consider the one dimensional motion of a particle which is restrained by reflecting walls. Here we shall assume a square well potential with infinitely high sides as shown in Fig. 1.11. The particle has a mass m and its position x at any instant is given by $0 < x < a$, *i.e.* particle is bound by impentrable walls of width a. The potential energy V of the particle is infinite on both sides of the box. The potential energy V of the particle can be assumed to be zero between $x = 0$ and $x = a$.

In terms of the boundary conditions imposed by the problem, the potential function is

$$V = 0 \text{ for } 0 < x < a$$

$$V = \infty \text{ for } x \le 0$$

$$V = \infty \text{ for } x \ge a \qquad \textbf{(1.71)}$$

The particle cannot exist outside the box and so its wave function ψ is 0 for $x \le 0$ and $x \ge a$. Our task is to find what ψ is within the box, viz., between $x = 0$ and $x = a$.

Within the box, the Schrödinger's equation becomes

$$\frac{d^2\psi}{d^2x} + \frac{8\pi^2 m}{h^2} E\psi = 0 \qquad \textbf{(1.72)}$$

Putting $\frac{8\pi^2 mE}{h^2} = k^2$ or $\frac{2mE}{\hbar^2} = k^2$, the equation becomes

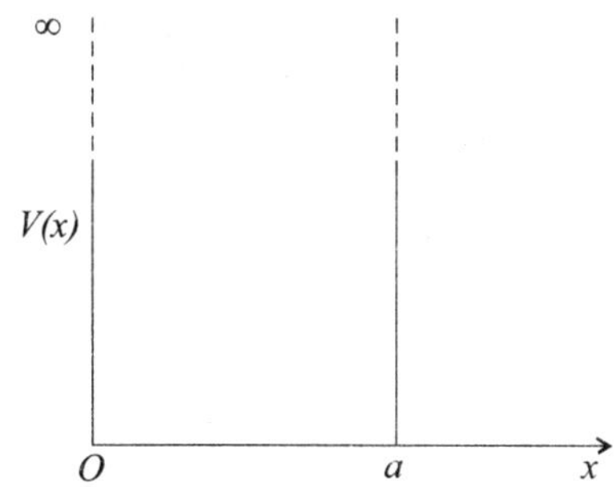

Fig. 1.11 One dimensional square well potential

$$\frac{d^2\psi}{dx^2} + k^2\psi = 0 .$$

The general solution of this equation is

$$\psi = A\sin kx + B\cos kx \qquad \textbf{(1.73)}$$

The boundary conditions can be used to evaluate the constants A and B in equation (1.73).

$\psi = 0$ at $x = 0$, hence $B = 0$

$\psi = 0$ at $x = a$, hence $0 = A\sin ka$

Since $A \ne 0$, $\therefore$ $\sin ka = 0 \Rightarrow ka = n\pi$ where n is an integer

$$\Rightarrow \qquad k = \frac{n\pi}{a} . \qquad \textbf{(1.74)}$$

$$\text{Thus } \psi_n(x) = A\sin\frac{n\pi x}{a}, \qquad \text{where } n = 1,2,3,\ldots \qquad \textbf{(1.75)}$$

To calculate constant A, we use normalization condition *i.e.*,

$$\int_0^a |\psi|^2\, dx = 1$$

$$\Rightarrow \qquad A^2 \int_0^a \sin^2\frac{n\pi x}{a}\, dx = 1$$

$$\Rightarrow \qquad \frac{A^2}{2}\int_0^a \left(1 - \cos\frac{2n\pi x}{a}\right) dx = 1$$

$$\Rightarrow \qquad \frac{A^2}{2}\left[x - \frac{a}{2n\pi}\sin\frac{2n\pi x}{a}\right]_0^a = 1$$

$$\Rightarrow \qquad \frac{A^2}{2}[a - 0] = 1$$

$$\Rightarrow \qquad A = \sqrt{\frac{2}{a}}$$

$$\text{Thus wave function } \psi_n\left(\vec{x}\right) = \sqrt{\frac{2}{a}}\sin\frac{n\pi x}{a} \qquad \textbf{(1.76)}$$

This equation (1.76) gives the eigenfunction of a particle, which is confined in one dimensional potential box and the eigenvalues are given by equation (1.74)

$$k = \sqrt{\frac{8\pi^2 mE}{h^2}} = \frac{n\pi}{a}$$

$$\frac{8\pi^2 mE}{h^2} = \frac{n^2\pi^2}{a^2}$$

$$\Rightarrow \quad E_n = \frac{n^2 h^2}{8ma^2}$$

$$E_n = \frac{n^2\pi^2\hbar^2}{2ma^2} \quad \left(\because \hbar = \frac{h}{2\pi}\right) \tag{1.77}$$

Here $n = 1, 2, 3,...$

Thus, corresponding to all integral values of n, there is an infinite sequence of discrete energy levels as shown in Fig. 1.12. Thus all values of energy are not allowed *i.e.* the energy is quantized.

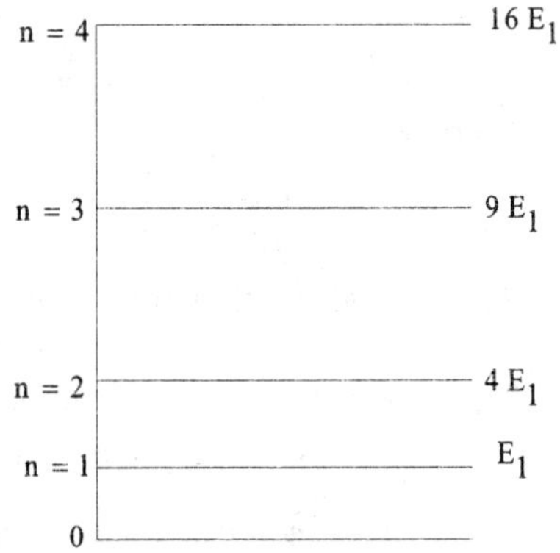

Fig. 1.12 Energy Level in One Dimensional Box

For each energy eigenvalue there is just one eigenfunction, so that the functions are non-degenerate. The ground state energy (n = 1) is $E_1 = \frac{h^2}{8ma^2} = \frac{\pi^2\hbar^2}{2ma^2}$. The uncertainty in the position of the particle is of the order of 'a' because the particle is constrained to remain in the potential well. This position uncertainty corresponds to a momentum uncertainty of the order of $(\hbar / a)$ which implies a minimum energy of the order of $\frac{1}{2m}\left(\frac{\hbar^2}{a^2}\right) = \frac{\hbar^2}{2ma^2}$. The ground state energy obtained above is definitely more than this values in agreement with the uncertainty relation.

Another feature of importance is the nodes of the eigenfunction (positions where eigenfunctions are zero). The number of nodes depends on the energy eigenvalue E or the frequency of waves associated with the particle. Higher is the value of E or the frequency, the more is the number of nodes. This feature is quite analogous to that observed in the case of stationary waves formed by vibrating strings.

The normalized wave functions ψ_1, ψ_2 and ψ_3 together with the probability densities $|\psi_1|^2$, $|\psi_2|^2$ and $|\psi_3|^2$ are plotted in Fig. (1.13).

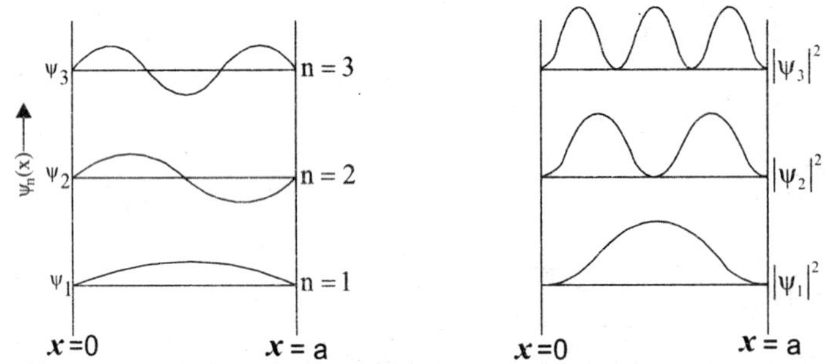

Fig. 1.13 Wave functions and probability density of a particle inside an infinite potential well

According to classical mechanics, a particle moving with a velocity v spends a time dx/v in travelling a distance dx and hence the probability distribution is constant. Quantum mechanically the distribution is entirely different. $|\psi_n|^2 = 0$ at $x = 0$ and $x = +a$. At a particular point, the probability of the particle being present is different for different quantum numbers. For example $|\psi_1|^2$ has its maximum value in the middle. Thus a particle in the lowest energy level of $n = 1$ is most likely to be in the middle while a particle in the next higher state of $n = 2$ is never there.

Example 1.33: An electron is trapped in infinitely deep cubical potential well of width 1 Å. What is its first excitation energy? (Given $m_e = 9.1 \times 10^{-31}\,\text{kg}, h = 6.62 \times 0^{-34}$ JS) **[Raj. B.Tech. I Sem. 2008]**

Solution: Given $a = 1\,\text{Å} = 1 \times 10^{-10}\,\text{m}$

The energy eigenvalue for infinite deep potential well

$$E_n = \frac{n^2\pi^2\hbar^2}{2ma^2} = \frac{n^2h^2}{8ma^2}$$

Put $n = 2$ for first excitation energy

$$E = \frac{4\,h^2}{8ma^2} = \frac{4 \times \left(6.62 \times 10^{-34}\right)^2}{8 \times 9.1 \times 10^{-31} \times \left(1 \times 10^{-10}\right)^2}$$

$$= \frac{175.29}{72^{-8}} \times 10^{-17}$$

$$\Rightarrow \quad E = 2.407 \times 10^{-17}\ \text{J}$$

Since $\quad 1\ \text{eV} = 1.6 \times 10^{-19}$ Joule

$$\therefore \quad E = \frac{2.407 \times 10^{-17}}{1.6 \times 10^{-19}}\text{eV}$$

$$= 1.5043 \times 10^2\,\text{eV}$$

$$E = 150.43\ \text{eV}$$

Ans.

Example 1.34: Considering one dimensional case, find the probability that a particle in a box of width 'a' can be found between $x = 0$ and $x = \frac{a}{n}$ when it is in the nth state. **[Raj. B.E. I 2004]**

Solution: The wave function of a particle in one dimensional box is given by

$$\psi_n = \sqrt{\frac{2}{a}}\sin\left(\frac{n\pi}{a}\right)x$$

Therefore the probability

$$P = \int_0^{a/n} |\psi_n|^2\,dx$$

$$= \frac{2}{a}\int_0^{a/n} \sin^2\left(\frac{n\pi}{a}\right)x$$

$$= \frac{2}{a}\int_0^{a/n}\left[\frac{1-\cos\left(\frac{2n\pi}{a}\right)x}{2}\right]dx$$

$$= \frac{2}{a}\times\frac{1}{2}\left[x - \frac{a}{2n\pi}\sin\left(\frac{2n\pi x}{a}\right)\right]_0^{a/n}$$

$$= \frac{1}{a}\left[\frac{a}{n} - 0\right]$$

$$P = \frac{1}{n}$$ **Ans.**

Example 1.35: Calculate the permitted energy levels of an electron in a box 1 Å wide.

Solution: The permitted electron energies are given as

$$E_n = \frac{n^2h^2}{8mL^2}$$

Here m = mass of electron = 9.1×10^{-31} kg,

L = 1 Å $= 10^{-10}$ m, $h = 6.626\times10^{-34}$ J-s

$$\therefore \quad E_n = \frac{n^2\left(6.626\times10^{-34}\right)^2}{8\times9.1\times10^{-31}\times\left(10^{-10}\right)^2}$$

$$= 6.03\times10^{-18}\,n^2\text{J}$$

$$= \frac{6\times10^{-18}}{1.6\times10^{-19}}n^2\text{eV} = 37.6n^2\text{eV} \approx 38n^2\text{eV}$$

The minimum energy, the electron can have, is $E_1 = 38$ eV$(n=1)$

The other values of energy are

$$E_2 = 4E_1 = 152 \text{ eV}$$

$$E_3 = 9E_1 = 342\,\text{eV} \text{ and so on.}$$ **Ans.**

Example 1.36: A particle is moving in a one dimensional box of infinite height of width 10 Å. Calculate the probability of finding the particle within an interval of 1 Å at the centre of the box, when it is in its state of least energy .

Solution: The wave function of the particle in the ground state (n = 1) is

$$\psi_1 = \left(\frac{2}{a}\right)^{1/2}\sin\frac{\pi x}{a}$$

The probability of finding the particle in unit interval at the centre of the box $\left(x = \frac{a}{2}\right)$ is

$$P = \psi_1^2 = \left[\left(\frac{2}{a}\right)^{1/2} \sin\frac{\pi(a/2)}{a}\right]^2$$

$$= \frac{2}{a}\sin^2\frac{\pi}{2} = \frac{2}{a}$$

∴ The probability of finding the particle within an interval of Δx at the centre of the box

$$W = |\psi_1|^2 \Delta x = \frac{2}{a}\Delta x$$

Here $a = 10\times10^{-10}\,\text{m}$, $\Delta x = 10^{-10}\,\text{m}$

∴ $$W = \frac{2\times10^{-10}}{10\times10^{-10}} = 0.2$$ **Ans.**

Example 1.37: For a particle in a one dimensional box of length L, calculate the probability of finding it in the region $B \le x \le B+b$. Compare this with the value expected for a classical system.

Solution: The total probability of finding the particle within domain $B \le x \le B+b$ is given by

$$P_b = \int_B^{B+b}\left(\sqrt{\frac{2}{L}}\sin\frac{n\pi x}{L}\right)^2 dx = \frac{2}{L}\int_B^{B+b}\frac{1}{2}\left(1-\cos\frac{2n\pi x}{L}\right)dx$$

$$= \frac{b}{L} - \frac{1}{2n\pi}\left[\sin\frac{2n\pi x}{L}\right]_B^{B+b}$$ **Ans.**

For $n\to\infty$, $P_b \to \frac{b}{L}$ **Ans.**

Which is the value expected for a classical system.

Example 1.38: In a long chain organic molecule of length 5Å, electron may be treated as free to move along the length. Deduce the zero point energy, the energy gap between the first two energy states of the electron and also the wavelength of absorption line arising from this transition.

(MREC-2002)

Solution: Zero point energy is given by

$$E_1 = \frac{h^2}{8ma^2}$$

Here $h = 6.626\times10^{-34}$ J-s, $m = 9.1\times10^{-31}\,\text{kg}$, $a = 5\text{Å} = 5\times10^{-10}\,\text{m}$

∴ $$E_1 = \frac{(6.626\times10^{-34})^2}{8\times9.1\times10^{-31}\times(5\times10^{-10})^2} = 2.4\times10^{-19}\,\text{J}$$ **Ans.**

Energy of electron in the second state

$$E_2 = 4E_1 = 4\times2.4\times10^{-19}\,\text{J} = 9.6\times10^{-19}\,\text{J}$$

Energy gap $\Delta E = E_2 - E_1 = 7.2\times10^{-19}\,\text{J}$

$$= \frac{7.2\times10^{-19}}{1.6\times10^{-19}} = 4.5\,\text{eV}$$ **Ans.**

Absorption wavelength is given by

$$\Delta E = h\nu = \frac{hc}{\lambda}$$

$$\Rightarrow \quad \lambda = \frac{hc}{\Delta E} = \frac{6.626 \times 10^{-34} \times 3 \times 10^{8} \times 10^{10}}{7.2 \times 10^{-19}} \text{Å}$$

$$= 2.8 \times 10^{3} \text{Å}$$

Ans.

Example 1.39: An electron is confined to move between two rigid walls separated by 10^{-9} m. Find the de Broglie wavelengths representing the first three allowed energy states of the electron and the corresponding energies.

Solution: The distance between the walls (a) must be an integral multiple of $\frac{\lambda}{2}$.

$$\therefore \quad a = \frac{n\lambda}{2}$$

$$\Rightarrow \quad \lambda = \frac{2a}{n}$$

Here $a = 10^{-9}$ m

$$\therefore \quad \lambda = \frac{2 \times 10^{-9}}{n} m = \frac{20}{n} \text{Å} \text{ where } n = 1, 2, 3, ...$$

Hence $\lambda_1 = 20$ Å; $\lambda_2 = \frac{20}{2} = 10$ Å; $\lambda_3 = 6.667$ Å **Ans.**

The permitted electron energies are given by

$$E_n = \frac{n^2 h^2}{8ma^2} = \frac{n^2 \times \left(6.626 \times 10^{-34}\right)^2}{8 \times \left(9.1 \times 10^{-31}\right) \times \left(10^{-9}\right)^2} \text{J}$$

$$= \frac{n^2 \times \left(6.626 \times 10^{-34}\right)^2}{8 \times \left(9.1 \times 10^{-31}\right) \times \left(10^{-9}\right)^2 \times 1.6 \times 10^{-19}} \text{eV}$$

$$= 0.38\, n^2 \text{ eV}$$

$\therefore$ $E_1 = 0.38$ eV; $E_2 = 4 \times 0.38 = 1.52$ eV; $E_3 = 9 \times 0.38 = 3.42$ eV **Ans.**

Example 1.40: A particle is moving in a one dimensional potential box of infinite height. What is the probability of finding the particle in a small interval Δx, at the centre of the box when it is in the energy state, next to the least energy state?

Solution: The wave function in the second state is given by

$$\psi_2 = \sqrt{\frac{2}{L}} \sin \frac{2\pi x}{L}$$

At the centre of the box, $x = \frac{L}{2}$

$$\therefore \quad \psi_2 = 0$$

$\therefore$ Probability of finding the particle in an interval Δx at the centre of the box is

$$P = \psi_2^{\ 2} \Delta x = 0$$

Ans.

Example 1.41: The size of the nucleus is 10^{-14} m. Treating it as a one dimensional box, show why electron does not exist in the nucleus ?

Solution: The zero-point energy is given by

$$E_1 = \frac{\pi^2 \hbar^2}{2ma^2}$$

Here $\hbar = 1.054 \times 10^{-34}$ J-s, $m = 9.1 \times 10^{-31}$ kg,

$a = 10^{-14}$ m

$$\therefore \quad E_1 = \frac{(3.14)^2 \times (1.054 \times 10^{-34})^2}{2 \times 9.1 \times 10^{-31} \times (10^{-14})^2} \text{ J}$$

$$= 6.0 \times 10^{-10} \text{ J}$$

The Coulombian atttraction enenrgy on the other hand is

$$V = -\frac{Ke^2}{r} = -\frac{(-1.6 \times 10^{-19})^2 \times 9 \times 10^9}{10^{-14}}$$

$$= -2.3 \times 10^{-14} \text{ J}$$

The Coulombian attraction energy is too weak for the nucleus to hold an electron. **Ans.**

Example 1.42: Find the lowest energy of an electron confined to move in a one-dimensional potential box of length 1Å. **(Raj. B.E.I. 2001)**

Solution: The zero-point energy is given by

$$E_1 = \frac{\pi^2 \hbar^2}{2ma^2}$$

Here $\hbar = 1.054 \times 10^{-34}$ J-s, $m = 9.1 \times 10^{-31}$ kg, $a = 10^{-10}$ m

$$E_1 = \frac{(3.14)^2 \times (1.054 \times 10^{-34})^2}{2 \times 9.1 \times 10^{-31} \times (10^{-20})}$$

$$= 6 \times 10^{-18} \text{ J}$$

$$= \frac{6 \times 10^{-18}}{1.6 \times 10^{-19}} = 37.5 \text{ eV}$$

Ans.

Example 1.43: A particle of mass $m = 1 \times 10^{-29}$ kg is confined to move in a box of length 2 Å whose ends are at $x = 0$ and $x = 2$ Å. What is the probability of finding the particle between 1.6000 Å and 1.6001 Å. Calculate for $n = 1$ and $n = 2$.

Solution: $$P = \frac{2}{a} \int_{1.6}^{1.6001} \sin^2 \frac{n\pi x}{a} dx$$

$$\Rightarrow \qquad P = \frac{1}{a}\left[x - \frac{a\sin\frac{2n\pi x}{a}}{2n\pi}\right]_{1.6}^{1.6001}$$

For $n = 1$

$$P = \frac{1}{2.0}\left[0.0001 - \left\{\frac{2\sin\frac{2\pi\times 1.6001}{2} - 2\sin\frac{2\pi\times 1.6000}{2}}{2\pi}\right\}\right]$$

$$= 3.45\times 10^{-5}$$

Similary for $n = 2$, we obtain

$$P = 9.05\times 10^{-5}$$

Ans.

QUESTIONS AND PROBLEMS

QUESTIONS

1. What is Compton effect? Give the theory of Compton scattering of X-rays by electrons. Which is the analogous effect in visible light ?
2. Why are X-rays best suited for the study of Compton effect ?

   Ans. The Compton shift is given by

   $$\Delta\lambda = \frac{h}{m_0 c}(1-\cos\theta) = 0.0243\ (1-\cos\theta)$$

   The maximum value of $(1-\cos\theta)$ is when $\theta = 180°$

   So, $\Delta\lambda \sim 0.05\text{Å}$

   This means that Compton effect can be detected only for those radiations whose wave–length is of the order of few Å. For example, for $\lambda = 1\,\text{Å}$ (wavelength of X-rays)

   $$\frac{\Delta\lambda}{\lambda}\times 100 = \frac{0.05}{1}\times 100 = 5\%$$

   Which can be easily detected. For $\lambda = 5000\text{Å}$ (wavelength of light)

   $$\frac{\Delta\lambda}{\lambda}\times 100 = \frac{0.05}{5000}\times 100 = 0.001\%$$

   Which cannot be detected. Hence X-rays are best suited for the study of compton effect.
3. Explain Compton scattering. Derive an expression for the compton shift.
   (MREC B.E. I 2002, Raj. B.E. I Back 2007, RTU B. Tech. II Sem. 2009, Back 2012)
4. What is compton effect? Obtain an expression for compton shift. Give its experimental verification. **(RTU B.Tech. I Sem. 2008, 2012)**

5. (*a*) Explain why compton effect is not observed experimentally for visible rays. Explain Comption scattering.

   (*b*) State Heisenberg's uncertainty principle. Explain its validity by any thought experiment. **(RTU B.Tech. Back, 2010)**

6. Explain the compton effect and its physical significance. **(Raj. B.E. I 2005)**

7. Derive the Schrödinger time dependent equation. Explain the following: **(RTU B.Tech. Back, 2011)**

   (*i*) Hamiltonian

   (*ii*) Physical significance of wavefunction ψ. **(RTU B.Tech. I Sem. 2009)**

   (*iii*) Normalized and orthogonalized wavefunctions. **(Raj. B.E.I 2002)**

8. Write down Schrödinger's equation for a free particle of mass m trapped in one dimensional box of side a. Solve it for energy eigenvalues and eigenfunction. Find lowest energy of the following states : (i) non-degenerate, (ii) doubley degenerate, (iii) triply degenerate. **(Raj. B.E.I 2001, RTU B.Tech. I 2007)**

9. Write down Schrödinger's time dependent and time independent wave equations. Give physical significance of wavefunction. **(Raj. B.E. I 2001, RTU B.Tech. II Sem. 2008, I Sem. 2010, 2012 & Back 2010)**

10. Write short notes on the following:

    (*i*) Physical interpretation of wavefunction and its properties. **(RTU B.Tech. I Sem. 2012)**

    (*ii*) Compton effect.

    (*iii*) Heisenberg's uncertainty principle. **(Raj. B.E. I 2004)**

    (*iv*) Determine the expectation value of position of a particle in one dimensional box. **(RTU B.Tech. I Sem. 2010)**

11. Write down the Schrödinger's equation for a particle enclosed in one dimensional box of size 'a'. Solve it for eigenvalues and eigenfunctions. **(Raj. B.E. I 1999, 2003, 2005, RTU B.Tech. I Sem. 2012, Back 2011)**

12. Discuss briefly any of the Physical experiments which provided evidence for quantum theory. What is de Broglie's hypothesis ? Explain the wave nature of matter. **(RTU B. Tech I Sem. 2009)**

13. Derive Schrödinger's time independent wave equation. What is the physical significance of wave function ψ used in this equation ? **(RTU B. Tech. II Sem. 2009)**

14. (*i*) What is the difference between compton effect and photoelectric effect?

    (*ii*) Explain how compton scattering confirms the quantum nature of radiation.

    (*iii*) Give reasons why the compton shift is detectable only in X-ray region and not in visible range of light. **(RTU. B.Tech. I Sem. 2010)**

PROBLEMS

1. Calcium is illuminated with ultraviolet light of wavelength 250 nm. If the work function of calcium is 3.2 eV, find the maximum kinetic energy of the emitted electron. **(Ans. 1.773 eV)**

2. Calculate the compton wavelength from standard data. **(Ans. 2.42×10^{-12} m)**
3. Find the change in wavelength of an X-ray photon when it is scattered through an angle of 90° by a free electron. **(Ans. 2.42×10^{-12} m)**
4. An incident photon of wavelength 0.003 nm recoils at an angle of 60° after being scattered by a free electron. Find the energy of the recoiling electron. **(Ans. 1.9×10^{-14} J)**
5. The uncertainty in the location of a particle is equal to its de Broglie wavelength. Find the uncertainty in its velocity. **(Ans. $v/4\pi$)**
6. Calculate the wave number of 10 KeV neutrons. **(Ans. $2.19 \times 10^{9} m^{-1}$)**
7. The position of 1 KV electron is located within 10^{-10} m. What is the uncertainty in its momentum? **(Ans. 5.276×10^{-25} Kg -m/s)**
8. An electron has a speed of 600 m/s with an accuracy of 0.005%. Calculate the certainty with which the position of the electron can be located. **(Ans. 1.056×10^{-30} m)**
9. Compare the uncertainties in the velocity of an electron and proton confined to a 10^{-18}m box. **(Ans. 5.48×10^{-4})**
10. An electron is constrained to a one dimensional box of side 0.1 nm. Obtain the first four eigenvalues in electron-volt. **(Ans. 34.4 eV ; 137.6 eV ; 309.6 eV ; 550.4 eV)**
11. Calculate the Compton shift for an X-ray photon if it is scattered at an angle of 60° by a free electron. **(Ans. 1.213×10^{-12} m)**
12. An X-ray of wavelength 1 Å is scattered at an angle of 90° by a free electron initially at rest. Compute the wavelength of scattered photon. **(Ans. 1.0242×10^{-10} m)**
(RTU B. Tech I Sem. 2012)
13. The life time of an excited state of an atom is about 10^{-8} sec. Calculate the minimum uncertainty in the determination of the energy of excited state. **(Ans. 6.5×10^{-8} eV)**
14. What is the lowest energy that a neutron can have if confined to move along the edge of an impenetrable box of length 10^{-14}m? **(Ans. 2.056 MeV)**
(RTU B. Tech. II Sem. 2011)
15. A particle is moving in one dimensional potential box (of infinite height) of width 25 Å. Calculate the probability of finding the particle within an interval of 5 Å at the centre of the box when it is in the state of least energy. **(Ans. 0.4)**
16. The wave function of a particle in its ground state in one dimensional box of length L is given by $\psi = \sqrt{\frac{2}{L}} \sin\frac{\pi}{L}x$. Calculate probability of finding the particle within a interval of 1 Å at the centre of box of length L = 10 Å. **(Ans. 0.2)**
(Raj. B.Tech. II Sem. 2008)

17. Find the lowest energy of an electron confined to move in a one dimensional box of length 1 Å. **(Ans. 38 eV)**

(RTU I Sem. 2012)

REFERENCES

1. E. Merzbacher, Quantum Mechanics, 2nd Ed., New York, John Wiley & Sons, 1970.
2. A. Beiser, Concepts of Modern Physics, McGraw Hill Book Company, New York, 1995.
3. N.F. Mott and I.N. Sneddon, Wave Mechanics and its Applications, London, Oxford University Press 1948, reprinted, New York, Dover, 1963.
4. H.C. Ohanian, Modern Physics, Prentice Hall of India Private Limited, New Delhi, 1994.
5. D.Haliday , R. Resnik and K.S.Krane , Physics, Vol. II extended version, John Wiley & Sons, Singapore, 1992.
6. J. Bernstein. P.M. Fishbane and S. Gasiorowicz, Modern Physics, Pearson Education, New Delhi, 2001.
7. S.T. Thornton and A. Rex, Modern Physics for Scientists and Engineers, Saunders College Publishing, New York, 2000.
8. R.L. Lioff, Introductory Quantum Mechanics, Pearson Publications, New Delhi, 2003.
9. J. Singh, Modern Physics for Engineers, John Wiley, New York, 2003.
10. S. Gasiorowiz, Quantum Physics, John Wiley, New York, 2005.

2

APPLICATIONS OF SCHRÖDINGER'S EQUATION

2.1 INTRODUCTION

We have already noted that Schrödinger was influenced by the matter–wave postulate of *de Broglie*. In order to describe the behaviour of a particle as a wave, Schrödinger proposed an equation in 1926 to be known after his name as *Schrödinger's Wave Equation*. The wave function ψ, which is a solution of Schrödinger's equation, is expected to describe the dynamic behaviour or dynamical state of a particle. Schrödinger's equation has been found to be a fundamental equation in quantum mechanics just as Newton's equation $\vec{F} = \frac{d\vec{p}}{dt}$ in classical Mechanics or Maxwell's equations in electromagnetism.

Erwin Rudolf Josef Alexander Schrödinger (12 August 1887 – 4 January 1961) was an Austrian physicist who achieved fame for his contributions to quantum mechanics. In January 1926, Schrödinger published in the Annalen der Physik the paper Quantisation as an Eigenvalue Problem on wave mechanics and what is now known as the Schrödinger equation, for which he received the Nobel Prize in 1933.

Schrödinger gave two types of wave equations namely time dependent and time independent. The time dependent Schrödinger's equation is given as

$$i\hbar \frac{\partial \psi(\vec{r},t)}{\partial t} = -\frac{\hbar^2}{2m}\nabla^2\psi(\vec{r},t) + V(\vec{r},t)\psi(\vec{r},t) \qquad \textbf{(2.1)}$$

and time independent Schrödinger's equation is given as

$$\nabla^2\psi(\vec{r}) + \frac{2m}{\hbar^2}\left[E - V(\vec{r})\right]\psi(\vec{r}) = 0 \qquad \textbf{(2.2)}$$

where the symbols carry their usual meaning.

Schrödinger's equation is extremely useful for investigating various quantum mechanical problems; viz particle in one and three dimensional box, one dimensional harmonic oscillator, step potential and potential barrier. In the following subsections, we discuss two quantum mechanical problems; particle in three dimensional box and potential barrier.

2.2 THE PARTICLE IN A THREE DIMENSIONAL BOX

Let us consider a particle which is enclosed inside a rectangular box having edges a, b and c in length as shown in Fig. 2.1. The potential function $V(x, y, z)$ is having a constant value of zero in the regions given below:

$$V(x, y, z) = 0 \quad \text{for} \quad 0 < x < a$$
$$V(x, y, z) = 0 \quad \text{for} \quad 0 < y < b$$
$$V(x, y, z) = 0 \quad \text{for} \quad 0 < z < c \qquad (2.3)$$

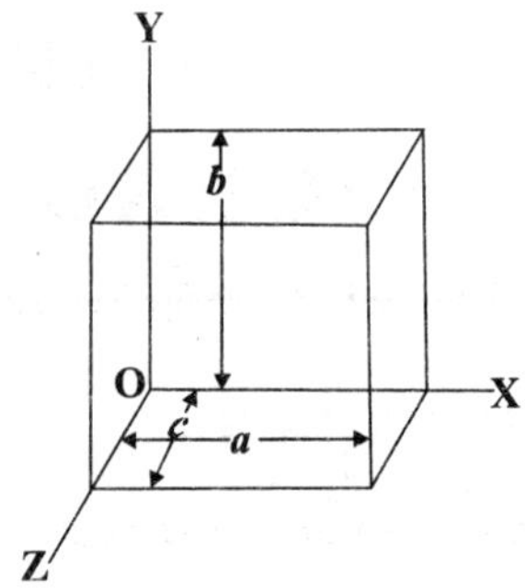

Fig. 2.1 Three dimensional box

The potential outside the box is infinite. The Schrödinger time independent wave equation inside the box for such a particle may be written as

$$\frac{\partial^2 \psi}{\partial x^2} + \frac{\partial^2 \psi}{\partial y^2} + \frac{\partial^2 \psi}{\partial z^2} + \frac{2m}{\hbar^2} E\psi = 0 \qquad (2.4)$$

To separate out the above equation let us put

$$\psi(x, y, z) = X(x)\ Y(y)\ Z(z) = XYZ \qquad (2.5)$$

Substituting the value of ψ from Eq. (2.5), we get

$$YZ\frac{\partial^2 X}{\partial x^2} + ZX\frac{\partial^2 Y}{\partial y^2} + XY\frac{\partial^2 Z}{\partial z^2} + \frac{2m}{\hbar^2} EXYZ = 0$$

Dividing this equation by XYZ, we get

$$\frac{1}{X}\frac{\partial^2 X}{\partial x^2} + \frac{1}{Y}\frac{\partial^2 Y}{\partial y^2} + \frac{1}{Z}\frac{\partial^2 Z}{\partial z^2} + \frac{2m}{\hbar^2} E = 0$$

$$\Rightarrow \quad \frac{1}{X}\frac{\partial^2 X}{\partial x^2} + \frac{1}{Y}\frac{\partial^2 Y}{\partial y^2} + \frac{1}{Z}\frac{\partial^2 Z}{\partial z^2} = -\frac{2m}{\hbar^2} E \qquad (2.6)$$

For the given energy of the particle, the term $\left(-\frac{2mE}{\hbar^2}\right)$ is constant and each term on the left side is a function of one variable only. If we allow only one of these (x or y or z) to vary at a time, the sum of the three terms is still equal to the constant on the right hand side. This means that each of the three terms on the left is itself a constant and independent of the other variables present in it. Let us represent each term on left side of equation, in succession, equal to the constants $-\alpha_1^2$, $-\alpha_2^2$ and $-\alpha_3^2$. These have a minus sign because the term on right side of equation has minus sign. We, therefore, write

$$\frac{1}{X}\frac{\partial^2 X}{\partial x^2} = -\alpha_1^2 \qquad (2.7)$$

$$\Rightarrow \quad \frac{\partial^2 X}{\partial x^2} + \alpha_1^2 X = 0 \qquad (2.8)$$

$$\frac{1}{Y}\frac{\partial^2 Y}{\partial y^2} = -\alpha_2^2 \qquad (2.9)$$

$$\Rightarrow \quad \frac{\partial^2 Y}{\partial y^2} + \alpha_2^2 Y = 0 \tag{2.10}$$

$$\frac{1}{Z}\frac{\partial^2 Z}{\partial z^2} = -\alpha_3^2 \tag{2.11}$$

$$\Rightarrow \quad \frac{\partial^2 Z}{\partial z^2} + \alpha_3^2 Z = 0 \tag{2.12}$$

On substituting equations (2.7), (2.9) and (2.11) in equation (2.6), we obtain

$$\alpha_1^2 + \alpha_2^2 + \alpha_3^2 = \frac{2mE}{\hbar^2} \tag{2.13}$$

The solutions of equations (2.8), (2.10) and (2.12) are given by

$$X = A_1 \cos\alpha_1 x + B_1 \sin\alpha_1 x \tag{2.14}$$

$$Y = A_2 \cos\alpha_2 y + B_2 \sin\alpha_2 y \tag{2.15}$$

$$Z = A_3 \cos\alpha_3 z + B_3 \sin\alpha_3 z \tag{2.16}$$

In the above equations A_1, A_2, A_3 and B_1, B_2, B_3 are constants. It is possible to obtain the values of these constants by applying the boundary conditions. As ψ vanishes at the surfaces (boundaries) of infinite potential, it means that

$\psi = 0$ when $x = 0,\ y = 0,\ z = 0$ and when $x = a,\ y = b,\ z = c$

If these boundary conditions are applied, then we have

$$A_1 = A_2 = A_3 = 0 \tag{2.17}$$

Also $\quad B_1 \neq 0 \ \therefore\ \sin\alpha_1 a = 0, i.e.,\ \alpha_1 a = n_x\pi$ or $\alpha_1 = \dfrac{n_x\pi}{a}$ **(2.18)**

$B_2 \neq 0 \ \therefore\ \sin\alpha_2 b = 0, i.e.,\ \alpha_2 b = n_y\pi$ or $\alpha_2 = \dfrac{n_y\pi}{b}$ **(2.19)**

$B_3 \neq 0 \ \therefore\ \sin\alpha_3 c = 0, i.e.,\ \alpha_3 c = n_z\pi$ or $\alpha_3 = \dfrac{n_z\pi}{c}$ **(2.20)**

On substituting equations (2.17 –2.20) in equations (2.14), (2.15) and (2.16), we obtain

$$X = B_1 \sin\frac{n_x\pi}{a}x \tag{2.21}$$

where $n_x = 1, 2, 3 \ldots$

$$Y = B_2 \sin\frac{n_y\pi}{b}y \tag{2.22}$$

where $n_y = 1, 2, 3 \ldots$

and $\quad Z = B_3 \sin\dfrac{n_z\pi}{c}z$ **(2.23)**

where $n_z = 1, 2, 3, \ldots$

Thus the complete wave function is given by

$$\psi_{n_x,n_y,n_z}^{(x,y,z)} = B_1 B_2 B_3 \sin\frac{n_x\pi}{a}x \sin\frac{n_y\pi}{b}y \sin\frac{n_z\pi}{c}z$$

$$\Rightarrow \quad \psi_{n_x,n_y,n_z}^{(x,y,z)} = K \sin\frac{n_x\pi}{a}x \sin\frac{n_y\pi}{b}y \sin\frac{n_z\pi}{c}z \tag{2.24}$$

In the above equation, K is termed as normalization constant. It is possible to obtain the value of K by using the normalization condition, *i.e.*,

$$\int_0^a\int_0^b\int_0^c \psi\psi^* \, dx\, dy\, dz = 1 \tag{2.25}$$

$$\Rightarrow \quad K^2 \int_0^a\int_0^b\int_0^c \sin^2\frac{n_x\pi x}{a} \sin^2\frac{n_y\pi y}{b} \sin^2\frac{n_z\pi z}{c} \, dx\, dy\, dz = 1 \tag{2.26}$$

But $$\int_0^a \sin^2\frac{n_x\pi x}{a} dx = \int_0^a \left[\frac{1}{2} - \frac{1}{2}\cos\frac{n_x\pi x}{a}\right] dx = \frac{a}{2}$$

Similarly $$\int_0^b \sin^2\frac{n_y\pi y}{b} = \frac{b}{2}$$

and $$\int_0^c \sin^2\frac{n_z\pi z}{c} = \frac{c}{2}$$

Thus equation (2.26) becomes

$$K^2 \frac{a}{2}.\frac{b}{2}.\frac{c}{2} = 1$$

$$\Rightarrow \quad K = \frac{2\sqrt{2}}{\sqrt{abc}} \tag{2.27}$$

Therefore, the complete wave function becomes

$$\psi_{n_x,n_y,n_z}^{(x,y,z)} = \frac{2\sqrt{2}}{\sqrt{abc}} \sin\frac{n_x\pi x}{a} \sin\frac{n_y\pi y}{b} \sin\frac{n_z\pi z}{c} \tag{2.28}$$

From equation (2.13), we have

$$\alpha_1^2 + \alpha_2^2 + \alpha_3^2 = \frac{2mE}{\hbar^2} \tag{2.29}$$

On substituting the value of α_1, α_2 and α_3 from equations (2.18 –2.20) into equation (2.29), we obtain

$$\Rightarrow \quad \frac{n_x^2\pi^2}{a^2} + \frac{n_y^2\pi^2}{b^2} + \frac{n_z^2\pi^2}{c^2} = \frac{2mE}{\hbar^2}$$

$$\Rightarrow \quad E_{n_x n_y n_z} = \frac{\pi^2 \hbar^2}{2m}\left[\frac{n_x^{\ 2}}{a^2} + \frac{n_y^{\ 2}}{b^2} + \frac{n_z^{\ 2}}{c^2}\right] \tag{2.30}$$

If we consider a box that is cubical in shape such that $a = b = c$, energy can be expressed as

$$E = \frac{\pi^2\hbar^2}{2ma^2}\left[n_x^{\ 2} + n_y^{\ 2} + n_z^{\ 2}\right] = \frac{h^2}{8ma^2}\left[n_x^{\ 2} + n_y^{\ 2} + n_z^{\ 2}\right]$$

$$\Rightarrow \quad E = E_1\left[n_x^{\ 2} + n_y^{\ 2} + n_z^{\ 2}\right] \tag{2.31}$$

where $$E_1 = \frac{\pi^2\hbar^2}{2ma^2} = \frac{h^2}{8ma^2} \tag{2.32}$$

2.2.1 Degeneracy of Energy Levels

In the lowest quantum state (1,1,1), in which n_x, n_y and n_z respectively are equal to unity, it is seen that $E = \frac{3h^2}{8ma^2} = 3E_1$. There is only one set of *quantum numbers* that gives this energy state, and this level is said to be *non-degenerate.*

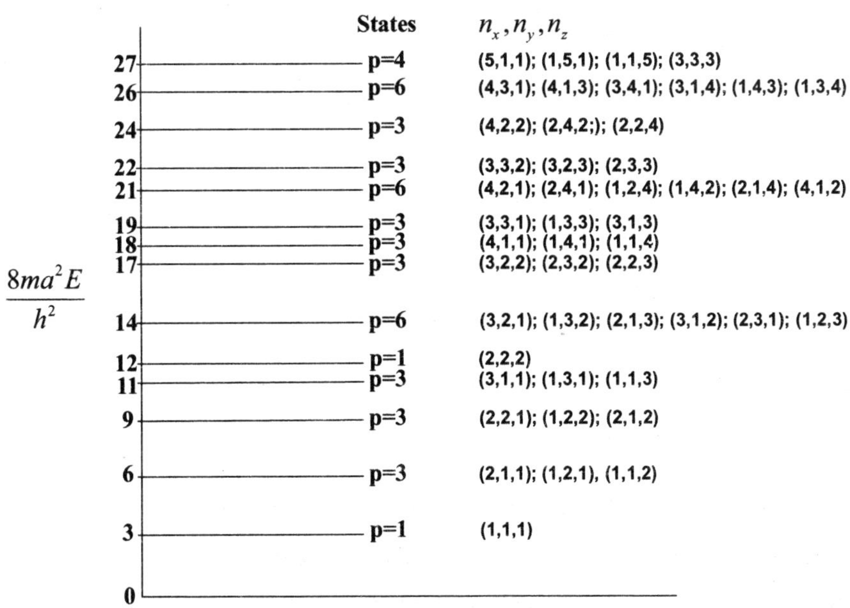

Fig. 2.2 Energy levels, degree of degeneracy and quantum numbers of a particle in a cubical box

If we now consider the second energy state as shown in Fig. 2.2, it is seen that there are three sets (1,1,2), (1,2,1) and (2,1,1) of the quantum number n_{x^2}, n_{y^2} and n_z that will give the same energy level $F = \frac{6h^2}{8ma^2}$, such a level is said to be degenerate, and in this particular case the level is triply *degenerate.*

For a cubical box, it can be concluded from the Fig. 2.2, that virtually all the energy levels are degenerate to some degree.

Example 2.1: A free particle is confined in a cubical box of side a. Write the eigenvalues and eigenfunctions for an energy state represented by $n_x + n_y + n_z = 4$. What is the order of degeneracy in this case? **[RTU B.Tech. I sem. 2007]**

Solution: Given $n_x + n_y + n_z = 4$

for this case, there are three possible combination of n_x, n_y and n_z as (2, 1, 1); (1, 2, 1); (1, 1, 2)

The eigenvalues and eigenfunctions for 3-D cubical box of side a, are given by

$$E_n = \frac{\pi^2 \hbar^2}{2ma^2}\left(n_x^2 + n_y^2 + n_z^2\right)$$

and $$\psi(x, y, z) = \sqrt{\frac{8}{a^3}} \sin\left(\frac{n_x \pi}{a}\right) x \sin\left(\frac{n_y \pi}{a}\right) y \sin\left(\frac{n_z \pi}{a}\right) z$$

There are three eigenfunctions corresponding to a single energy value $E = \frac{6\pi^2\hbar^2}{2ma^2}$. Hence, the order of degeneracy is 3.

The correseponding eigenfunctions are

$$\psi(2, 1, 1) = \sqrt{\frac{8}{a^3}} \sin\left(\frac{2\pi}{a}\right) x \sin\left(\frac{\pi}{a}\right) y \sin\left(\frac{\pi}{a}\right) z$$

$$\psi(1,2,1) = \sqrt{\frac{8}{a^3}} \sin\left(\frac{\pi}{a}\right) x \sin\left(\frac{2\pi}{a}\right) y \sin\left(\frac{\pi}{a}\right) z$$

and $$\psi(1,1,2) = \sqrt{\frac{8}{a^3}} \sin\left(\frac{\pi}{a}\right) x \sin\left(\frac{\pi}{a}\right) y \sin\left(\frac{2\pi}{a}\right) z$$ **Ans.**

Example 2.2: Answer the following questions with respect to a particular cubical box of side a.

(*i*) Is $n_x = n_y = n_z = 1$ state degenerate?

(*ii*) What shall happen to the degeneracies for $n_x + n_y + n_z = 4$ if the box is not cubical but rectangular parallelopiped with sides a, b and c such that $a = b \neq c$?

Solution: The eigenvalue of a particle in three dimensional box of sides a, b and c is

$$E_n = \frac{\pi^2 \hbar^2}{2m}\left(\frac{n_x^2}{a^2} + \frac{n_y^2}{b^2} + \frac{n_z^2}{c^2}\right)$$

and the eigenfunction for that particle is

$$\psi(x,y,z)=\sqrt{\frac{8}{abc}}\sin\left(\frac{n_x\pi}{a}\right)x\sin\left(\frac{n_y\pi}{b}\right)y\sin\left(\frac{n_z\pi}{c}\right)z$$

(*i*) for a cubical box $a = b = c$ and for $n_x = n_y = n_z = 1$, there is only one eigenfunction corresponding to eigenvalue $\frac{3\pi^2\hbar^2}{2ma^2}$, so it is non-degenerate.

(*ii*) Given that $n_x + n_y + n_z = 4$ and $a = b \neq c$. Therefore there are three states (2, 1, 1), (1, 2, 1) and (1, 1, 2).

$$E_{2,1,1}=\frac{\pi^2\hbar^2}{2m}\left[\frac{4}{a^2}+\frac{1}{a^2}+\frac{1}{c^2}\right]=\frac{\pi^2\hbar^2}{2m}\left[\frac{5}{a^2}+\frac{1}{c^2}\right]$$

$$E_{1,2,1}=\frac{\pi^2\hbar^2}{2m}\left[\frac{1}{a^2}+\frac{4}{a^2}+\frac{1}{c^2}\right]=\frac{\pi^2\hbar^2}{2m}\left[\frac{5}{a^2}+\frac{1}{c^2}\right]$$

$$E_{1,1,2}=\frac{\pi^2\hbar^2}{2m}\left[\frac{1}{a^2}+\frac{1}{a^2}+\frac{4}{c^2}\right]=\frac{\pi^2\hbar^2}{2m}\left[\frac{2}{a^2}+\frac{4}{c^2}\right]$$

Thus the energy $\frac{\pi^2\hbar^2}{2m}\left(\frac{5}{a^2}+\frac{1}{c^2}\right)$ is doubly degenerate corresponding to eigenfunctions $\psi_{2,1,1}$ and $\psi_{1,2,1}$ while energy $\frac{\pi^2\hbar^2}{2m}\left(\frac{2}{a^2}+\frac{4}{c^2}\right)$ is non-degenerate.

2.3 THE POTENTIAL BARRIER (TUNNEL EFFECT)

Let us consider the motion of a particle in the x-direction, under the action of a potental which is zero everywhere except over a certain region AB where it has a cosntant value V_0. Such a region AB is called a square *potential barrier* (Fig. 2.3).

According to classical mechanics a particle travelling in region I (on the left of the barrier) will overcome the barrier if its initial kinetic energy E is greater than the barrier height V_0. If its energy is smaller than V_0, it will be totally reflected and the particle can never be found in region III (on the right of the barrier). Quantum mechanical calculations predict that the particle can penetrate through the barrier even when its energy is less than V_0. This phenomenon is known as *tunneling* and the effect is known as *'Tunnel Effect'*.

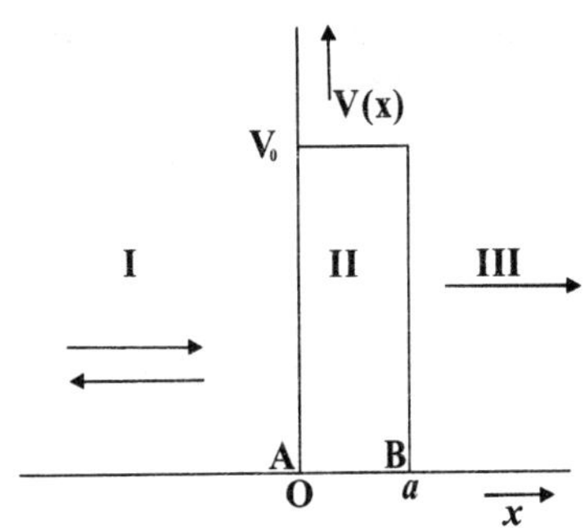

Fig. 2.3 Potential barrier

Consider a particle of mass m and energy E in the region I. The potential field can be represented as

$$V(x) = 0 \quad \text{for } x < 0 \quad \text{Region I}$$

$$V(x) = V_0 \quad \text{for } 0 \le x \le a \quad \text{Region II}$$

$$V(x) = 0 \quad \text{for } x > a \quad \text{Region III} \tag{2.33}$$

The situation represents the encounter between a free particle and a potential barrier. The particle can be turned back by the barrier or it can penetrate through. Thus in the region I the desired wave function has to represent a particle moving to the right (incident particle) and a particle moving to the left (reflected particle). In the region III the wave function has to represent only a particle moving to the right (transmited particle).

In the force free region a particle moving in a definite direction (x-direction) with a definite energy E has a definite momentum p_x. However its position is completely uncertain and it can be represented by a travelling wave. In the region I and III where $V(x) = 0$, the wave equation is given by

$$\frac{\partial^2 \psi}{\partial x^2} + \frac{2m}{\hbar^2} E\psi = 0$$

$$\Rightarrow \quad \frac{\partial^2 \psi}{\partial x^2} + k^2 \psi = 0 \tag{2.34}$$

where $\quad k^2 = \dfrac{2mE}{\hbar^2}$

The solutions of this equation are

$$\psi_1(x) = Ae^{ikx} + Be^{-ikx} \quad \text{for } x < 0 \quad \text{(Region I)} \tag{2.35}$$

and

$$\psi_3(x) = Ce^{ikx} \quad \text{for } x > a \quad \text{(Region III)} \tag{2.36}$$

For $x > a$, e^{-ikx} term has been omitted as there is no wave travelling in-x direction.

The physical meaning of the coefficients A, B and C can be understood in terms of probability current density. If v is the speed of a particle with propagation constant k, then the flux of particles in the incident beam is given by f_l = probability density of particles in the incident beam × particle velocity.

$$\Rightarrow \quad f_l = \left(Ae^{ikx} A^* e^{-ikx}\right) v$$

$$= (AA^*) v = |A|^2 v$$

Similarly the flux of particles in the reflected beam and transmitted beam may be written as $f_2 = |B|^2 v$ and $f_3 = |C|^2 v$, respectively. Therefore reflection coefficient (R) and transmission coefficient (T) are defined as $R = \dfrac{|B|^2 v}{|A|^2 v} = \left|\dfrac{B}{A}\right|^2$ and $T = \dfrac{|C|^2 v}{|A|^2 v} = \left|\dfrac{C}{A}\right|^2$ respectively. Thus the absolute values of A, B and C are not important only their ratios are of more significance.

In the region II $\left(0 \le x \le a \text{ and } V(x) = V_0\right)$, the solution of Schrödinger wave equation will depend on the energy of particle, *i.e.*, $E < V_0$ or $E > V_0$. Therefore both cases are discussed separately.

Case I When $E < V_0$ **(Tunnel Effect)**

In this case the wave equation is

$$\frac{\partial^2 \psi_2}{\partial x^2} - \frac{2m}{\hbar^2}(V_0 - E)\psi_2 = 0$$

$$\Rightarrow \qquad \frac{\partial^2 \psi_2}{\partial x^2} - \alpha^2 \psi_2 = 0 \tag{2.37}$$

where $\alpha = \left[\frac{2m(V_0 - E)}{\hbar^2}\right]^{1/2}$

The solution of this wave equation is

$$\psi_2(x) = Fe^{\alpha x} + Ge^{-\alpha x} \tag{2.38}$$

Now we will use the boundary condition to calculate the reflection coefficient and transmission coefficient.

$$\text{At } x = 0,\ \psi_1 = \psi_2 \text{ and } \frac{\partial \psi_1}{\partial x} = \frac{\partial \psi_2}{\partial x} \tag{2.39}$$

Applying this boundary condition on equations (2.35) and (2.38)

$$A + B = F + G \tag{2.40}$$

$$i\,k\,(A - B) = \alpha\,(F - G) \tag{2.41}$$

$$\text{Similarly} \quad x = a,\ \psi_2 = \psi_3,\ \text{and } \frac{\partial \psi_2}{\partial x} = \frac{\partial \psi_3}{\partial x} \tag{2.42}$$

using this boundary condition on equations (2.36) and (2.38)

$$Ce^{ika} = Fe^{\alpha a} + Ge^{-\alpha a} \tag{2.43}$$

$$ikCe^{ika} = \alpha\left(Fe^{\alpha a} - Ge^{-\alpha a}\right) \tag{2.44}$$

From equations (2.40) and (2.41)

$$2A = F\left(1 + \frac{\alpha}{ik}\right) + G\left(1 - \frac{\alpha}{ik}\right) = F\left(1 - \frac{i\alpha}{k}\right) + G\left(1 + \frac{i\alpha}{k}\right) \tag{2.45}$$

and

$$2B = F\left(1 - \frac{\alpha}{ik}\right) + G\left(1 + \frac{\alpha}{ik}\right) = F\left(1 + \frac{i\alpha}{k}\right) + G\left(1 - \frac{i\alpha}{k}\right) \tag{2.46}$$

$$\Rightarrow \qquad \frac{B}{A} = \frac{(k + i\alpha)F + G(k - i\alpha)}{(k - i\alpha)F + G(k + i\alpha)}$$

$$=\frac{(k+i\alpha)+\frac{G}{F}(k-i\alpha)}{(k-i\alpha)+\frac{G}{F}(k+i\alpha)} \tag{2.47}$$

Now dividing equation (2.44) by equation (2.43)

$$ik=\frac{\alpha\left(Fe^{\alpha a}-Ge^{-\alpha a}\right)}{Fe^{\alpha a}+Ge^{-\alpha a}}$$

$$\Rightarrow \quad ikFe^{\alpha a}+ikGe^{-\alpha a}=\alpha Fe^{\alpha a}-\alpha Ge^{-\alpha a}$$

$$\Rightarrow \quad (ik-\alpha)Fe^{\alpha a}=-(ik+\alpha)Ge^{-\alpha a}$$

$$\Rightarrow \quad \frac{G}{F}=-\left(\frac{ik-\alpha}{ik+\alpha}\right)e^{2\alpha a}$$

$$=-\left(\frac{-k-i\alpha}{-k+i\alpha}\right)e^{2\alpha a}$$

$$\frac{G}{F}=-\left(\frac{k+i\alpha}{k-i\alpha}\right)e^{2\alpha a} \tag{2.48}$$

Put this value in equation (2.47), we get

$$\frac{B}{A}=\frac{(k+i\alpha)+(k-i\alpha)\left[-\left(\frac{k+i\alpha}{k-i\alpha}\right)e^{2\alpha a}\right]}{(k-i\alpha)+(k+i\alpha)\left[-\left(\frac{k+i\alpha}{k-i\alpha}\right)e^{2\alpha a}\right]}$$

$$=\frac{(k+i\alpha)\left[1-e^{2\alpha a}\right](k-i\alpha)}{(k-i\alpha)^2-(k+i\alpha)^2e^{2\alpha a}}$$

$$=\frac{\left(k^2+\alpha^2\right)\left[1-e^{2\alpha a}\right]}{\left(k^2-2i\alpha k-\alpha^2\right)-\left(k^2+2i\alpha k-\alpha^2\right)e^{2\alpha a}}$$

$$=\frac{\left(k^2+\alpha^2\right)\left[1-e^{2\alpha a}\right]}{\left(k^2-\alpha^2\right)\left[1-e^{2\alpha a}\right]-2i\alpha k\left[1+e^{2\alpha a}\right]}$$

$$\frac{B}{A}=\frac{\left(k^2+\alpha^2\right)\left[e^{2\alpha a}-1\right]}{\left(k^2-\alpha^2\right)\left[e^{2\alpha a}-1\right]+2i\alpha k\left[e^{2\alpha a}+1\right]}$$

Since $\sinh x=\frac{e^x-e^{-x}}{2}$ and $\cosh x=\frac{e^x+e^{-x}}{2}$

$$\therefore \qquad \frac{B}{A}=\frac{\left(k^2+\alpha^2\right)\left[e^{\alpha a}-e^{-\alpha a}\right]}{\left(k^2-\alpha^2\right)\left[e^{\alpha a}-e^{-\alpha a}\right]+2i\alpha k\left[e^{\alpha a}+e^{-\alpha a}\right]}$$

$$\frac{B}{A}=\frac{\left(k^2+\alpha^2\right)\sinh\alpha a}{\left(k^2-\alpha^2\right)\sinh\alpha a+2i\alpha k\cosh\alpha a} \tag{2.49}$$

Now equation (2.43) dividing by equation (2.45), we get

$$\frac{Ce^{ika}}{2A}=\frac{Fe^{\alpha a}+Ge^{-\alpha a}}{F\left(1-\frac{i\alpha}{k}\right)+G\left(1+\frac{i\alpha}{k}\right)}$$

$$\Rightarrow \qquad \frac{C}{A}=\frac{2k\left[e^{\alpha a}+\frac{G}{F}e^{-\alpha a}\right]}{(k-i\alpha)+\frac{G}{F}(k+i\alpha)}$$

Put the value of $\frac{G}{F}$ using equation (2.48)

$$\frac{C}{A}=\frac{2k\left[e^{\alpha a}+e^{-\alpha a}\left\{-\left(\frac{k+i\alpha}{k-i\alpha}\right)e^{2\alpha a}\right\}\right]e^{-ika}}{(k-i\alpha)+(k+i\alpha)\left\{-\left(\frac{k+i\alpha}{k-i\alpha}\right)e^{2\alpha a}\right\}}$$

$$=\frac{2ke^{\alpha a}\left[(k-i\alpha)-(k+i\alpha)\right]e^{-ika}}{(k-i\alpha)^2-(k+i\alpha)^2e^{2\alpha a}}$$

$$=-\frac{4i\alpha ke^{\alpha a}e^{-ika}}{\left(k^2-\alpha^2-2i\alpha k\right)-\left(k^2-\alpha^2+2i\alpha k\right)e^{2\alpha a}}$$

$$=\frac{-4i\alpha ke^{\alpha a}e^{-ika}}{\left(k^2-\alpha^2\right)\left(1-e^{2\alpha a}\right)-2i\alpha k\left(e^{2\alpha a}+1\right)}$$

$$=\frac{4i\alpha ke^{\alpha a}e^{-ika}}{\left(k^2-\alpha^2\right)\left(e^{2\alpha a}-1\right)+2i\alpha k\left(e^{2\alpha a}+1\right)}$$

$$=\frac{2i\alpha ke^{-ika}}{\left(k^2-\alpha^2\right)\left(\frac{e^{\alpha a}-e^{-\alpha a}}{2}\right)+2i\alpha k\left(\frac{e^{\alpha a}+e^{-\alpha a}}{2}\right)}$$

$$\frac{C}{A} = \frac{2i\alpha k e^{-ika}}{\left(k^2 - \alpha^2\right)\sinh \alpha a + 2i\alpha k \cosh \alpha a} \tag{2.50}$$

Now reflection coefficient $R = \left|\frac{B}{A}\right|^2 = \left(\frac{B}{A}\right)\left(\frac{B}{A}\right)^*$

From equation (2.49)

$$R = \frac{\left(k^2 + \alpha^2\right)^2 \sinh^2 \alpha a}{\left[\left(k^2 - \alpha^2\right)\sinh \alpha a + 2i\alpha k \cosh \alpha a\right]\left[\left(k^2 - \alpha^2\right)\sinh \alpha a - 2i\alpha k \cosh \alpha a\right]}$$

$$= \frac{\left(k^2 + \alpha^2\right)^2 \sinh^2 \alpha a}{\left(k^4 + \alpha^4 - 2k^2\alpha^2\right)\sinh^2 \alpha a + 4k^2\alpha^2 \cosh^2 \alpha a}$$

$$= \frac{\left(k^2 + \alpha^2\right)^2 \sinh^2 \alpha a}{\left(k^4 + \alpha^4 - 2k^2\alpha^2\right)\sinh^2 \alpha a + 4k^2\alpha^2 \left(1 + \sinh^2 \alpha a\right)}$$

$$R = \frac{\left(k^2 + \alpha^2\right)^2 \sinh^2 \alpha a}{\left(k^2 + \alpha^2\right)^2 \sinh^2 \alpha a + 4k^2\alpha^2} \tag{2.51}$$

$$R = \frac{1}{\left[1 + \frac{4k^2\alpha^2}{\left(k^2 + \alpha^2\right)^2 \sinh^2 \alpha a}\right]} \tag{2.52}$$

As we know that $k^2 = \frac{2mE}{\hbar^2}$ and $\alpha^2 = \frac{2m}{\hbar^2}\left(V_0 - E\right)$

$$R = \left[1 + \frac{4E\left(V_0 - E\right)}{\left(V_0\right)^2 \sinh^2 \alpha a}\right]^{-1} \tag{2.53}$$

and transmission coefficient $T = \left|\frac{C}{A}\right|^2 = \left(\frac{C}{A}\right)\left(\frac{C}{A}\right)^*$

From equation (2.50)

$$T = \frac{4\alpha^2 k^2}{\left(k^2 + \alpha^2\right)^2 \sinh^2 \alpha a + 4\alpha^2 k^2} \tag{2.54}$$

$$= \left[1 + \frac{\left(k^2 + \alpha^2\right)^2 \sinh^2 \alpha a}{4\alpha^2 k^2}\right]^{-1}$$

$$T = \left[1 + \frac{(V_0)^2 \sinh^2 \alpha a}{4E(V_0 - E)}\right]^{-1} \tag{2.55}$$

Thus $T \neq 0$, this implies that the particle has finite probability of penetrating the potential barrier.. This penetration is due to the wave nature of matter and is known as tunnel effect or quantum tunneling. This effect provides explanation for the following phenomena:

(*i*) the electrical break down of insulators

(*ii*) the switching action of a tunnel diode

(*iii*) the emission of α-particles from a radioactive nuclei

(*iv*) the emission of electrons from a cold metallic surface, *i.e.*,electron emission from a metal under the action of an electric field.

(*v*) the reverse break down of semiconductor diodes.

From equations (2.51) and (2.54) it can be proved that $R + T = 1$ **(2.56)**

The transmission coefficient is maximum when E approaches V_0. So αa is very small, *i.e.*,

$$E \to V_0,\ \alpha a \to 0 \ \Rightarrow \sinh \alpha a \to \alpha a$$

Using equation (2.55)

$$T = \left|\frac{C}{A}\right|^2 = \left[1 + \frac{(V_0)^2 \alpha^2 a^2}{4E(V_0 - E)}\right]^{-1}_{E \to V_0}$$

$$= \left[1 + \frac{(V_0)^2 2m(V_0 - E)a^2}{\hbar^2 4E(V_0 - E)}\right]^{-1}_{E \to V_0}$$

$$= \left[1 + \frac{m(V_0)^2 a^2}{2\hbar^2 E}\right]^{-1}_{E \to V_0}$$

$$T = \left[1 + \frac{m(V_0)a^2}{2\hbar^2}\right]^{-1} \tag{2.57}$$

From equation (2.57) it is clear that the width a or the height of potential barrier (V_0) or both increases, the transmission probability of particle decreases.

If $\alpha a >> 1$, then $\sinh \alpha a \approx \frac{1}{2} e^{\alpha a}$

Put this value in equation (2.55), the transmission coefficient becomes

$$T \approx \left[1+\frac{(V_0)^2 e^{2\alpha a}}{16E(V_0 - E)}\right]^{-1}$$

$$\Rightarrow \qquad T \approx \frac{16E(V_0 - E)}{(V_0)^2} e^{-2\alpha a} \qquad \textbf{(2.58)}$$

$\Rightarrow$ As αa increases, T decreases exponentially.

In the case of $E < V_0$, the wave function will be of the form as depicted in Fig. (2.4).

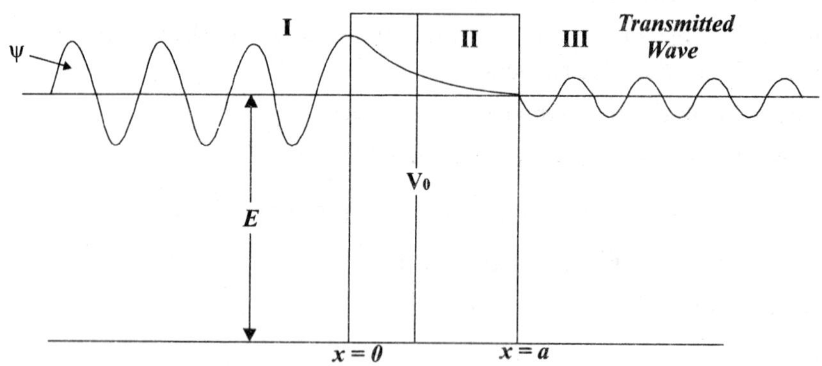

Fig. 2.4 Barrier penetration

Case II When $E > V_0$ the Schrödinger equations for all three regions

$$\frac{\partial^2 \psi_1}{\partial x^2} + \frac{2m}{\hbar^2} E\psi_1 = 0$$

or $$\frac{\partial^2 \psi_1}{\partial x^2} + k^2\psi_1 = 0 \qquad x < 0 \qquad \textbf{(2.59)}$$

$$\frac{\partial^2 \psi_2}{\partial x^2} + k'^2 \psi_2 = 0 \qquad 0 \le x \le a \qquad \textbf{(2.60)}$$

and $$\frac{\partial^2 \psi_3}{\partial x^2} + k^2\psi_3 = 0 \qquad x > a \qquad \textbf{(2.61)}$$

where $k^2 = \frac{2mE}{\hbar^2}$ and $k'^2 = \frac{2m}{\hbar^2}(E - V_0)$ **(2.62)**

As we know that $\alpha^2 = \frac{2m}{\hbar^2}(V_0 - E)$ in previous case $E < V_0$

Therefore $\alpha^2 = -\frac{2m}{\hbar^2}(E - V_0)$

$$\alpha^2 = i^2 k'^2$$

$\Rightarrow$ $\alpha = ik'$ $\therefore$ $\sinh \alpha a = i \sin k'a$ **(2.63)**

Thus the value of reflection coefficient (R) and transmission coefficient (T) can be calculated by putting $\alpha = ik'$ in equations (2.52) and (2.54).

$$R = \left[1 + \frac{4k^2 k'^2}{\left(k^2 - k'^2\right)^2 \sin^2 k'a}\right]^{-1} \quad \text{and} \quad T = \left[1 + \frac{\left(k^2 - k'^2\right)^2 \sin^2 k'a}{4k^2 k'^2}\right]^{-1} \tag{2.64}$$

Put the values of k and k' then R and T become

$$R = \left[1 + \frac{4E\left(E - V_0\right)}{\left(V_0\right)^2 \sin^2 k'a}\right]^{-1} \tag{2.65}$$

and $$T = \left[1 + \frac{\left(V_0\right)^2 \sin^2 k'a}{4E\left(E - V_0\right)}\right]^{-1}$$

(2.66)

From equation (2.64) it can be proved that $R + T = 1$. Keeping $E > V_0$, as E approaches its lower limit V_0, k' becomes small so that $\sin^2 k'a \approx k'^2 a^2 = \frac{2m\left(E - V_0\right)}{\hbar^2} a^2$

In this limiting condition

$$T = \left|\frac{C}{A}\right|^2 = \left[1 + \frac{\left(V_0\right)^2 2m\left(E - V_0\right)a^2}{4E\hbar^2\left(E - V_0\right)}\right]^{-1}_{E \to V_0}$$

$$T = \left[1 + \frac{mV_0 a^2}{2\hbar^2}\right]^{-1} \tag{2.67}$$

i.e., the transmission probability of wave packet is less than one, even when the energy of the particle is greater than the height of potential barrier. When $k'a = \pi, 2\pi, 3\pi, ... = n\pi$ then $\sin k'a = 0$ and $T = 1$, *i.e.*, there is perfect transmission.

Since $$\lambda' = \frac{2\pi}{k'}$$

$\therefore$ $$\frac{2\pi a}{\lambda'} = n\pi$$

$\Rightarrow$ $$a = \frac{n\lambda'}{2}$$

Hence when the barrier thickness is equal to an integral multiple of half wavelength, the barrier does not obstruct the motion of particle, *i.e.*, $T = 1$.

As E increases (with $E > V_0$) the transmission coefficient oscillates and for large values of E it approcaches unity (Fig. 2.5).

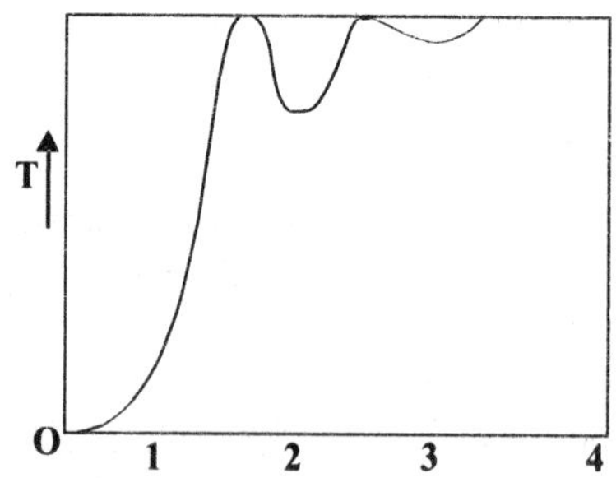

Fig. 2.5 Oscillations of transmission coefficient

Example 2.3: The potential barrier problem is a good approximation to the problem of an electron trapped inside but near the surface of metal. Calculate the probability of transmission that a 1.0 eV electron will penetrate a potential barrier of 4.0 eV when the barrier width is 2.0 Å.

[RTU B.Tech. II sem. 2012]

Solution: The transmission coefficient is given by

$$T \approx \frac{16E(V_0 - E)}{V_0^2} \exp\left[\frac{-2a}{\hbar}\sqrt{2m(V_0 - E)}\right]$$

Here $E = 1$ eV, $V_0 = 4$ eV, $a = 2$ Å $= 2 \times 10^{-10}$ m

$\hbar = 1.054 \times 10^{-34}$ J-s, $m = 9.1 \times 10^{-31}$ kg

$(V_0 - E) = (4 - 1)\text{eV} = 3 \times 1.6 \times 10^{-19}$ J

$$\therefore \quad T = 16\left(\frac{1}{4}\right)\left(1 - \frac{1}{4}\right)\exp\left[\frac{-2 \times 2 \times 10^{-10} \times \sqrt{2 \times 9.1 \times 10^{-31} \times 3 \times 1.6 \times 10^{-19}}}{1.054 \times 10^{-34}}\right]$$

$\sim 0.087 = 8.7\%$ **Ans.**

Thus, only about 8 electrons of energy 1.0 eV out of every hundred, penetrate the barrier.

Example 2.4: Consider an electron whose total energy is 5 eV approaching a barrier whose height is 6 eV and width is 7 Å. Find out the de Broglie wavelength of incident electron and probability of transmission through the barrier. (Mass of electron $m = 9.1 \times 10^{-31}$ kg, Planck's constant $= 6.6 \times 10^{-34}$ J-s).

(Raj. B.E.I. 2000)

Solution: The de Broglie wavelength of an electron is given by

$$\lambda = \frac{12.27}{\sqrt{V}} \text{Å}$$

Here $V = 5$ Volt (as electron energy is 5 eV)

$$\therefore \quad \lambda = \frac{12.27}{\sqrt{5}} = 5.48 \text{ Å}$$ **Ans.**

The transmission coefficient is given by

$$T = \frac{16E(V_0 - E)}{V_0^2} \exp\left[\frac{-2a}{\hbar}\sqrt{2m(V_0 - E)}\right]$$

Here $V_0 = 6$ eV, $a = 7 \times 10^{-10}$ m, $\hbar = 1.054 \times 10^{-34}$ J-s

$E = 5$ eV

$$\therefore \quad T = \frac{16\times5\times(6-5)}{36}\exp\left[\frac{2\times7\times10^{-10}}{1.054\times10^{-34}}\sqrt{2\times9.1\times10^{-31}\times1\times1.6\times10^{-19}}\right]$$

$$\approx 1.69\times10^{-3}$$
$$= 0.169\ \%$$

Ans.

Example 2.5: Electrons of energy 2 eV are incident on a potential barrier of height 5 eV and width 5Å . Find tunnel probability for these electrons. **[RTU B.Tech. I Sem. 2008]**

Solution: Given $E = 2\text{ eV} = 2\times1.6\times10^{-19}\text{ J}$

$V_0 = 5\text{ eV} = 5\times1.6\times10^{-19}\text{ J}$

and $a = 5\text{ Å} = 5\times10^{-10}\text{ m}$

The tunnel probability is given by

$$T = \frac{16E(V_0-E)}{V_0^2}e^{-2\alpha a}$$

where

$$\alpha = \left[\frac{2m}{\hbar^2}(V_0-E)\right]^{1/2}$$

$$= \left[\frac{2\times9.1\times10^{-31}\times(5-2)\times1.6\times10^{-19}}{(1.054\times10^{-34})^2}\right]^{1/2}$$

$$= \left(\frac{29.12\times3}{1.11}\right)^{1/2}\times10^9$$

$$= (26.23\times3)^{1/2}\times10^9$$

$$= 5.12\times\sqrt{3}\times10^9$$

$$= 5.12\times1.732\times10^9$$

$$= 8.8628\times10^9$$

Now

$$T = \frac{16\times2\times(5-2)}{25}\exp\left[-2\times8.867\times10^9\times5\times10^{-10}\right]$$

$$= \frac{96}{25}\exp[-8.867]$$

$$= 3.84\exp(-8.867)$$

$$= 3.84\times1.42\times10^{-4}$$

$$T = 5.452\times10^{-4}$$
$$= 0.054\ \%$$

Ans.

Example 2.6: A beam of electrons is incident on a potential barrier of height 5 eV and width 0.3 nm. What should be the energy of the electrons so that $\left(\frac{1}{3}\right)$ of them are able to penetrate through the barrier ?

Solution: Given $V_0 = 6$ eV and $a = 0.3$ nm $= 3\times10^{-10}$ m

The probability of transmission is given by

$$T = \frac{16E(V_0 - E)}{V_0^2} e^{-2\alpha a} = T_0 e^{-2\alpha a}$$

where $T_0 = \frac{16E(V_0 - E)}{V_0^2}$ maximum transmission probability

and $$\alpha = \sqrt{\frac{2m}{\hbar^2}(V_0 - E)}$$

Given $$T = \frac{T_0}{3}$$

$\therefore$ $$\frac{T_0}{3} = T_0 e^{-2\alpha a}$$

$\Rightarrow$ $$e^{2\alpha a} = 3$$

$\Rightarrow$ $$2\alpha a = \ln 3$$

$\Rightarrow$ $$(2\alpha a)^2 = (\ln 3)^2$$

$$4.\frac{2m(V_0 - E)}{\hbar^2} a^2 = (\ln 3)^2$$

$$(V_0 - E) = \frac{\hbar^2 (\ln 3)^2}{8ma^2}$$

$$(V_0 - E) = \frac{(1.054\times10^{-34})^2 (1.098)^2}{8\times9.1\times10^{-31}\times(3\times10^{-10})^2}$$

$$V_0 - E = 0.201\times10^{-19}\text{ J}$$

$$V_0 - E = \frac{0.201\times10^{-19}}{1.6\times10^{-19}}\text{eV} = 0.125\text{ eV}$$

$\Rightarrow$ $$E = (V_0 - 0.125)\text{ eV} = 6 - 0.125$$

$$E = 5.875\text{ eV}$$ **Ans.**

2.4 THEORY OF α-DECAY

The theory of α-decay was explained by George Gamow in 1928 on the basis of quantum mechanical barrier penetration. There are certain nuclei which emit α-particles and get converted into new nuclei with atomic number less by two and mass number less by four. This process is known as α-decay and is depicted as follows:

$$_{Z}X^{A} \rightarrow {}_{Z-2}Y^{A-4} + {}_{2}\text{He}^{4}$$

When any α-particle is present inside the nucleus, the energy of α-particle cannot be greater than the height of the potential barrier which is existing at the surface of the nucleus. It implies that according to classical mechanics it is never possible for the α-particle to come out of the nucleus.

The first difficulty involves the energy of α- particle in the field of the nucleus. It was concluded from Rutherford's scattering experiments that the force experienced by α-particles is a Coulomb force which acts at very small distance $\left(=10^{-14}\text{m}\right)$ from the centre of the nucleus. After this direct distance, Coulomb's law breaks down and the boundary of the nucleus starts. Thus, there is only Coulomb potential outside the nucleus which is depicted in Fig. 2.6. As soon as the α-particle enters the nucleus ($0 < r < a$), it will be under the influence of very strong attractive nuclear forces. These forces are represented by a potential well as depicted in Fig. 2.6.

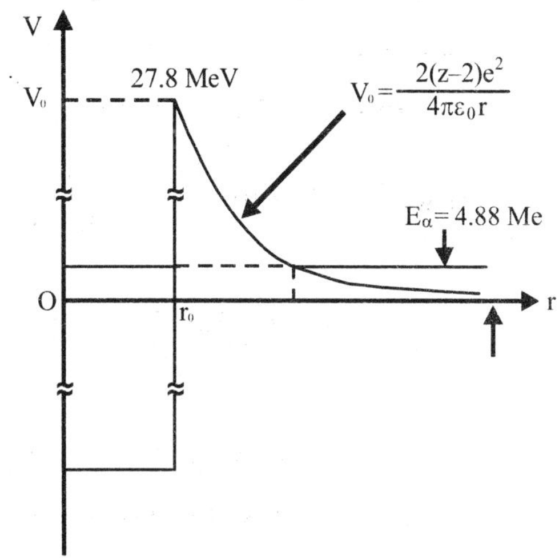

Fig. 2.6 Potential for the α-particle in the nucleus

If the radius of the radium nucleus is 9.1×10^{-15} m, the Coulomb potential outside the nucleus is 27.8 MeV. However, the energy of α-particle emitted by a radium nucleus is 4.88 MeV only. Now the question arises: How can a α-particle of 4.88 MeV energy go through a potential barrier of 27.8 MeV ?

The second difficulty is concerned with the law of casuality. According to the radioactive decay law, the amaount of radium left after 1620 years should be half. After several half lives, a small fraction of radium should be left. Now the question arises. *Why are only some radium atoms decaying in the first few years and why are some others surviving for thousands of years ?*

When quantum mechanics was applied to the problem of α-decay, the difficulties of classical mechanics as described above disappear. According to quantum mechanics, every particle is assigned a wave aspect. If this fact is taken into consideration, then the electrostatic potential barrier, although very high, cannot completely rule out the passage of wave through it. There always exists certain probability of the particle to pentrate through the barrier, how small the energy might be. It is possible to obtain some quantitative features of the theory from the expression of the transmission coefficient.

$$T = \frac{16E\left(V_0 - E\right)}{V_0^2}\exp\left[\frac{-2a}{\hbar}\sqrt{2m\left(V_0 - E\right)}\right] \tag{2.68}$$

Now an attempt is being made to calculate the half life time in case of uranium nucleus which is having radius of about 10^{-14} m. There is an evidence that the α-particle moves back and forth freely with an average speed of 10^7 m/s. Thus the α-particle will strike the barrier $10^7/10^{-14} = 10^{21}$ times per second. The probability that it penetrates the barrier is equal to the transmittance (T).

Hence, the probability (P) that α-particle leaks out from nucleus in one second, which is called decay constant (λ), *i.e.*,

$$\lambda = 10^{21}\, T \text{ per sec}$$

$$= 10^{21} \times \frac{16E(V_0 - E)}{V_0^{\,2}} \exp\left[\frac{-2a}{\hbar}\sqrt{2m(V_0 - E)}\right] \qquad \textbf{(2.69)}$$

For uranium, E = 4.2 MeV and V_0 = 30 MeV.

Mass of α-particle $= 4 \times 1.67 \times 10^{-27}$ kg.

$\hbar = 1.054 \times 10^{-34}$ J-s, $a = 2 \times 10^{-14}$m

Thus, $$\lambda = 10^{21} \times 16 \times \frac{(4.2\text{ MeV}) \times (30 - 4.2)\text{MeV}}{(30\text{ MeV})^2} \exp\left[\frac{-2a}{\hbar}\sqrt{2m(V_0 - E)}\right]$$

$$\Rightarrow \lambda = 10^{21} \times \frac{16 \times 4.2 \times 25.8}{30 \times 30} \exp\left[\frac{-2\sqrt{(2 \times 4 \times 1.67 \times 10^{-27} \times 25.8 \times 1.6 \times 10^{-13})} \times 2 \times 10^{-14}}{1.054 \times 10^{-34}}\right]$$

$$= 1.9 \times 10^{21} e^{(\approx -89)} = 1.9 \times 2.22 \times 10^{21} \times 10^{-39}$$

$$\approx 4.218 \times 10^{-18} \text{ per sec}$$

$$= 1.328 \times 10^{-10} \text{ per year}$$

Therefore half-life time of nucleus is

$$T_{1/2} = \frac{0.693}{\lambda} = 5.21 \times 10^{9} \text{ years} \qquad \textbf{(2.70)}$$

But experimentally, $T_{1/2}$ had been found to be 10^{19} years. The discrepancy has been ascribed to the fact that the barrier is not rectangular. This result explains why most energetic particles are decaying in first few years and why least energetic particles are surviving for thousands of years.

Example 2.7: Find the height of the potential barrier for α-particles emitted from radon $\left({}_{86}Rn^{222}\right)$ assuming that the effective nuclear radius is given by $r_0 = 1.5 \times 10^{-15} A^{1/3}$m.

Solution: $$V_0 = \frac{2(Z-2)e^2}{4\pi\varepsilon_0 r_0}$$

Here $Z = 86$, $e = 1.6 \times 10^{-19} C$

$$\frac{1}{4\pi\varepsilon_0} = 9 \times 10^{9}, \quad r_0 = 1.5 \times 10^{-15}(222)^{1/3}$$

$$\therefore \quad V_0 = \frac{2 \times (86-2) \times \left(1.6 \times 10^{-19}\right)^2 \times 9 \times 10^{9}}{1.5 \times 10^{-15} \times (222)^{1/2} \times 1.6 \times 10^{-19}}$$

$$= 26.62 \times 10^{6}\text{eV}$$

$$= 26.62 \text{ MeV}$$

Ans.

Example 2.8: Find the width of the potential barrier for an α-particle emitted with kinetic energy 5.5 MeV from $\left({}_{86}\text{Rn}^{222}\right)$. How many collisions does the α-particle make per second on the barrier. Assume that it collides on both sides of the sphere to cover the distance $2R$ during each collision, R being the radius of the nucleus.

Solution: The width of the potential barrier due to the nucleus is given by $a = r_1 - r_0$ where r_0 is the effective nuclear radius, and r_1 is the distance from the centre of the nucleus at which the potential energy of the α-particle is equal to the kinetic energy of the emitted α-particle.

$$E = \frac{2(Z-2)e^2}{4\pi\varepsilon_0 r_1}$$

$$\Rightarrow \qquad r_1 = \frac{2(Z-2)e^2}{4\pi\varepsilon_0 E}$$

Here $r_0 = 1.5\times10^{-15}A^{1/3} = 1.5\times10^{-15}(222)^{1/3}$

$$= 9.08\times10^{-15}\,\text{m}$$

$$r_1 = \frac{2\times84\times\left(1.6\times10^{-19}\right)^2\times9\times10^9}{5.5\times1.6\times10^{-13}}$$

$$= 43.98\times10^{-15}\,\text{m}$$

$$\therefore \qquad a = r_1 - r_0 = (43.98 - 9.08)\times10^{-15}\,\text{m}$$

$$= 34.90\times10^{-15}\,\text{m}$$ **Ans.**

The velocity of electron at kinetic energy 5.5 MeV

$$5.5\text{ MeV} + m_0c^2 = mc^2$$

$$\Rightarrow \qquad mc^2 = 5.5 + 0.505 \approx 6\text{ MeV}$$

$$\frac{m_0c^2}{\sqrt{1-\frac{v^2}{c^2}}} = 6$$

$$\sqrt{1-\frac{v^2}{c^2}} = \frac{0.5}{6} = \frac{1}{12}$$

$$\Rightarrow \qquad \frac{v^2}{c^2} = \frac{143}{144}$$

$$v = c\sqrt{\frac{143}{144}} = 0.996c$$

$$v = 2.98\times10^8\text{ m/s}$$

Therefore number of collisions per second is

$$n = \frac{v}{2r_0}$$

$$= \frac{2.98 \times 10^8}{2 \times 9.08 \times 10^{-15}}$$

$$= 1.64 \times 10^{22}$$ **Ans.**

2.5 SOMMERFELD'S FREE ELECTRON GAS MODEL OF SOLIDS (METALS)

According to free electron theory, the solid crystal is a composition of positive ion cores and free electrons which are unaffected by the periodic potential of the ion cores. This model was first developed by Drude and Lorentz in 1900 to explain the electrical and thermal conductivity of metals. They assumed the free electrons in a metal to behave like molecules of a perfect gas obeying Maxwell Boltzmann Statistics of electron distribution. This theory provides the correct explanation of Ohm's Law, the variation of electrical resistance with temperature at ordinary temperature and Wiedmann-Franz Law (the ratio of electrical conductivity to the thermal conductivity) but fails to explain many experimentally observed facts. Like the variation of electrical conductivity of a metal with temperature, heat capacity and paramagnetic susceptibility of the conduction electrons, the large variation in mean free path (λ) values *etc.* could not be explained by Drude-Lorentz theory. In 1928 Sommerfeld modified the free electron model by taking into account the quantum statistics of Fermi-Dirac in place of the classical statistics (Maxwell-Boltzmann Statistics) and the Pauli's exclusion principle.

Arnold Johannes Wilhelm Sommerfeld
(5 December 1868 – 26 April 1951) was a German theoretical Physicist who pioneered developments in atomic and quantum physics. He introduced the 3rd quantum number (azimuthal quantum number) and the 4th quantum number (spin quantum number). He also introduced the fine-structure constant, and pioneered X-ray wave theory. In 1927 Sommerfeld applied Fermi-Dirac statistics to the Drude model of electrons in metals.

2.5.1 Postulates of Sommerfeld Model

This model is based on following postulates.

(*a*) In this model, the valence electrons of atoms are assumed to be the free electrons. Since the potential inside the metal is taken as constant so that the force acting on the valence electron is zero.

(*b*) Since electron emission from metals does not occur at room tamperature, the potential energy of these free electrons is assumed to be less than that of free electrons outside the metal. Thus the motion of a free electron in a solid is treated as the motion of free particle in a three dimensional box.

(*c*) The free electron gas in a solid obeys Fermi-Dirac Statistics. According to this , the probability $F(E)$ of an electron occupying a given energy level is given by

$$F(E) = \frac{1}{1 + e^{(E-E_f)/k_B T}}$$

where E_f is the highest energy state for the electrons to occupy at absolute zero ($T = 0\ K$). It is called Fermi energy.

(*d*) The electrons obey Pauli's exclusion principle. According to this, not more than two electrons may occupy any orbital state.

Thus Sommerfeld treated the solid as a three dimensional potential box and calculated the density of energy states which is defined as the number of energy states per unit energy range per unit volume of a solid.

2.5.2 Density of Energy States and Fermi Energy

According to Sommerfeld, the metal (solid) is in the form of a cube of side a and electrons are moving inside the solid with zero potential energy. The wave function for a free particle inside a three dimensional cubical box is given by

$$\psi = \left(\frac{2}{a}\right)^{3/2} \sin\frac{n_x\pi}{a}x \sin\frac{n_y\pi}{a}y \sin\frac{n_z\pi}{a}z \tag{2.71}$$

The possible energy levels are given by

$$E = \frac{\pi^2\hbar^2}{2ma^2}\left(n_x^2 + n_y^2 + n_z^2\right) = \frac{\pi^2\hbar^2}{2ma^2}n^2 \tag{2.72}$$

where n_x, n_y and n_z are positive integers and represent the coordinates in a three dimensional space [Fig. 2.7] and $n^2 = n_x^2 + n_y^2 + n_z^2$.

To find the number of states with energy lying between zero and E, let us construct a space of points represented by the values of n_x, n_y and n_z. Each point in this space with integer values of the coordinates represent an energy state.

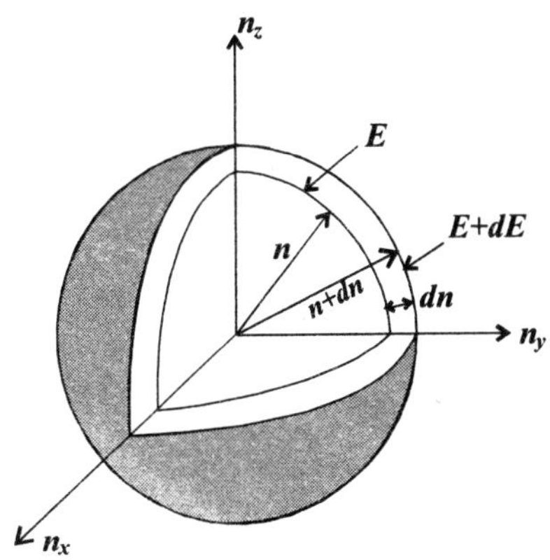

Fig. 2.7 Representative space to calculate density of states

Hence, the number of states in any volume is just equal to the numerical value of the volume. Therefore the number of available states within a sphere of radius n, *i.e.*, energy E is given by $= \frac{1}{8}\left(\frac{4}{3}\pi n^3\right)$ *

Similarly the number of states within a sphere of radius $(n + dn)$, *i.e.*, energy $(E + dE)$ is

$$= \frac{1}{8}\left(\frac{4\pi}{3}(n+dn)^3\right)$$

Thus the number of energy states having energy values between E an $E + dE$ is given by

$$dN(E) = \frac{1}{8}\left(\frac{4\pi}{3}(n+dn)^3\right) - \frac{1}{8}\left(\frac{4\pi}{3}n^3\right)$$

$$= \frac{\pi}{6}\left[n^3 + dn^3 + 3n^2dn + 3ndn^2 - n^3\right]$$

Since dn is very small therefore we neglect dn^2 and dn^3 terms.

$$dN(E) \approx \frac{\pi}{6} \times 3n^2dn \approx \frac{\pi}{2}n^2dn$$

As we know that the Pauli's exclusion principle permits two electrons in each state, so that the number of energy levels actually available are

$$dN(E) \approx 2 \times \frac{\pi}{2}n^2dn$$

$$dN(E) \approx \pi n^2 dn \qquad \textbf{(2.73)}$$

From equation (2.72)

$$E = \frac{\pi^2\hbar^2}{2ma^2}n^2$$

$$\Rightarrow \qquad n^2 = \frac{2ma^2}{\pi^2\hbar^2}E \text{ and } n = \left[\frac{2ma^2}{\pi^2\hbar^2}E\right]^{1/2}$$

Differentiating the n^2 term, we get

$$2ndn = \frac{2ma^2}{\pi^2\hbar^2}dE$$

Put the value of n in equation (2.73)

$$dN(E) = \pi n(ndn)$$

$$= \pi\left[\frac{2ma^2}{\pi^2\hbar^2}E\right]^{1/2}\frac{ma^2}{\pi^2\hbar^2}dE$$

$$= \frac{\pi}{2}\left[\frac{2ma^2}{\pi^2\hbar^2}\right]^{3/2}E^{1/2}dE$$

---

* The factor $\frac{1}{8}$ accounts for the fact that only positive integers are allowable and thus only one octant of the sphere alone be considered.

$$= \frac{\pi}{2} a^3 \left[\frac{2m}{\pi^2 \hbar^2}\right]^{3/2} E^{1/2} dE$$

$$dN(E) = \frac{\pi}{2} V \left[\frac{2m}{\pi^2 \hbar^2}\right]^{3/2} E^{1/2} dE \quad (2.74)$$

where $V = a^3$ is the volume of the potential box. Therefore the density of states is given by

$$Z(E) = \frac{dN(E)}{dE} \frac{1}{V} = \frac{\pi}{2} \left[\frac{2m}{\pi^2 \hbar^2}\right]^{3/2} E^{1/2} \quad (2.75)$$

Here $Z(E)$ is the number of states per unit energy range at the energy E per unit volume. The function $Z(E)$ is plotted in Fig. (2.8). The area of the strip of width dE gives the corresponding number $d\,N(E) = Z(E)\,dE$ of the states in such an energy range. The area under the curve from $E = 0$ to $E = E_f$ gives the total number of states in that energy interval.

The total number of electrons per unit volume $n(E)dE$ in a given interval dE will be obtained by multiplying the number of energy states in the interval by the Fermi distribution function.

$$n(E)\,dE = Z(E)\,dE\,F(E) = \frac{\pi}{2} \left[\frac{2m}{\pi^2 \hbar^2}\right]^{3/2} E^{1/2} \frac{d(E)}{1 + e^{(E-E_f)/k_B T}}$$

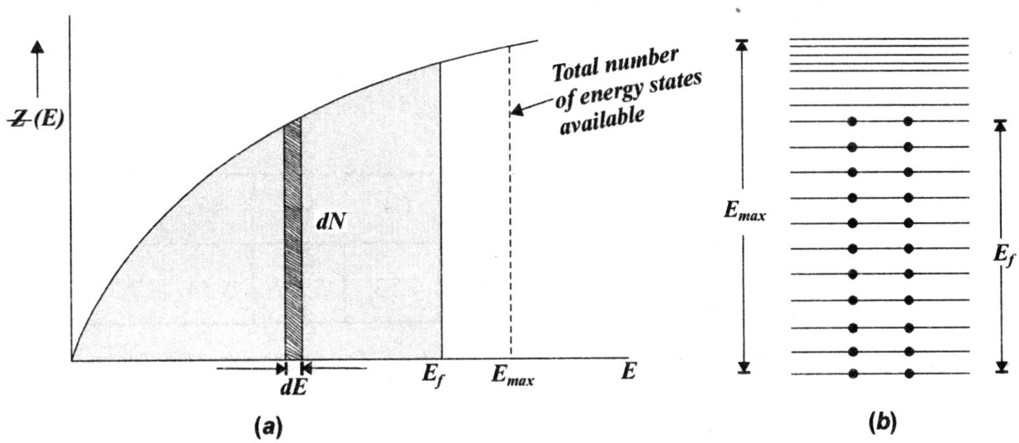

Fig. 2.8 (*a*) Density of energy levels in a solid (*b*) Distribution of free electrons among energy states

At absolute zero temperature all states of lowest energy are filled with electrons up to a certain energy level called Fermi level (E_f) and those above E_f are empty, *i.e.*, $F(E) = 1$ at $T = 0\,K$ and $E < E_f$. Hence the number of free electrons per unit volume occupied up to the Fermi energy level is given by

$$n_f = \int_0^{E_f} Z(E)dE$$

$$= \int_0^{E_f} \frac{\pi}{2}\left[\frac{2m}{\pi^2\hbar^2}\right]^{3/2} E^{1/2} dE$$

$$= \frac{\pi}{2}\left(\frac{2m}{\pi^2\hbar^2}\right)^{3/2} \frac{2}{3} E_f^{\ 3/2}$$

$$n_f = \frac{\pi}{3}\left(\frac{2m}{\pi^2\hbar^2}\right)^{3/2} E_f^{\ 3/2} \quad \textbf{(2.76)}$$

thus Fermi energy

$$E_f = \frac{\pi^2\hbar^2}{2m}\left(\frac{3n_f}{\pi}\right)^{2/3} \quad \textbf{(2.77)}$$

$$\Rightarrow \quad E_f = \frac{(3.14)^2 (1.054\times10^{-34})^2 \times 3^{2/3}}{2\times 9.1\times 10^{-31} \times (3.14)^{2/3}} n_f^{\ 2/3}$$

$$E_f = 5.838\times10^{-38} n_f^{\ 2/3} \text{J}$$

$$E_f = \frac{5.838\times10^{-38}}{1.6\times10^{-19}} n_f^{\ 2/3} \text{ eV}$$

$$E_f = 3.65\times10^{-19} n_f^{\ 2/3} \text{ eV} \quad \textbf{(2.78)}$$

This is expression for Fermi energy. When the Fermi energy is equal to the energy bandwidths, the band is fully occupied. The following table gives the calculated values of the Fermi energies of a number of metals.

Metal	Li	Na	K	Rb	Cu	Cs	Ag	Au	Mg	Al
E_f ,ev	4.72	3.12	2.14	1.82	4.07	1.53	5.51	5.54	7.3	11.9

2.5.3 The Fermi Distribution Function and Effect of Temperature on it

This function tells us that how many electrons of a real material distribute among various available energy states. The probability $F(E)$ for the occupation of a particular energy level E by an electron is given by

$$F(E) = \frac{1}{1+e^{(E-E_f)/k_BT}} \quad \textbf{(2.79)}$$

Here $F(E)$ = Fermi distribution function

E = Energy of the level whose occupancy is being considered

E_f = Fermi energy level

k_B = Boltzmann constant

and T = Absolute temperature

Case I

At absolute zero temperature ($T = 0\ K$) and for $E < E_f$, the function attains the value 1, when $E > E_f$, $F(E)$ will be zero. This means that at absolute zero, all the quantum states below E_f are filled while all the quantum states having energies greater than E_f are empty. Thus the Fermi energy level is the highest energy state for the electrons to occupy at absolute zero.

Case II

At $T > 0\ K$ and $E = E_f$, $F(E) = \frac{1}{2}$. Therefore Fermi level is that level at which the probability of occupation is $\frac{1}{2}$.

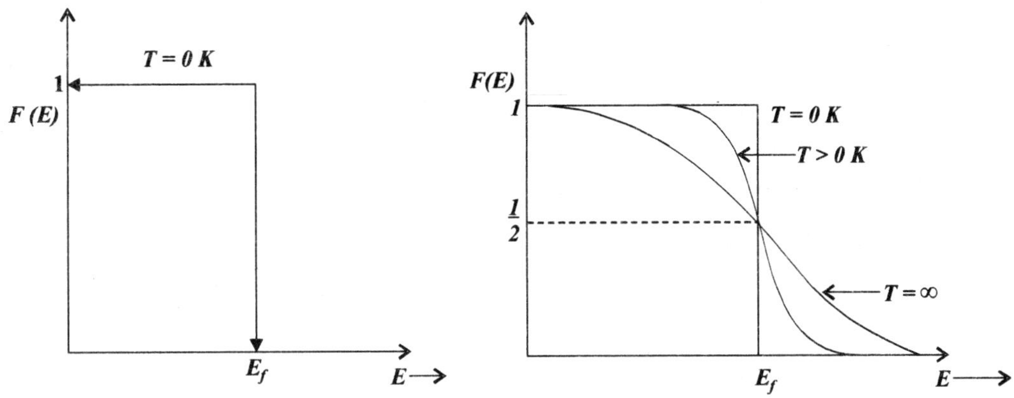

Fig. 2.9 The Fermi distribution function at different temperatures

2.5.4 Mean Energy of Electron Gas at Absolute Zero

As we know that the Fermi energy at absolute zero is given by

$$E_f = \frac{\pi^2 \hbar^2}{2m}\left(\frac{3n_f}{\pi}\right)^{2/3} \tag{2.80}$$

To calculate the average electron energy at 0 K, first obtain the total energy U_0 at 0 K, which is given by the following equation.

$$U_0 = \int_0^{E_f} Z(E)dE\, V\, F(E) \quad E \text{ with } F(E) = 1 \tag{2.81}$$

Using equation (2.75)

$$U_0 = \frac{\pi}{2}\left(\frac{2m}{\pi^2\hbar^2}\right)^{3/2} V \int_0^{E_f} E.E^{1/2}dE$$

$$U_0 = \frac{\pi}{2}\left(\frac{2m}{\pi^2\hbar^2}\right)^{3/2} V \int_0^{E_f} E^{3/2}dE$$

$$U_0 = \frac{\pi}{2} V \left(\frac{2m}{\pi^2\hbar^2}\right)^{3/2} \frac{2}{5} E_f^{\,5/2}$$

$$U_0 = \frac{\pi}{5} V \left(\frac{2m}{\pi^2\hbar^2}\right)^{3/2} E_f^{\,5/2} \tag{2.82}$$

The total number of electrons in a cube of volume V is given by

$$N = n_f \times V$$

From equation (2.76)

$$N = \frac{\pi}{3}\left(\frac{2m}{\pi^2\hbar^2}\right)^{3/2} E_f^{\,3/2} V \tag{2.83}$$

Thus the mean energy of the electron at 0 K is

$$\overline{E_0} = \frac{U_0}{N}$$

$$= \frac{\frac{\pi}{5} V \left(\frac{2m}{\pi^2\hbar^2}\right)^{3/2} E_f^{\,5/2}}{\frac{\pi}{3} V \left(\frac{2m}{\pi^2\hbar^2}\right)^{3/2} E_f^{\,3/2}}$$

$$\overline{E_0} = \frac{3}{5} E_f$$

Thus the mean energy of the electron at 0 K is 60% of the Fermi energy.

Example 2.9: There are 2.54×10^{22} free electrons per cm^3 in sodium. Calculate its Fermi energy, Fermi velocity and Fermi temperature.

($h = 6.63 \times 10^{-34}$ J-s, $m = 9.11 \times 10^{-31}$ kg, $k = 1.38 \times 10^{23}$ J/K, 1 eV $= 1.6 \times 10^{-19}$ Joule)

Solution: The Fermi energy of a metal is given by

$$E_f = \frac{h^2}{2m}\left(\frac{3N}{8\pi V}\right)^{2/3}$$

where m is the mass of electron, N is the number of free electrons in a volume V of the metal.

Here $\frac{N}{V} = 2.54 \times 10^{22}\,/\text{cm}^3 = 2.54 \times 10^{28}\,/\text{m}^3$

$$\therefore\ E_f = \frac{(6.63 \times 10^{-34})^2}{2 \times 9.1 \times 10^{-31}} \times \left(\frac{3 \times 2.54 \times 10^{28}}{8 \times 3.14}\right)^{2/3}$$

$$= 5.0 \times 10^{-19}\,\text{J} = 3.1\ \text{eV}$$ **Ans.**

This is the maximum kinetic energy of the free electron at absolute zero. If v_F is the Fermi velocity, then

$$E_F = \frac{1}{2}mv^2{}_F = 5 \times 10^{-19}\,\text{J}$$

$$\therefore\quad v_F = \left(\frac{2 \times 5 \times 10^{-19}}{9.11 \times 10^{-31}}\right)^{1/2} = 1.05 \times 10^6\ \text{m/s}$$

The Fermi temperature is defined by

$$T_F = \frac{E_F}{k} = \frac{5 \times 10^{-19}}{1.38 \times 10^{-23}} = 3.6 \times 10^4\,\text{K}$$ **Ans.**

Example 2.10: The Fermi energy in silver is 5.51eV

(*a*) What is the average energy of the free electrons in silver at 0 K?

(*b*) At what temperature a classical free particle will have this kinetic energy?

($k = 1.38 \times 10^{-23}$ J/K and $1\ \text{eV} = 1.6 \times 10^{-19}\,\text{J}$)

Solution: (*a*) The average energy of an electron is electron gas at 0 K is

$$\overline{E_0} = \frac{3}{5}E_F$$

Here $E_F = 5.51$ eV

$$\therefore\quad \overline{E_0} = \frac{3}{5} \times 5.51 = 3.306\ \text{eV}$$ **Ans.**

(*b*) The kinetic energy of a classical particle at $T\,K$ is $\frac{3}{2}kT$. Then, if the particle has an energy 3.306 eV, we have

$$\overline{E_0} = \frac{3}{2}kT$$

$$\Rightarrow \quad T = \frac{2\overline{E_0}}{3k} = \frac{2 \times 3.306 \times 1.6 \times 10^{-19}}{3 \times 1.38 \times 10^{-23}}$$

$$= 2.25 \times 10^4 \text{ K}$$ **Ans.**

Example 2.11: The Fermi energy for lithium is 4.72 eV at absolute zero. Calculate the number of conduction electrons per unit volume in lithium. ($h = 6.63 \times 10^{-34}$ J-s , $m = 9.11 \times 10^{-31}$ kg.)

Solution: The number of conduction electrons per unit volume in a metal is given by

$$\frac{N}{V} = \frac{8\pi}{3}\left(\frac{2m}{h^2}\right)^{3/2} E_F^{3/2}$$

Here $E_F = 4.72 \times 1.6 \times 10^{-19}$ J

$$\frac{N}{V} = \frac{8 \times 3.14}{3} \times \left(\frac{2 \times 9.11 \times 10^{-31}}{\left(6.63 \times 10^{-34}\right)^2}\right)^{3/2} \times \left(4.72 \times 1.6 \times 10^{-19}\right)^{3/2}$$

$$= 4.64 \times 10^{28} \text{ /m}^3$$ **Ans.**

Example 2.12: For lithium, the Fermi energy is 4.70 eV and the density of electrons is 4.6×10^{28} /m^3. Find the electron density for a metal with Fermi energy 2.35 eV.

Solution: The electron density for a metal with Fermi energy E_F is given by

$$\frac{N}{V} = n = \frac{8\pi}{3}\left(\frac{2m}{h^2}\right)^{3/2} E_F^{3/2}$$

$$\Rightarrow \quad n \propto E_F^{3/2}$$

$$\therefore \quad \frac{n'}{n} = \left(\frac{E_F'}{E_F}\right)^{3/2}$$

$$\therefore \quad n' = 4.6 \times 10^{28} \times \left(\frac{2.35}{4.70}\right)^{3/2} = 1.6 \times 10^{28} \text{ /m}^3$$ **Ans.**

Example 2.13: Calculate the Fermi energy in eV for silver at 0 *K*, given that the density of silver = 10500 kg/m^3, atomic weight = 107.9, and it has one conduction electron per atom.

[RTU B.Tech. II sem. 2010]

Solution: The number of atoms per unit volume is

$$= \frac{N}{V} = \frac{N_A \rho}{M}$$

where ρ = Density of element

M = Atomic weight of element

and N_A = Avogadro's number represents atoms per mole

$\therefore$ Number of electrons per unit volume

$$n_e = \frac{6.023 \times 10^{26} \times 10500}{107.9}$$

$$n_e = 5.86 \times 10^{28} \text{ electrons/m}^3$$

The Fermi energy for silver is

$$E_f = \frac{h^2}{2m}\left(\frac{3n_e}{8\pi}\right)^{2/3}$$

$$E_f = 3.65 \times 10^{-19}(n_e)^{2/3}$$

$$= 3.65 \times 10^{-19} \times 10^{18}(58.6)^{2/3}$$

$$= 0.365 \times (3.88)^2$$

$E_f = 5.49$ eV **Ans.**

Example 2.14: Calculate the number of states lying in an energy interval of 0.02 electrons volt above the Fermi energy for sodium crystal of unit volume (E_f = 3.22 eV for sodium).

Solution: Given E_f = 3.22 eV

$\therefore$ E_1 = 3.22 eV, E_2 = 3.24 eV

The number of energy states in unit volume having energies lying between E_1 and E_2 is given by

$$N(E) = \int_{E_1}^{E_2} \frac{\pi}{2}\left(\frac{2m}{\pi^2 \hbar^2}\right)^{3/2} E^{1/2} dE$$

$$= \frac{\pi}{2}\left(\frac{8m}{h^2}\right)^{3/2} \cdot \int_{E_1}^{E_2} E^{1/2} dE$$

$$= \frac{\pi}{2}\left(\frac{8m}{h^2}\right)^{3/2} \frac{2}{3}\left[E^{3/2}\right]_{E_1}^{E_2}$$

$$= \frac{\pi}{3}\left(\frac{8m}{h^2}\right)^{3/2}\left[E_2^{3/2} - E_1^{3/2}\right]$$

$$= \frac{3.14}{3}\left[\frac{8\times 9.1\times 10^{-31}}{\left(6.625\times 10^{-34}\right)^2}\right]^{3/2}\left[\left(3.24\times 1.6\times 10^{-19}\right)^{3/2}-\left(3.22\times 1.6\times 10^{-19}\right)^{3/2}\right]$$

$$= \frac{1.04\times 19.58\times 10^{-45}}{290.77\times 10^{-102}}\left[0.373-0.369\right]\times 10^{-27}$$

$$= \frac{20.36\times 10^{30}\times 4\times 10^{-3}}{290.77}$$

$$= \frac{814.40}{290.77}\times 10^{26}$$

$$N(E) = 2.80\times 10^{26}$$ **Ans.**

Example 2.15: At what temperature we can expect a 10% probability the electrons in silver have an energy which is 1 % above the Fermi energy (E_f = 5.5 eV for silver).

Solution: Given E_f = 5.5 eV and $\frac{E_f}{100} = \frac{5.5}{100} = 0.055\,\text{eV}$

$\therefore \quad E - E_f = 0.055$ eV

From the Fermi Dirac distribution function

$$F(E) = \frac{1}{1+e^{(E-E_f)/k_BT}}$$

$$\therefore \quad \frac{10}{100} = \frac{1}{1+e^{\frac{0.055}{k_BT}}}$$

$$\frac{1}{10} = \frac{1}{1+e^x}, \quad \text{where } x = \frac{0.055}{k_BT}$$

$$\Rightarrow \quad 0.1 + 0.1e^x = 1$$

$$\Rightarrow \quad e^x = 9$$

$$\Rightarrow \quad x = 2.303\times \log_{10}^{9}$$

$$x = 2.197$$

$$\therefore \quad \frac{0.055}{k_BT} = 2.197$$

$$\Rightarrow \quad k_BT = \frac{0.055}{2.197}$$

$$T = \frac{0.055 \times 1.6 \times 10^{-19}}{2.197 \times k_B} \text{ Joule}$$

$$= \frac{0.055 \times 1.6 \times 10^{-19}}{2.197 \times 1.38 \times 10^{-23}}$$

$$T = 290 \text{ K}$$

Ans.

Example 2.16: Calculate Fermi energy in copper assuming that each copper atom contributes one free electron to the gas. Given density of copper $= 8.94 \times 10^3 \text{ kg/m}^3$ and atomic mass is 63.5 *u*.

[Raj. B.E.I 2001]

Solution: The volume of the given copper crystal is

$$V = \frac{\text{mass}}{\text{density}} = \frac{63.5 \times 10^{-3}}{8.94 \times 10^3} = 7.103 \times 10^{-6} \text{m}^3$$

Number of electrons per unit volume

$$n = \frac{6.02 \times 10^{23}}{7.103 \times 10^{-6}} = 8.47 \times 10^{28} /\text{m}^3$$

Now the Fermi energy of copper metal

$$E_F = \frac{h^2}{2m}\left(\frac{3N}{8\pi V}\right)^{2/3}$$

$$= \frac{\left(6.63 \times 10^{-34}\right)^2}{2 \times 9.11 \times 10^{-31}} \times \left(\frac{3 \times 8.47 \times 10^{28}}{8 \times 3.14}\right)^{2/3}$$

$$= 2.412 \times 10^{-37} \times 4.68 \times 10^{18}$$

$$= 11.3 \times 10^{-19} \text{J} = \frac{11.3 \times 10^{-19}}{1.6 \times 10^{-19}}$$

$$= 7.05 \text{ eV}$$

Ans.

Example 2.17: Calculate Fermi energy for copper, given that number of conduction electron per unit valume to be $n = 8.49 \times 10^{28} \text{ m}^{-3}$. **[RTU B.Tech .I sem. 2008]**

Solution: The Fermi energy

$$E_f = \left(3.65 \times 10^{-19}\right) n^{2/3} \text{ eV}$$

$$= 3.65 \times 10^{-19} \left(8.49 \times 10^{28}\right)^{2/3} \text{ eV}$$

$$= 3.65 \times 10^{-19} \left(84.9 \times 10^{27}\right)^{2/3} \text{ eV}$$

$$= 3.65 \times 19.32 \times 10^{-1} \text{ eV}$$

$$= 70.51 \times 10^{-1}$$

$$E_f = 7.051 \text{ eV}$$ **Ans.**

Example 2.18: For potassium, fermi-energy is 2.14 eV and the density of electrons is $1.4 \times 10^{28}/\text{m}^3$. Find the electron density of a metal for which fermi energy is 4.72 eV.

[Raj. B.E. I 2006, RTU B.Tech. II Sem. 2012]

Solution: The electron density of a metal

$$n = \frac{\pi}{3}\left(\frac{2m}{\lambda^2 \hbar^2}\right)^{3/2} E_f^{3/2}$$

$$\Rightarrow \quad n \,\alpha\, E_f^{3/2}$$

$$\Rightarrow \quad \frac{n_2}{n_1} = \left(\frac{E_{f_2}}{E_{f_1}}\right)^{3/2}$$

$$\Rightarrow \quad n_2 = n_1 \left(\frac{E_{f_2}}{E_{f_1}}\right)^{3/2}$$

$$= 1.4 \times 10^{28} \left(\frac{4.72}{2.14}\right)^{3/2}$$

$$= 1.4 \times 10^{28} (2.20)^{3/2}$$

$$= 1.4 \times 10^{28} \times 3.27$$

$$n_2 = 4.578 \times 10^{28} \text{ per m}^3$$ **Ans.**

Example 2.19: Given that electron density in silver to be $5.8 \times 10^{28}\,\text{m}^{-3}$. Calculate the Fermi energy for free electron gas. What is speed of electrons at this energy?

[Raj. B.E. I 2004, RTU B. Tech. Back 2010 & II Sem. 2011]

Solution: Given $n = 5.8 \times 10^{28}/\text{m}^3$

The Fermi energy

$$E_f = 3.65 \times 10^{-19} n^{2/3} \text{ eV}$$

$$= 3.65 \times 10^{-19} \left(5.8 \times 10^{28}\right)^{2/3}$$

$$= 3.65 \times 10^{-19} (58)^{2/3} 10^{18}$$

$$= 3.65 \times 14.98 \times 10^{-1}$$

$$= 54.67 \times 10^{-1}$$

$$E_f = 5.467 \text{ eV}$$ **Ans.**

Since $$E_f = \frac{1}{2} m v_f^2$$

$\rightarrow$ $$v_f = \sqrt{\frac{2E_f}{m}} = \sqrt{\frac{2 \times 5.467 \times 1.6 \times 10^{-19}}{9.1 \times 10^{-31}}}$$

$$= \sqrt{\frac{17.49}{9.1}} \times 10^6$$

$$v_f = \sqrt{1.92} \times 10^6$$

$$v_f = 1.39 \times 10^6 \text{ m/s}$$ **Ans.**

QUESTIONS

1. Explain the meaning of the terms degeneracy and tunnelling. **[Raj. B.E.I. 1998, Back 2010]**
2. Write down Schrödinger's equation for a free particle of mass m trapped in one dimensional box of side a. Solve it for energy eigenvalues and eigenfunctions. How do solutions modify if particles were in three dimensional cubical box of side a? Find lowest energy of the following states: (i) non-degenerate, (ii) doubly degenerate, (iii) triply degenerate (iv) six fold degenerate for 3-D cubical box. **[Raj. B.E.I. 2001]**
3. Discuss the degeneracy of energy states. Solve the Schrödinger's equation for a free electron in 3-Dimensional box and find the eigenvalues and eigenfunctions of free particle. **[RTU B.Tech. II Sem. 2010 & 2011]**
4. What is tunnel effect? Write down Schrödinger's equations for potential barrier problem and steps to find out the transmission coefficient of a particle having energy less than the height of potential barrier.
5. Obtain an expression for the density of states for free electron gas in metal and hence find expression for the Fermi energy. **[Raj. B.E.I. 2001]**
6. What is density of energy states in metals? Derive an expression for the density of energy states and hence obtain Fermi energy of metals. **[MREC. B.E.I. 2002]**
7. Wrtie short notes on the following:
   (*a*) Fermi energy and density of states **[Raj. B.E.I. 2002]**
   (*b*) Quantum mechanical tunnelling **[MREC. B.E.I. 2002]**
   (*c*) Degeneracy **[Raj. B.E.I. 2003]**
8. Obtain an expression for the "density of states" for a Fermi-gas and hence find expression for Fermi-energy of metal at 0 K having n electrons per unit volume. **[RTU B.Tech. I Sem. 2008]**

9. Write down Schrödinger's equation for a particle trapped in an infinitely deep cubical potential well of side 'a'. Write expression for its energy eigenvalues. What is degeneracy of second excited state? What will happen to above degeneracy if the well is a rectangular parallelopiped shape with sides $a = b \neq c$? **[RTU B.Tech. I Sem. 2008]**

10. Prove, according to quantum mechanics, that even with particle energy E less than the barrier height V, there is always a finite probability of transmission through the barrier.

11. What is potential barrier? Calculate the transmission probability for rectangular potential barrier of small width for $E < V_0$. **[RTU. B.Tech. I Sem. 2008 & II Sem. 2012, Back 2012]**

12. What do you mean by quantum mechanical tunnelling? Show that the tunning probability is given by the expression $T = \left[1 + \frac{V_0^2 \sinh^2 \alpha a}{4E(V_0 - E)}\right]^{-1}$ where V_0= height of the rectangular potential barrier. Draw graphs showing variation of T with particle energy E and barrier width a. **[RTU B.Tech. II Sem. 2008]**

13. Write down basic postulates of Sommerfeld's free electron gas model. Show that the number of energy states per unit energy is given by $g(E) = \frac{4\pi V}{h^3}(2m)^{3/2} E^{1/2}$.What do you mean by Fermi energy level? **[RTU B.Tech. II Sem. 2008, 2010]**

14. With help of suitable diagram explain the phenomenon of quantum mechanical tunneling in α-decay process. **[RTU B.Tech. II Sem. 2008]**

15. Using free electron model, calculate density of energy states for free electron in the energy range from E to $E + dE$. **[Raj. B.E.I. 2003, RTU B.Tech. Back June, 2010]**

16. What is density of energy states in metals ? Derive an expression for the density of density of energy states and hence obtain Fermi energy of metals. **[MREC, B.E. I 2002, RTU B.Tech. II Sem. 2008]**

17. What is tunnel effect? Discuss the Alpha decay as an example of tunnel effect. Prove that tunnelling increases on decreasing the height and width of the barrier. **[RTU B.Tech. II Sem. 2010]**

18. Give the basic assumptions made in free electron model of solid. **[RTU B.Tech. I 2007]**

19. What are the postulates of Sommerfeld's free electron gas model? What do you mean by Fermi energy level ? **[RTU B. Tech. Back Exam. 2010]**

20. Define the Fermi energy. Write down the expression for the Fermi-Dirac distribution law. Also derive an expression for the Fermi energy for the system of particles. **[RTU B.Tech. II Sem. 2011]**

PROBLEMS

1. What is the lowest energy that a neutron can have if confined to move along the edge of an impenetrable box of length 10^{-14} m? **(Ans. 2.056 MeV)**

2. Determine the degree of degeneracy of the energy level $\left(38\dfrac{h^2}{8ma^2}\right)$ of a particle in a cubical potential box of side *a*. **(Nine fold)**

3. The electrons in a cubical box of a metal are subjected to the influence of a magnetic field such that the length increases by *da* while the width and thickness remain the same. Show that the energy difference between the states (3, 1, 1) and (1, 3, 1) is $\left(\dfrac{h^2}{8ma^2}\right)\left(\dfrac{16da}{a}\right)$ in the new position.

4. Calculate the energy of the electron in the energy state immediately after the lowest energy level confined in a cubical box of side 0.1 nm. Also find the temperature at which the average energy of the molecules of a perfect gas would be equal to the energy of the electron in the above said level. **(225.8 eV and 17.4×10⁵ K)**

5. The Fermi energy of sodium is 3 eV. Calculate the difference in energy between the neighbouring levels at the highest energy state in a cubical box of side 1cm. Given $n_x = n_y = n_z$.

   [**Hint:** The energy level just below Fermi level [1.2×10^{-7}eV] is obtaining using the quantum numbers $(n_x - 1)$, n_x and n_x in this problem.]

6. When 1 eV electrons are incident on a potential barrier of 8 eV, what fraction of electrons penetrate the barrier if it is 5 Å wide? **(Ans. 2.17×10^{-6})**

7. A particle is moving in one dimensional potential box (of infinite height) of width 25 Å. Calculate the probability of finding the particle within an interval of 5 Å at the centre of the box when it is in the state of least energy. **(Ans. 0.4)**

8. Electrons with energy of 2 eV are incident on a barrier 5 eV high and 5 Å wide. Find its transmission probability. **(Ans. 1.4×10^{-4})**

9. A stream of electrons, each of energy E = 3 eV is incident on a potential barrier of height V_0 = 4 eV. The width of the barrier is 2 nm. Calculate the percentage transmission of the beam through this barrier. **(1.029×10^{-2})**

10. An α-particle having energy 10 MeV approaches a potential barrier of height equal to 30 MeV. Determine the width of the potential barrier if the transmission coefficient is 2×10^3.

    [Given: Mass of α-particle is $4\times1.6\times10^{-27}$ kg, $\hbar = 1.054\times10^{-34}$ Js.] **(1.95×10^{-15}m)**

11. An electron in a metal encounters a barrier layer of height 6 eV and thickness 0.5 nm. If the energy of electron is 5 eV. What is the probability of tunnelling through the barrier? **(1.33%)**

12. An α-particle is trapped in a nucleus whose radius is $r_0 = 1.4\times10^{-15}$ m. What is the probability that α-particle will escape from the nucleus if its energy is 2 MeV? The potential barrier at the surface of the nucleus is 4 MeV. **(0.124)**

13. Calculate the temperature at which there is 1 % probability that a state, with an energy 0.5 eV above the Fermi energy, will be occupied by an electron. **(1262 K)**

14. The density of zinc is 7.13×10^3 kg/m^3 and its atomic weight is 65.4. Calculate the Fermi energy in zinc. Also calculate the mean energy at zero kelvin. The effective mass of electron in zinc is 0.85 m_e. **(9.4 eV and 5.7 eV)**

15. Show that the wavelength associated with an electron having an energy equal to Fermi energy is given by $2\left(\frac{\pi}{3n}\right)^{1/3}$.

REFERENCES

1. E. Merzbacher, Quantum Mechanics, 2nd Ed., New York, John Wiley & Sons, 1970.
2. A. Beiser, Concepts of Modern Physics, McGraw Hill Book Company, New York, 1995.
3. N.F. Mott and I.N. Sneddon, Wave Mechanics and its Applications, London, Oxford University Press, Reprinted, New York, 1963.
4. Charles Kittel, Introduction to Solid State Physics–VII ed., John Wiley & Sons (Asia) Pvt. Ltd. Singapore, 2004.
5. H.C. Ohanian, Modern Physics, Prentice Hall of India Private Ltd., New Delhi, 1994.
6. D. Haliday, R. Rensik and K.S.Krane, Physics, Vol. II Extended Version, John Wiley & Sons, Singapore, 1992.
7. S.O. Pillai, Solid State Physics–VII ed., New Age International Publishers, New Delhi, 2005.
8. J. Bernstein, P.M. Fishbane and S. Gasiorowicz, Modern Physics, Pearson Education, New Delhi, 2001.
9. S.T. Thornton and A. Rex, Modern Physics for Scientists and Engineers, Saunders College Publishing, New York, 2000.
10. R.L. Liboff, Introductory Quantum Mechanics, Pearson Publication, New Delhi, 2003.
11. A. J. Dekkar, Solid State Physics, MacMillan India Ltd. New Delhi, 1985.
12. J. Singh, Modern Physics for Engineers, John Wiley, New York, 2003.
13. S. Gasiorowicz, Quantum Physics, John Wiley, New York, 2005.

3

COHERENCE AND OPTICAL FIBRES

3.1 COHERENCE

Coherence is one of the most important concept in optics and is related to the phase correlation between the light waves at different points in space at a given time or different times at a given point.

If the phase difference between the waves from two sources reaching any point in space does not change with time, rather it remains constant then these sources of light are known as '*Coherent sources of light*'. This property of the waves is known as coherence.

A wave which appears to be a pure sine wave for an infinitely large period of time or in an infinitely extended space is said to be a perfectly coherent wave. Thus there are two different criteria of coherence: the criterion of time and criterion of space. This gives rise to temporal coherence and spatial coherence.The temporal coherence is related to the finite bandwidth of the source, whereas the spatial coherence is related to the finite size of the source.

3.1.1 Spatial Coherence

The ultimate source of light is an atom. All conventional light sources are an aggregate of atomic electromagnetic oscillators operating independently of each other. The atoms are continuously pumped to the excited states from where they fall back spontaneously to their ground states, emitting photons. There is no phase correlation between these de-excitation processes and the photons are emitted at random. The light from such a source is emitted in the form of independent waves which interfere constructively or destructively in a random fashion and the resultant disturbance produced varies from point to point and from instant to instant. We observe only the time averaged effect at any point in space. If the time t in which we take an obervation is less than a characteristic time τ_c, the mean life time of an excited state ($\approx 10^{-8}$ sec), then during the observation the different atomic sources will continue to keep a constant phase relationship. Hence it will be possible to observe a correlation between the waves reaching two different points in space.

The definite phase correlationship between the waves reaching at two different points in space at a given time is defined as spatial coherence.

3.1.2 Temporal Coherence

We now consider the correlation in the wave trains reaching a given point in space at different times. If our source is a monochromatic point source operating continuously (infinite life time) then the disturbance at different times will be completely correlated. But with a source of mean life time τ_c the radiation emitted can be considered to be small wave trains of duration τ_c. The different wave trains will have no phase correlationship between them. In such a case no definite correlation can be observed at the given point at different times, when the difference in times is much greater than τ_c.

The phase correlation in wave trains at different times at a given point is defined as temporal coherence.

3.2 DETAILED CONCEPT OF TEMPORAL AND SPATIAL COHERENCE

(Coherence Length, Coherence Time and Q-Factor)

3.2.1 Temporal Coherence and Monochromaticity of the Source

In temporal coherent wave, there is definite relationship between the phase of wave at a given instant of time. Its frequency spectrum is as shown in Fig. 3.2. The oscillating electric field E of a perfectly coherent light wave would have a constant amplitude of vibration at any time, while its phase would vary linearly with time. As a function of time, the field would appear as shown in Fig. 3.1.

Fig. 3.1 Perfectly coherent light wave

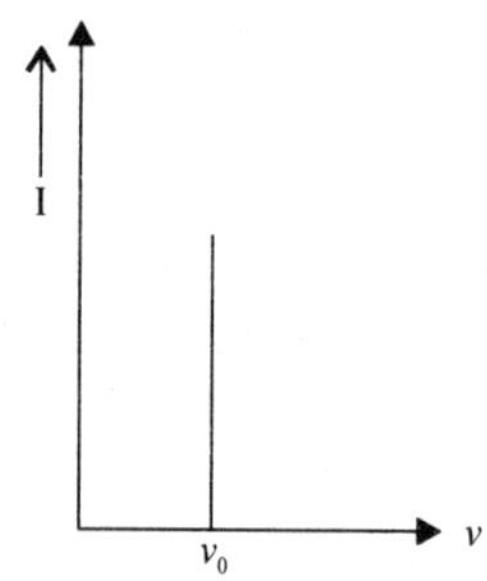

Fig. 3.2

However, no actual source of light produces an ideal sinusoidal field for all values of time. When an excited atom returns to the initial state, it emits pulse of short duration of the order of 10^{-10} second. Thus the field remains sinusoidal for time-intervals of the order of 10^{-10} second, after which the phase changes abruptly. The field due to an actual light source appears as shown in Fig. 3.3. Its frequency spectrum is as shown in Fig. 3.4.

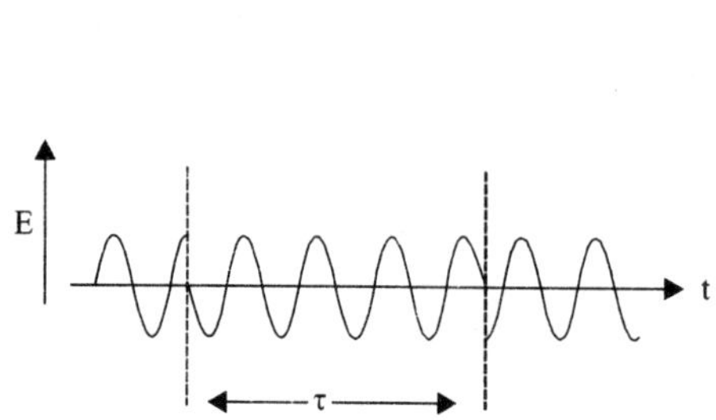

Fig. 3.3 Incoherent light wave

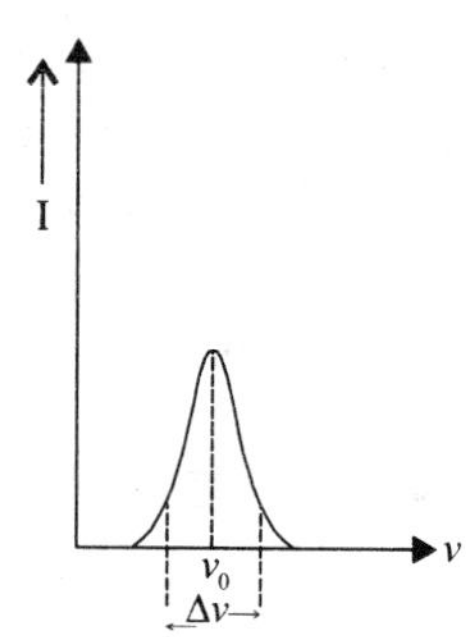

Fig. 3.4

The average time interval for which the field remains sinusoidal (*i.e.*, definite phase relationship exists) is known as '*coherence time*' or '*temporal coherence*' of the light beam, and is denoted by τ. The distance L for which the field is sinusoidal is given by

$$L = \tau c \tag{3.1}$$

where c is the speed of light. L is called the '*coherence length*' of the light beam.

Coherence length in terms of frequency

For monochromatic light, the frequency range is zero, *i.e.*, $\Delta\nu = 0$. But the frequency range of waves emitted by an actual light source is greater than zero. The reason of this is that every spectral line has a finite wavelength spread, *e.g.*, for cadmium red line ($\lambda = 6438$ Å), $\Delta\lambda = 0.01$ Å and for sodium D_2 – line ($\lambda = 5896$ Å) $\Delta\lambda = 0.06$ Å. The measure of the finite purity factor of spectral lines (Q) is expressed as

$$Q = \frac{\lambda}{\Delta\lambda} \tag{3.2}$$

The time of coherence is of the order of the reciprocal of frequency range i.e.,

$$\tau = \frac{1}{\Delta\nu} \tag{3.3}$$

From Eqs. (3.1) and (3.3)

$$\text{Coherence length } L = \tau c = \frac{c}{\Delta\nu} \tag{3.4}$$

For a monochromatic light source $\Delta\nu = 0$. Therefore coherence length will be infinity. Such a source of light is known as perfectly coherent source.

Coherence length in terms of wavelength

$$\text{For light waves } \nu = \frac{c}{\lambda} \tag{3.5}$$

Differentiating, we get

$$\Delta\nu = -\frac{c}{\lambda^2}\Delta\lambda$$

$$\Rightarrow |\Delta\nu| = \frac{c\Delta\lambda}{\lambda^2} \tag{3.6}$$

$$\text{Coherence time } \tau = \frac{\lambda^2}{c\Delta\lambda} \tag{3.7}$$

$$\text{From Eq. (3.1), coherence length } L = \tau c = \frac{\lambda^2}{\Delta\lambda} \tag{3.8}$$

From Eqs. (3.2) and (3.8)

$$\text{Coherence length } L = Q\lambda = \text{purity factor} \times \text{wavelength} \tag{3.9}$$

Thus, the coherent length is inversely proportional to the width of the spectral line. *i.e.,* the width of the spectral line is a measurement of coherent length.

3.2.2 Spatial Coherence and Size of the Source

In spatial coherent wave, there is definite relationship between the phase of the wave at a given point and at another point at a certain distance away. Let us consider light waves originating from point source S (Fig. 3.5). Let X and Y be two points lying on a line joining them with S. The phase relationship between X and Y depends on the distance XY and on temporal coherence of the beam. If $XY << L$ (Coherence length), there will be a definite phase relation between X and Y, *i.e.*, there will be high coherence between X and Y. On the other hand if $XY >> L$, there will be no coherence between X and Y.

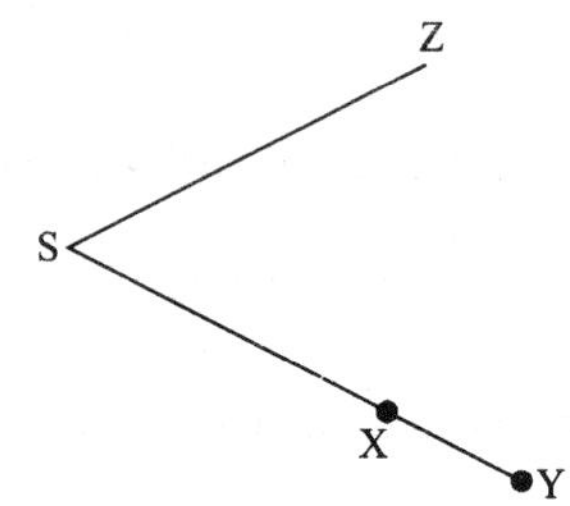

Fig. 3.5 Spatial coherence

Let us now consider points X and Z which are equidistant from S. If the source S is a true point source, then the waves reach X and Z in exactly same phase, *i.e.*, the two points will have perfect spatial coherence. If, however, the source S is an extended source, the points X and Z will no longer remain in coherence.

This may be demonstrated by Young's double slit experiment illustrated in Fig. 3.6. Actual source of light is not a point source, rather it is extended. Its size can be reduced by placing a narrow slit in front of it, but it cannot be made zero.

Let the width of the slit be e. There are two slits S_1 and S_2 which are symmetrically placed in front of S as shown in the figure. Let the distance between S_1 and S_2 be y. The maximum value of y for which the sources S_1 and S_2 remain coherent is defined as '*spatial coherence length* (L_W)'.

$$\therefore \qquad (L_W) = y_{max} \qquad \textbf{(3.10)}$$

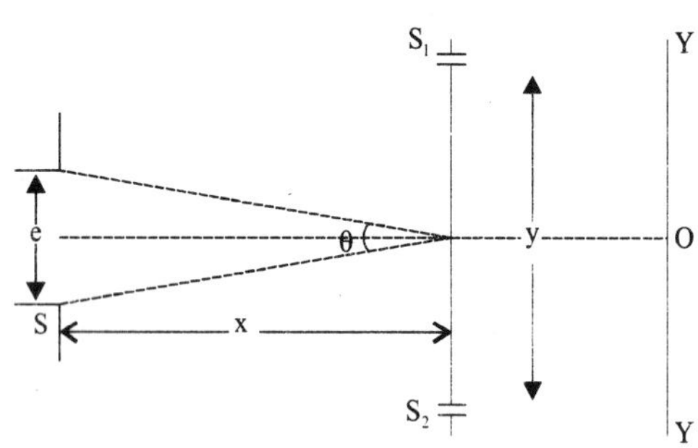

Fig. 3.6 Transverse spatial coherence

The angular spread of the light waves emitted by each excited atom situated in the slit of width e is $\frac{\lambda}{e}$, hence if $\angle S_1SS_2 < \frac{\lambda}{e}$, then S_1 and S_2 will lie on the same wavefront emitted by the source S and hence S_1 and S_2 will be coherent. If the distane of S from the plane S_1S_2 is x, then

$$\angle S_1SS_2 \approx \frac{y}{x}$$

Therefore, for S_1 and S_2 to be coherent

$$\frac{y}{x} < \frac{\lambda}{e} \tag{3.11}$$

So that the spatial coherence length

$$L_W = y_{max} = \frac{x\lambda}{e} \tag{3.12}$$

But $\frac{e}{x}$ is the angle (θ) subtended by the slit S at the point of observation in the plane of S_1 and S_2, *i.e.*,

$$\theta = \frac{e}{x} \tag{3.13}$$

∴ Spatial coherence length from Eqs. (3.12) and (3.13) is

$$L_W = \frac{\lambda}{\theta} \tag{3.14}$$

For circular aperture S

$$L_W = \frac{1.22\lambda}{\theta} \tag{3.15}$$

From the Eqs. (3.12) and (3.14) it is observed that spatial coherence length can be increased by decreasing e or θ.

3.3 VISIBILITY AS A MEASURE OF COHERENCE

The quality of fringes produced by an interferometric system can be described quantitatively using the visibility (V). It is given by

$$V = \frac{I_{max} - I_{min}}{I_{max} + I_{min}} \tag{3.16}$$

where I_{max} and I_{min} are the maximum and minimum values of intensity in the fringe pattern. The concept of visibility was first formulated by Michelson to measure contrast of interference fringes. For two coherent waves of equal amplitude or intensity, the fringe visibility is unity since $I_{min} = 0$.

The visibility is a dimensionless quantity, which indicates the extent of coherence of a light wave. In the laboratory, the observed value of visibility of the fringes is less than unity even with the waves of equal amplitude. Therefore only partially coherent wave is a physical reality. The ability of light waves to produce interference is measured by the degree of coherence of light waves. Higher the degree of coherence higher is the probability that the waves produce a contrast interference pattern. Light is considered to be highly coherent when the degree of coherence exceeds 0.88, partially coherent for values less than 0.88 but more than nearly zero values, and incoherent for nearly zero and zero values.

If the waves are partially coherent but have the same intensity, the fringe visibility becomes a measure of the degree of coherence.

To explain this, let us consider two interfering waves of same intensity I_0. Each of them consists of coherent part I_c and incoherent part I_{inc}. If c be the degree of coherence then the coherent part and incoherent part can be written as

$$I_c = cI_0 \text{ and } I_{inc} = (1-c)I_0 \tag{3.17}$$

When superimposed, the coherent part shall interfere adding thereby their amplitudes whereas for incoherent parts, the intensities are simply added. Thus

$$I_{max} = (a_1 + a_2)_c^2 + (I_1 + I_2)_{inc}$$

$$I_{max} = 4cI_0 + 2(1-c)I_0$$

and

$$I_{min} = (a_1 - a_2)_c^2 + (I_1 + I_2)_{inc}$$

$$I_{min} = 0 + 2(1-c)I_0$$

Thus

$$\text{visibility} = \frac{I_{max} - I_{min}}{I_{max} + I_{min}}$$

$$= \frac{4cI_0 + 2(1-c)I_0 - 2(1-c)I_0}{4cI_0 + 2(1-c)I_0 + 2(1-c)I_0}$$

$$\Rightarrow \quad V = \frac{4cI_0}{4I_0} = c \tag{3.18}$$

Therefore, the fringe visibility is a measure of degree of coherence between waves of equal intensities. If the intensities of waves are unequal, the visibility is given by

$$V = \frac{2\sqrt{I_1 I_2}}{I_1 + I_2}|c| \tag{3.19}$$

Example 3.1: If light of wavelength 660 nm has wave trains $20\,\lambda$ long. What are its coherence length and coherence time?

Solution: The coherence length is

$$L = 20\,\lambda = 20 \times 660 \times 10^{-9} = 13.2 \times 10^{-6} \text{ m}$$ **Ans.**

The coherence time is

$$\tau = \frac{L}{c} = \frac{13.2 \times 10^{-6}}{3 \times 10^8} = 4.4 \times 10^{-14}\,\text{s}$$ **Ans.**

Example 3.2: Determine the coherence time and coherence length for white light. (Given that $\nu_{violet} = 7.7 \times 10^{14}$ Hz and $\nu_{red} = 3.85 \times 10^{14}$ Hz)

Solution: Frequency bandwidth

$$\Delta\nu = \nu_{violet} - \nu_{red}$$

$$= (7.7 - 3.85) \times 10^{14} = 3.85 \times 10^{14}\,\text{Hz}$$

$\therefore$ Coherence time $\tau = \dfrac{1}{\Delta\nu}$

$$\Rightarrow \quad \tau = \frac{1}{3.85 \times 10^{14}} = 2.597 \times 10^{-15}\ \text{sec.}$$ **Ans.**

Coherence length $L = c\tau$

$$\Rightarrow \quad L = 3 \times 10^{8} \times 2.597 \times 10^{-15} = 7.79 \times 10^{-7}\ \text{m.}$$ **Ans.**

Example 3.3: The band-width of light source of wavelength 6×10^{-7} m is 10^{-11} m. Determine its coherence length.

Solution: $L = \dfrac{\lambda^2}{\Delta\lambda} = \dfrac{36 \times 10^{-14}}{10^{-11}} = 36 \times 10^{-3}\,\text{m} = 36\ \text{mm}$ **Ans.**

Example 3.4: In Michelson interferometer experiment with He-Ne laser, fringes are visible upto path difference spread of 8 m. Determine the lower limits of the following : (*i*) Coherence length (*ii*) Coherence time (*iii*) Spectral line width (*iv*) Purity factor (Given : $\lambda = 11.5 \times 10^{-7}\,\text{m}$)

Solution: (*i*) Coherence length L = 8 m **Ans.**

(*ii*) Coherence time $\tau = \dfrac{L}{c} = \dfrac{8}{3 \times 10^{8}} = 2.67 \times 10^{-8}$ sec **Ans.**

(*iii*) Spectral line width

$$\Delta\nu = \frac{1}{\tau} = \frac{1}{2.67 \times 10^{-8}} = 3.75 \times 10^{7}\,\text{Hz}$$ **Ans.**

(*iv*) Purity factor $Q = \dfrac{\nu}{\Delta\nu} = \dfrac{c}{\lambda\Delta\nu}$

$$\Rightarrow \quad Q = \frac{3 \times 10^{8}}{11.5 \times 10^{-7} \times 3.75 \times 10^{7}} = 6.96 \times 10^{6}$$ **Ans.**

Example 3.5: The frequency range of light emitted by a mercury lamp is $\Delta\nu = 10^{9}\,\text{Hz}$. Determine the coherence time and coherence length.

Solution: Coherence time $\tau \approx \dfrac{1}{\Delta\nu}$

$$\Rightarrow \quad \tau = \frac{1}{10^{9}} = 10^{-9}\,\text{sec}$$

Coherence length $L = \tau c = 10^{-9} \times 3 \times 10^{8} = 0.3$ m **Ans.**

Example 3.6: Michelson interferometer is illuminated by a source of light whose wavelength is $(6500 \pm 0.015)\text{Å}$. Initially the interferometer is set for zero optical path difference. Through what displacement the mirror of the interferometer should be moved so that the interference fringes may disappear?

Solution: If the optical path difference exceeds the coherence length, the interference fringes disappear. In Michelson interferometer optical path difference is twice the distance (d) moved by the mirror.

Given $\lambda = 6500\text{Å}$ and $\Delta\lambda = 0.015\text{Å}$

$$\therefore \quad L = 2d \text{ But } L = \frac{\lambda^2}{\Delta\lambda}$$

$$\therefore \quad d = \frac{\lambda^2}{2\Delta\lambda} = \frac{\left(6.5 \times 10^{-7}\right)^2}{2 \times 0.015 \times 10^{-10}} = 14.0\,\text{cm}$$

Ans.

Example 3.7: The coherence length of a LASER source is 110 m and wavelength of light used is 6600Å. Calculate the bandwidth and frequency spread.

Solution: Bandwidth $\Delta\lambda = \frac{\lambda^2}{L}$

$$\Rightarrow \quad \Delta\lambda = \frac{\left(6.6 \times 10^{-7}\right)^2}{110} = \frac{6.6 \times 6.6 \times 10^{-14}}{110}$$

$$\Rightarrow \quad \Delta\lambda = 3.96 \times 10^{-15}\,\text{mm}$$

Frequency spread $\Delta v = \frac{c}{L} = \frac{3 \times 10^8}{110} 2.73 \times 10^6\,\text{Hz}$ **Ans.**

Example 3.8: The figure shows a Young's double slit experimental arrangement. What should be the maximum distance between S_1 and S_2 so that distinct interference fringes may be obtained? The wavelength of light used is $6 \times 10^{-7}\,\text{m}$.

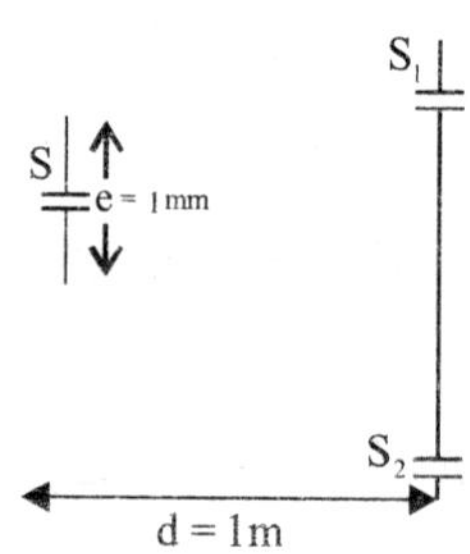

Solution: For distinct interference fringes the spatial coherence length S_1S_2 is

$$L_w = \frac{\lambda d}{e} = \frac{6 \times 10^{-7} \times 1}{10^{-3}}$$

$= 6 \times 10^{-4}\,\text{m} = 0.6$ mm. **Ans.**

Example 3.9: The coherent time for Na-light of wavelength 5890 Å is $\tau = 10^{-10}\,s$. What is the maximum thickness of film that could be measured using interference of Na-light?

[Raj. B.E.I. 2003]

Solution: The maximum thickness of film is equal to the coherence length, so

$$L_c = \tau c = 10^{-10} \times 3 \times 10^{10} = 3 \text{ cm}$$ **Ans.**

Example 3.10: Two lights having intensities in the ratio of 1 : 9 produce interference fringes of visibility 0.3. What information do we get about the degree of coherence?

Solution: Given $\dfrac{I_1}{I_2} = \dfrac{1}{9}$

$$\therefore \quad I_{max} = 1 + 9 + 2\sqrt{1 \times 9} = 10 + 2 \times 3 = 16$$

$$I_{min} = 1 + 9 - 2\sqrt{1 \times 9} = 10 - 2 \times 3 = 4$$

Visibility $$V = \frac{I_{max} - I_{min}}{I_{max} + I_{min}} = \frac{16 - 4}{16 + 4} = \frac{12}{20}$$

$V = 0.6$

Therefore the degree of coherence is 50% . **Ans.**

Example 3.11: The coherence length for sodium D_2 line is 2.5 cm. Deduce (*i*) the coherence time τ_1 (*ii*) the spectral width of the line and (*iii*) the purity factor.

Solution: The coherence time is given by

$$\tau = \frac{l}{c}$$

Here l = 2.5 cm and speed of light $c = 3 \times 10^8$ m/s = 3×10^{10} cm/s

$$\therefore \quad \tau = \frac{2.5}{3 \times 10^8} = 0.8 \times 10^{-10} \text{ s}$$ **Ans.**

The spectral width of the line is given by

$$\Delta\lambda = \frac{\lambda^2}{L}$$

For sodium D_2 line $\lambda = 5896 \text{A}^0 = 5.896 \times 10^{-5}$ cm

$$\therefore \quad \Delta\lambda = \frac{(5.896)^2 \times 10^{-10}}{2.5} = 13.90 \times 10^{-10} \text{ cm}$$ **Ans.**

The purity factor is given by

$$Q = \frac{\lambda}{\Delta\lambda} = \frac{5.896 \times 10^{-5}}{13.90 \times 10^{-10}} = 4.24 \times 10^4$$ **Ans.**

Example 3.12: With He-Ne laser, Michelson interferometer fringes remained clearly visible when the path difference was increased upto 8 m. Deduce the lower limits for (*i*) the coherence length (*ii*) coherence time (*iii*) spectral half width and (*iv*) Q of the line, (Given $\lambda = 1.15 \times 10^{-5}$ cm)

Solution: The coherence length must be greater than the maximum path difference.

Hence

$$L \geq 8m$$ **Ans.**

The coherence time is given by

$$\tau = \frac{L}{c} > \frac{8}{3 \times 10^8} = 2.7 \times 10^{-8} \text{ sec}$$ **Ans.**

Spectral half width is given by

$$\Delta\lambda = \frac{\lambda^2}{L}$$

Here $\lambda = 1.15 \times 10^{-5}$ cm

$$\therefore \quad \Delta\lambda = \frac{(1.15)^2 \times 10^{-10}}{800} = 1.65 \times 10^{-3} \text{ Å}$$ **Ans.**

Q-factor is given by

$$Q = \frac{L}{\lambda} = \frac{8}{1.15 \times 10^{-7}} = 7 \times 10^7$$ **Ans.**

Example 3.13: Imagine that we chop a continuous laser beam (assumed to be perfectly monochromatic $\lambda_0 = 6500$Å) into 0.5 ns pluses using some sort of shutter. Compute the coherence length, bandwidth and the linewidth.

Solution: The coherence length is given by

$$L = c \times \Delta t$$

Here $\Delta t = 5 \times 10^{-10}$ s, $c = 3 \times 10^8$ m/s

$$\therefore \quad L = 3 \times 10^8 \times 5 \times 10^{-10} = 0.15 \text{ m}$$ **Ans.**

Bandwidth is given by

$$\Delta \nu = \frac{1}{\Delta t} = \frac{1}{5 \times 10^{-10}} = 2 \times 10^9 \text{ Hz}$$ **Ans.**

Linewidth is given by

$$\Delta\lambda = \frac{\lambda^2}{c} \Delta \nu$$

Here $\lambda = 6.5 \times 10^{-7}$ m

$$\therefore \quad \Delta\lambda = \frac{(6.5 \times 10^{-7})^2 \times 2 \times 10^9}{3 \times 10^8} = 0.028 \text{ Å}$$ **Ans.**

Example 3.14: Light of wavelength 6000 Å has average wavetrain length of 20 waves. Determine its coherent time and Q-value. **[RTU B.Tech. I Sem. 2008]**

Solution: Given coherence length is a length of 20 waves, *i.e.*,

$$L = 20 \times 6000 \times 10^{-10} \text{ m}$$

$$= 12 \times 10^{-6} \text{ m}$$

$\therefore$ Coherence time $\tau = \dfrac{L}{c} = \dfrac{12 \times 10^{-6}}{3 \times 10^{8}}$

$\tau = 4 \times 10^{-14}$ sec **Ans.**

Again coherence length $L = \dfrac{\lambda^2}{\Delta\lambda} = \dfrac{\lambda}{\Delta\lambda}\lambda$

$$L = Q \times \lambda$$

$\therefore$ Q-factor $= \dfrac{\lambda}{\Delta\lambda} = \dfrac{L}{\lambda}$

$$= \frac{12 \times 10^{-6}}{6000 \times 10^{-7}} = \frac{12 \times 10^{-6}}{6 \times 10^{-7}}$$

$Q = 20$ **Ans.**

Example 3.15: A laser operates at wavelength of 6000 $\AA$ and its spectral line width Δv is 10^2 Hz. For this calculate

(*i*) coherence length (*ii*) Q-factor. **[RTU B.Tech. I 2007]**

Solution: $\lambda = 6000\text{Å} = 6000 \times 10^{-10}$ m and $\Delta v \doteq 100$ Hz

$\therefore$ Coherence length $L = c \times \tau = \dfrac{c}{\Delta v} = \dfrac{3 \times 10^8}{10^2} = 3 \times 10^6$ m **Ans.**

and Q-factor $= \dfrac{\lambda}{\Delta\lambda} = \dfrac{v}{\Delta v} = \dfrac{c}{\lambda \Delta v} = \dfrac{3 \times 10^8}{6 \times 10^{-7} \times 10^2}$

$Q = 5 \times 10^{12}$ **Ans.**

Example 3.16: The spectral spread of a red cadmium light of wavelength 694.3 nm is 0.001 nm. Calculate spectral purity factor, coherence length and coherence time.

[Raj. B.E.I. 2004, RTU B. Tech. Back 2011]

Solution: Given $\lambda = 694.3$ nm and $\Delta\lambda = 0.001$ nm

The spectral purity of the line (Q-factor)

$$= \frac{\lambda}{\Delta\lambda} = \frac{694.3}{0.001}$$

$= 6.943 \times 10^5$ **Ans.**

Coherence length $L = \lambda Q$

$= 694.3 \times 10^{-9} \times 6.943 \times 10^5$

$= 0.482$ m **Ans.**

Coherence time $= \dfrac{L}{c} = \dfrac{0.482}{3 \times 10^8}$

$= 1.6 \times 10^{-9}$ sec **Ans.**

Example 3.17: The coherence length for sodium light is 2.945×10^{-2} m. The wavelength for sodium light is 5890 Å. Calculate (*i*) Number of oscillation and (*ii*) Coherence time.

Solution: Given $L_e = 2.945\times10^{-2}\,\text{m}$

$$\lambda = 5890\text{Å} = 5890\times10^{-10}\,\text{m}$$

(*i*) The number of oscillations $= \dfrac{2L_c}{\lambda}$

$$= \frac{2\times2.95\times10^{-2}}{5890\times10^{-10}}$$

$$= \frac{5.89\times10^{-12}}{5890}$$

$$= 10^{15}$$ **Ans.**

(*ii*) Coherence time $= \dfrac{L_c}{c} = \dfrac{2.945\times10^{-2}}{3\times10^{8}}$

$$= 0.98\times10^{-10}\ \text{sec}$$ **Ans.**

Example 3.18: Calculate the temporal coherence length for mercury vapour lamp emitting in green portion of spectrum at wavelength 546.1 nm with emission bandwidth of $\Delta\nu = 6\times10^{8}\,\text{Hz}$.

[RTU B. Tech. II Sem. 2010]

Solution: Given $\lambda = 546.1\,\text{nm}$

and $\Delta\nu = 6\times10^{8}\,\text{Hz}$

The temporal coherence length

$$L_c = \frac{\lambda^2}{|\Delta\lambda|} = \frac{c}{\Delta\nu}$$

$$= \frac{3\times10^{8}}{6\times10^{8}} = \frac{1}{2}$$

$$L_c = 0.5\,\text{m}$$ **Ans.**

3.4 OPTICAL FIBRE

John Tyndall (August 2, 1820 – December 4, 1893) was a prominent 19th century Irish physicist. His initial scientific fame arose in the 1850s from his study of diamagnetism, which led to his appointment as successor to Michael Faraday at the Royal Institution in 1862. John Tyndall in 1854 demonstrated to the Royal Institution that a water jet can act as a light guide.

In today's information age, the need for fast, accurate and high volume communication is more important. Since optical frequencies are extremely large ($\approx 10^{15}$ Hz) as compared to the conventional radio waves ($\approx 10^{6}$ Hz) and microwaves ($\approx 10^{10}$ Hz) so a light beam acting as a carrier wave is capable of carrying for more information than radio waves and microwaves. Therefore in order to have an efficient and dependable com munication system, one would require a guiding medium in which the information carried by light waves could be transmitted. This guiding medium is known as optical fibre, which is a glass or plastic pipe as thin as human hair, that guides light waves from one place to another place in the same way that copper wire conducts electricity or pipes conduct water. Therefore it is also known as a optical waveguide.

3.4.1 Importance of Optical Fibres

1. Attenuation in a fibre is markedly lower than that in a coaxial cable, hence transmission over a wide range of distance is possible.
2. Optical fibres have smaller size and lighter weight.
3. Electromagnetic waves generated from electrical disturbances do not interfere with light signals in fibres.
4. The optical fibres are much more reliable because they can better withstand the environmental conditions such as pollution, humidity etc.
5. No physical electrical connection is required between the sender and the receiver.
6. Bandwidth of optical fibre is higher than that of an equivalent transmission line.
7. As fibres are very good dielectrics, isolation coating is not required.
8. Data rate is much higher in a fibre and hence much more information can be carried by each fibre.

3.4.2 Structure of Optical Fibre

Physically an optical fibre is a very thin and flexible medium, having a cylindrical shape consisting of three sections: (*i*) the core (*ii*) the cladding and (*iii*) the jacket.

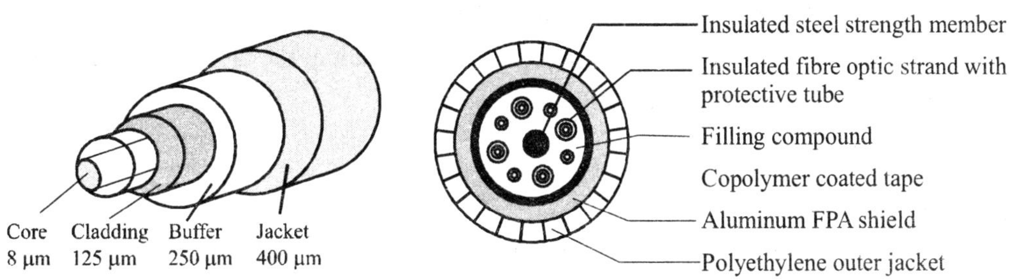

(*a*) (*b*)

Fig. 3.7 (a) Structure of optical fibre and **(b)** Cross sectional view of a fibre cable

Out of these, the core is the innermost section and is made of transparent dielectric material like glass or plastic. This is the actual fibre and has the remarkable property of conducting an optical beam. It is surrounded by its own cladding- a glass or plastic coating, that has optical properties which are very different from those of the core. The outer section is called the jacket made of plastic or polymer (Polyethylene) and other materials and is provided for protection against moisture, absorption, crushing and other environmental dangers.

The core acts like a continuous layer of two parallel mirrors. A singal is first enclosed into a light beam, which is then passed in between the two boundaries and propagated as a result of multiple internal reflections. The cross sectional view of a typical optical fibre cable is shown in Fig. 3.7 (*b*). Each optical fibre consists of a core surrounded by a cladding which in turn is coated with an insulating jacket. The fibres are individually buffered and strengthened. The fibres are wrapped with mylar film to bind the assembly. The assembly is then fitted in a corrugated aluminium sheath which acts as a shield. A jacket is put over the top.

3.5 PROPAGATION OF LIGHT WAVES THROUGH FIBRE

3.5.1 Mechanism

If light wave enters at one end of a fibre in proper conditions most of it is propagated down the length of the fibre and comes out from the other end of the fibre. There may be some loss due to a small fraction leakage through the side walls of the fibre. The reason of confining the light beam inside the fibre, is the total internal reflection of light waves by the inside surface of the fibre. Fibre obeys laws of reflection and refraction of light waves. The light which enters at one end of a fibre at a slight angle to the axis of the fibre, follows a zig-zag path due to series of reflections down the length of the fibre.

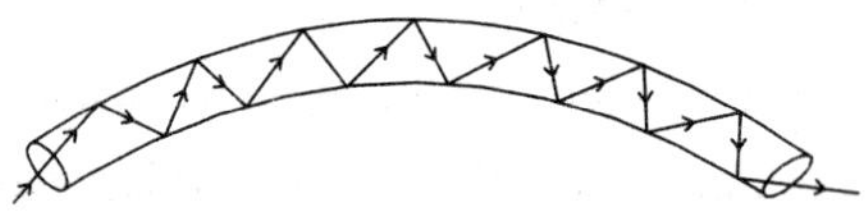

Fig. 3.8 Propagation of light in optical fibre

3.5.2 Conditions

Total internal reflection in the walls of the fibre can occur, if and only if, the following two conditions are satisfied:

(*i*) The glass around the centre of the fibre should have higher refractive index (μ_1) than that of the material (cladding) surrounding the fibre (μ_2), ($\mu_1 > \mu_2$).

(*ii*) The light should be incident at an angle θ (between the path of the ray and normal to the fibre wall) which should be greater than critical angle, θ_c

$$\sin\theta_c = \frac{\mu_2}{\mu_1} \tag{3.20}$$

Reflection, refraction and total internal reflection of light waves are shown in Fig. 3.9.

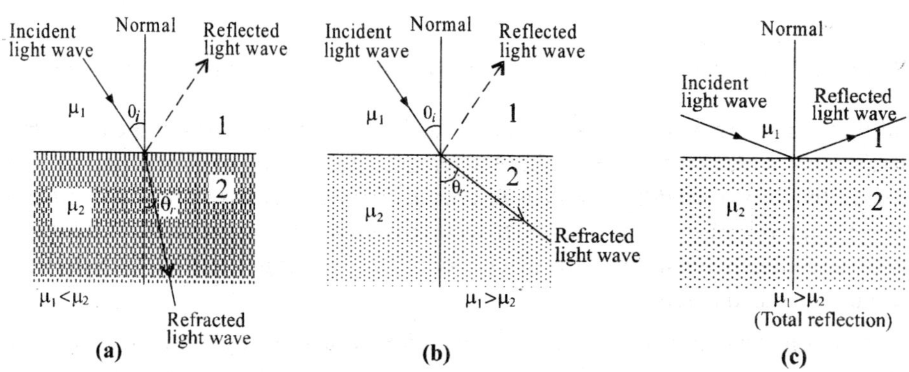

Fig. 3.9 Reflection, refraction and total Internal reflection

Let us now write down the conditions of reflection, refraction and total internal reflection.

1. In reflection angle of incidence (θ_i) is equal to the angle of reflection.
2. In refraction.
   (*i*) $\mu_1 \sin\theta_i = \mu_2 \sin\theta_r$ (θ_r= angle of refraction)
   (*ii*) The refracted wave should move towards the normal, if the light wave passes from optically rarer medium to optically denser medium. The refracted wave should move away from the normal, if the light wave travels from the optically denser to optically rarer medium.
3. The angle of incidence at which total reflection first occurs is called the critical angle θ_c for the two media. Light waves incident at angles greater than θ_c will be totally reflected.

3.6 TYPES OF OPTICAL FIBRE

There are two types of optical fibre: (*i*) Step index (SI) optical fibre and (*ii*) graded index (GRIN) optical fibre. This classification is based on the basic structure of an optical fibre and propagation of light wave through it.

3.6.1 Step Index (SI) Optical Fibre

When the refractive index of the fibre changes suddenly at the core-cladding boundary, then the fibre is known as the step index optical fibre.

Fig. 3.10 represents a longitudinal cross-section of a fibre. A light wave, which travels along the core and meets the cladding at angle greater than the critical angle θ_c, will be totally reflected. This reflected ray will then meet the opposite surface of the cladding, again at the same angle greater than θ_c and so is again totally reflected. Therefore, the light wave is propagated along the fibre core by a series of total internal reflections from the core cladding interface. Step index fibre can be either single mode or multimode.

(*a*) Multimode Step Index Fibre

Light energy emanating from any practical source, will have several paths with different angles of incidence at the core-cladding interface. It may also contain different colours with different frequencies (and so wavelengths). Then it is called *step-index multimode propagation* as shown in Fig. 3.10.

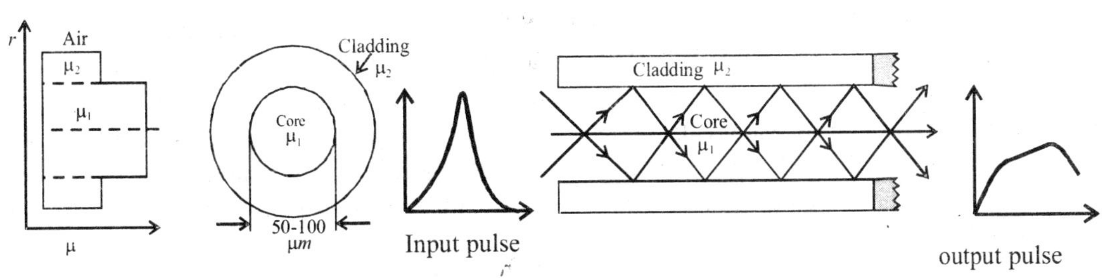

Fig. 3.10 Refractive index profile, core size and light ray propagation in a multimode step index optical fibre

Thus, the various light waves, travelling along the core, will have propagation paths of different lengths. Hence they will take different times to reach a given destination. Thus a distortion is produced which is called '*transit-time dispersion.*' As a result of this distortion, the variations of successive pulses of light may overlap into each other causing distortion of the information being carried.

(*b*) Single Mode Step Index Fibre

However, the above defect can be minimised by making the core diameter of the same order as the wavelength of light wave to be propagated. The resultant propagation is a single light wave along the axis only as shown in Fig. 3.11.

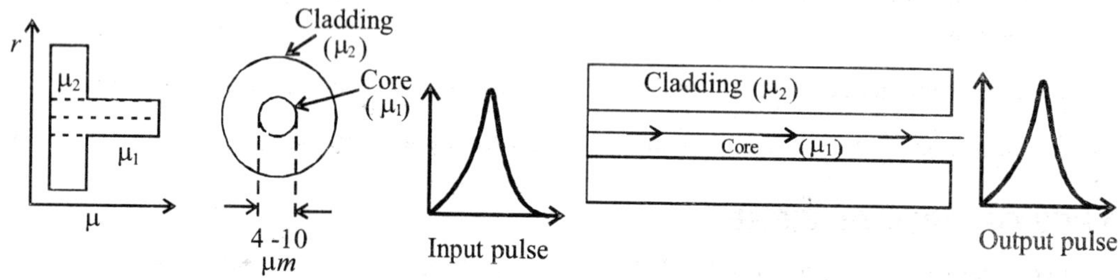

Fig. 3.11 Refractive index profile, core size and light ray propagation in a single mode step index optical fibre

This type of fibre is called a '*single mode step index fibre*'. These fibres are capable of transferring higher amount of data due to low fibre dispersion, so these fibres are most useful for large bandwidth and high capacity data applications.

3.6.2 Graded Index (GRIN) Optical Fibre

This fibre is also known as multimode graded index fibre. When the refractive index of the core changes gradually with the distance from the fibre axis, then the fibre is known as graded index optical fibre. It has a high value at the centre and falls off with increasing radial distance from the axis. The propagation of multiwave light energy in graded index fibre is shown in Fig. 3.12.

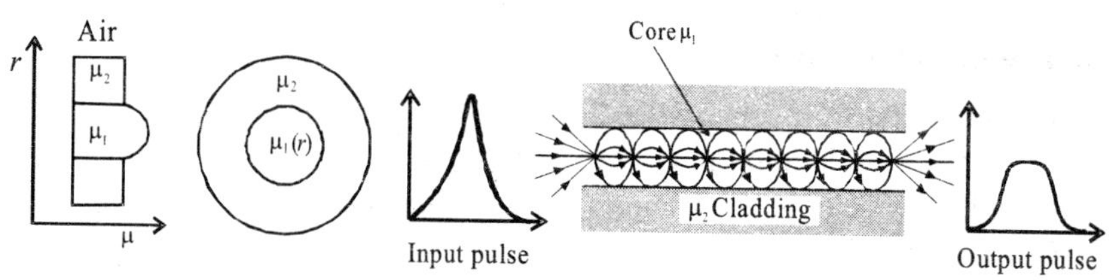

Fig. 3.12 Refractive index profile, core size and light ray propagation in a multimode graded index optical fibre

It is obvious that light waves with large angle of incidence travel more paths than those with smaller angles. But we know that the decrease of refractive index allows a higher velocity of propagation. Thus all waves will reach a given point along the fibre at virtually the same time. As a result the transit time dispersion is greately reduced. This type of light wave propagation is referred to as '*multi-mode graded index propagation*'.

3.7 ACCEPTANCE ANGLE AND ACCEPTANCE CONE OF A STEP INDEX FIBRE

We know that in a stepped index multimode fibre any light wave which travels along the core and meets the cladding at the critical angle of incidence θ_c will be totally reflected. This reflected wave will then meet the opposite surface of the cladding, again at the critical angle θ_c and so it will be again totally reflected. Any other light wave, which is meeting the core-cladding interface at or above the critical value θ_c will also be totally reflected and hence will propagate along the core. However, any light wave meeting the core-cladding interface at angle below θ_c will be absorbed by the cladding. In the reverse case ($\mu_1 < \mu_2$) a critical angle does not exist and the optical power loss is high. For a fibre for which $\mu_1 > \mu_2$, light impinging on the core within a critical angle θ_c is coupled into the fibre and will propagate. If the external incident angle is θ_0 corresponding to the critical angle of incidence θ_c at the core-cladding interface of the fibre, as shown in the Fig. 3.13 the light will stay in the fibre.

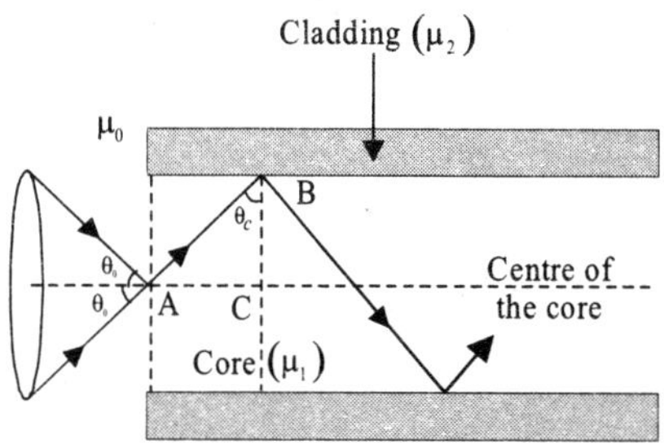

Fig. 3.13

So any light wave impinging on the core within this maximum external incident angle θ_0 is coupled into the fibre and will propagate. The angle ($2\theta_0$) is called the '*acceptance angle*'. It is unique for a particular fibre but different for different fibres and depends on the material and core diameter.

Let us now derive a mathematical relation for half the acceptance angle θ_0 from the Figs. 3.14 and 3.15. Let the refractive index of the medium outside the fibre be μ_0.

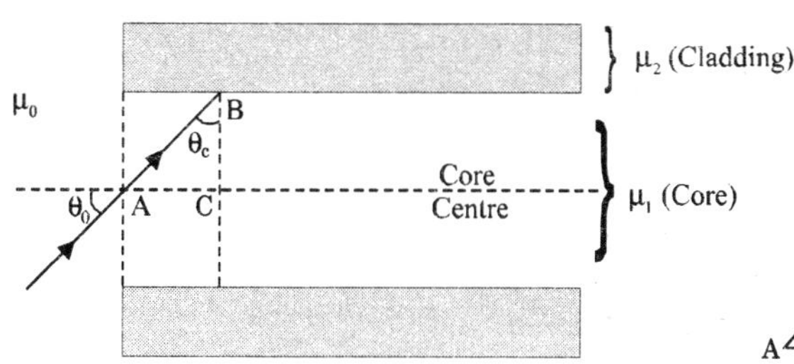

Fig. 3.14

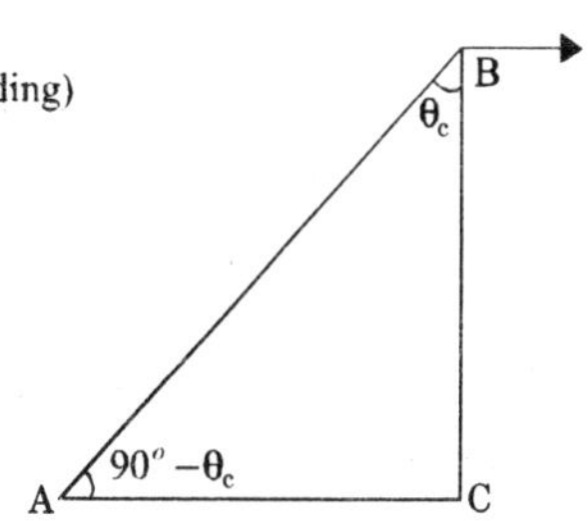

Fig. 3.15

From Figs. 3.14 and 3.15 and applying the law of refraciton at *A* and *B* we get

$$\mu_0 \sin\theta_0 = \mu_1 \sin(90° - \theta_c) \qquad \text{at } A \qquad (3.21)$$

$$\mu_1 \sin\theta_c = \mu_2 \sin 90° \text{ at } B \tag{3.22}$$

$$\sin\theta_c = \frac{\mu_2}{\mu_1} \text{ and } \mu_0 \sin\theta_0 = \mu_1 \cos\theta_c$$

$$\Rightarrow \quad \mu_0 \sin\theta_0 = \mu_1 \left(1 - \sin^2\theta_c\right)^{1/2}$$

$$= \mu_1 \left(1 - \frac{\mu_2^2}{\mu_1^2}\right)^{1/2}$$

$$= \sqrt{\mu_1^2 - \mu_2^2}$$

$$\Rightarrow \quad \sin\theta_0 = \frac{\left(\mu_1^2 - \mu_2^2\right)^{1/2}}{\mu_o}$$

$\therefore$ Half the acceptance angle is given by

$$\theta_0 = \sin^{-1}\frac{\left(\mu_1^2 - \mu_2^2\right)^{1/2}}{\mu_0} \tag{3.23}$$

If fibre surrounding medium is air i.e. $\mu_0 = 1$, then

$$\theta_0 = \sin^{-1}\sqrt{\mu_1^2 - \mu_2^2} \tag{3.24}$$

Thus the light, which travels within a cone defined by the *acceptance angle,* is trapped and guided. This is the fundamental property of light propagation in an optical fibre.

The cone defined by the angle of acceptance ($2\,\theta_0$) is defined as '*acceptance cone*'.

3.7.1 Numerical Aperture

Another important term associated with a fibre is the '*numerical aperture*'sometimes called the '*figure of merit*' for optical fibres. It is defined as μ_0 times the *sin* of the maximum angle at which light rays can enter the fibre and be conducted down the fibre, *i.e.,* numerical aperture of an optical fibre is defined as

$$NA = \mu_0 \sin\theta_0(\max) = \sqrt{\mu_1^2 - \mu_2^2} \tag{3.25}$$

If the fibre is surrounded by air $(\mu_0 = 1)$, then

$$NA = \sqrt{\mu_1^2 - \mu_2^2} \tag{3.26}$$

Generally μ_1 is only a few percentage greater than μ_2

$$\therefore NA = \sqrt{(\mu_1 + \mu_2)(\mu_1 - \mu_2)} \approx \sqrt{2\mu_1(\mu_1 - \mu_2)}$$

$$= \sqrt{2\mu_1^2 \frac{\mu_1 - \mu_2}{\mu_1}} = \mu_1\sqrt{2\Delta} \tag{3.27}$$

where $\mu_1 + \mu_2 \approx 2\mu_1$ and $\Delta = \frac{\mu_1 - \mu_2}{\mu_1}$

Δ is the fractional difference between the refractive indices of the core and the cladding. Fig. 3.16 shows the variation of *NA* with acceptance angle

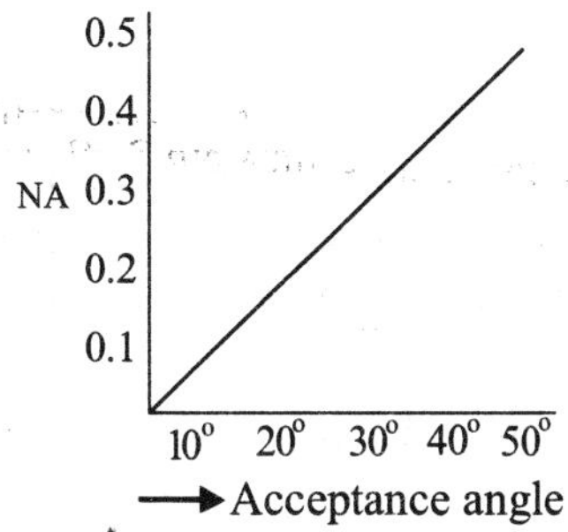

Fig. 3.16

It has been observed that numerical apertures for the fibres used in short distance communication are in the range of 0.4 to 0.5, whereas for long distance communications, numerical apertures are in the range of 0.1 to 0.3. It has been observed that smaller *NA* is not an advantage because it makes it harder to launch power into the fibre.

3.7.2 V-Number

The path along which a light wave propagates through an optical fibre depends on the angle of incidence. Since the angle of incidence through (θ) can vary, so the paths will also vary. These paths are called modes. It is revealed theoretically that the rays making only certain discrete angles θ are possible. If the core radius a or the refractive index difference between the core and the cladding, *i.e.*, $(\mu_1 - \mu_2)$ decreases then the possible value of θ also decreases.

The number of modes that can be propagated through the fibre may be related to an important parameter *V*, known as *normalized frequency* or *V-number.* It is given by the relation

$$V = \frac{2\pi}{\lambda} a\sqrt{\mu_1^2 - \mu_2^2} \tag{3.28}$$

where λ = operating wavelength,

a = the radius of the core

This is a dimensionless number that determines how many modes a fibre can support. As

$$\sqrt{\mu_1^2 - \mu_2^2} = NA$$

$$\therefore \quad V = \frac{2\pi}{\lambda} a(NA) \tag{3.29}$$

Again $\quad NA = \mu_1\sqrt{2\Delta}$

$$\therefore \quad V = \frac{2\pi a}{\lambda} \mu_1\sqrt{2\Delta} \tag{3.30}$$

The maximum number of modes N_m supported by a step index fibre is given by

$$N_m = \frac{V^2}{2} \tag{3.31}$$

and for a graded index fibre

$$N_m = \frac{V^2}{4} \tag{3.32}$$

Thus for a step index fibre, if $V = 10$, $N_m = 50$. If the value of V is less than 2.4045, then only one guided mode (as if there is only one discrete value of θ) is possible and the fibre is known as a single mode fibre (SMF). For $V > 2.4045$ the fibre can support many modes simultaneously and hence called multi-mode fibre (MMF).

The wavelength corresponding to the value of $V = 2.4045$ is called *cut-off wavelength* of the fibre and is given as $\lambda_c = \dfrac{\lambda V}{2.4045}$. **(3.33)**

Example 3.19: Compute the numerical aperture and the acceptance angle of an optical fibre from the following data. $\mu_{core} = 1.55$ and $\mu_{clad} = 1.50$

Solution: $NA = \mu_{core}\left(2\dfrac{\mu_{core} - \mu_{clad}}{\mu_{core}}\right)^{1/2}$

Here $\mu_{core} = 1.55$, $\mu_{clad} = 1.50$

$\therefore$ $NA = 1.55 \times \left(2 \times \dfrac{1.55 - 1.50}{1.55}\right)^{1/2} = 0.3937$ **Ans.**

Acceptance angle $\theta_0 = \sin^{-1}(NA) = \sin^{-1}(0.3937) = 23.2°$ **Ans.**

Example 3.20: Calculate the velocity of light in the optically active region of a fibre at 850 nm. Also compute the corresponding wavelength.

(Given: Refractive index of the substance at 850 nm = 3.6)

Solution: $v_{\text{substance}} = \dfrac{v_{vac}}{\mu}$

Here $\mu = 3.6$ and $v_{vac} = 3 \times 10^8$

$\therefore$ $v_{\text{substance}} = \dfrac{3 \times 10^8}{3.6} = 8.33 \times 10^7$ m/s **Ans.**

$\lambda = \dfrac{\lambda_{vac}}{\mu} = \dfrac{850}{3.6} = 236.11$ nm **Ans.**

Example 3.21: Calculate the refractive indices of the core and cladding materials of a fibre from the following data.

$NA = 0.22$ and $\Delta = 0.012$

Solution: $NA = \mu_{core}\sqrt{2\Delta}$

$$\therefore \quad \mu_{core} = \frac{NA}{\sqrt{2\Delta}}$$

Here $NA = 0.22$ and $\Delta = 0.012$

$$\therefore \quad \mu_{core} = \frac{0.22}{\sqrt{2 \times 0.012}} = 1.42$$ **Ans.**

Also $$\Delta = \frac{\mu_{core} - \mu_{clad}}{\mu_{core}}$$

$$0.012 = \frac{1.42 - \mu_{clad}}{1.42}$$

$$\Rightarrow \quad \mu_{clad} = 1.42 - 0.012 \times 1.42$$

$$\Rightarrow \quad \mu_{clad} = 1.40$$ **Ans.**

Example 3.22: A ruby laser emits 0.1 Joule pluses of light of wavelength 7200 Å. What is the minimum number of Cr^{3+} ions in the ruby?

Solution: Number of photons taking part in laser action

$$n = \frac{E\lambda}{hc}$$

Here $E = 0.1$ J, $\lambda = 7.2 \times 10^{-7}$ m, $h = 6.62 \times 10^{-34}$ J-s and $c = 3 \times 10^{8}$ m/s

$$\therefore \quad n = \frac{0.1 \times 7.2 \times 10^{-7}}{6.62 \times 10^{-34} \times 3 \times 10^{8}} = 3.625 \times 10^{17}$$

Since one Cr^{3+} ion emits one photon. Hence minimum number of Cr^{3+} ions in the ruby crystal is 3.625×10^{17} ions. **Ans.**

Example 3.23: Show that numerical aperture of a step index fibre is given by

$$NA = \frac{\sqrt{2\Delta}}{n_0} n_1$$

where symbols have their usual meanings. **[Raj. B.E. I 2003]**

Solution: Numerical aperture is defined as

$$NA = \frac{\sqrt{n_1^2 - n_2^2}}{n_0}$$

where n_1 = R.I. of core
n_2 = R.I. of cladding
n_0 = R.I. of surrounding medium

or $$NA = \frac{\sqrt{(n_1 + n_2)(n_1 - n_2)}}{n_0} \quad \text{but } n_1 = \frac{n_1 + n_2}{2}$$

so $$NA = \frac{\sqrt{2n_1(n_1 - n_2)}}{n_0} = \frac{\sqrt{2n_1^2\left(\frac{n_1 - n_2}{n_1}\right)}}{n_0}$$

or $$NA = \frac{\sqrt{2\Delta}}{n_0} n_1$$

where $\Delta = \frac{n_1 - n_2}{n_1}$ = fractional change in refractive index. **Ans.**

Example 3.24: The refractive index of core of an optical fibre is $n_1 = 1.45$ and the relative refractive index difference is 0.01. Find maximum angle of acceptance for this fibre.

[RTU B.Tech. I Sem. 2008]

Solution: Given $n_1 = 1.45$ and $n_1 - n_2 = 0.01$ then $\Delta = \frac{n_1 - n_2}{n_1} = 0.006$

If θ_0 is the maximum acceptance angle, then

$$NA = \sin\theta_0 = n_1\sqrt{2\Delta}$$

$$\Rightarrow \quad \theta_0 = \sin^{-1}\left(1.45 \times \sqrt{2 \times 0.006}\right)$$

$$\Rightarrow \quad \theta_0 = \sin^{-1}(1.45 \times 0.109)$$

$$= \sin^{-1}(0.158)$$

$$\therefore \quad \theta_0 = 9.13°$$ **Ans.**

Example 3.25: An optical fibre has refractive index of core to be 1.5 and the refractive index difference of core-cladding to be 0.01. Determine numerical aperture and maximum angle of acceptance.

[RTU B.Tech. II Sem. 2008]

Solution: Given $n_1 = 1.5$

and $n_1 - n_2 = 0.01$

then refractive index is

$$\Delta = \frac{n_1 - n_2}{n_1} = \frac{0.01}{1.5}$$

$$\Rightarrow \quad \Delta = 0.006$$

Since $NA = n_1\sqrt{2\Delta}$

$$= 1.5\sqrt{2 \times 0.006}$$

$$= 1.5\sqrt{0.012}$$

$\therefore \qquad NA = 0.164$ **Ans.**

and maximum angle of acceptance is

$$\theta_0 = \sin^{-1} NA$$

$$= \sin^{-1}(0.164)$$

$\therefore \qquad \theta_0 = 9.45°$ **Ans.**

Example 3.26: The refractive index of the core is 1.8 and the fractional refractive index between the core and the cladding is 1.5 %. Calculate (*a*) numerical aperture (*b*) acceptance angle (*c*) critical angle at the core-cladding interface (*d*) the velocity of light in the core and (*e*) the velocity of light in the cladding.

Solution: Given $\mu_1 = 1.8, \; \Delta = 1.5 \times 10^{-2}$

We have $NA = \mu_1 (2\Delta)^{1/2}$

$$= 1.8(2 \times 1.5 \times 10^{-2})^{1/2}$$

$NA = 0.312$ **Ans.**

$\therefore$ Acceptance angle (i_{max}) = $\sin^{-1}$ (NA)

$$= \sin^{-1} (0.312)$$

$= 18.18°$ **Ans.**

Also $\Delta = \dfrac{\mu_1 - \mu_2}{\mu_1} = 0.015$

$$\therefore \qquad \mu_1 - \mu_2 = 0.015\mu_1$$

$$\Rightarrow \qquad 1 - \frac{\mu_2}{\mu_1} = 0.015$$

$$\Rightarrow \qquad \frac{\mu_2}{\mu_1} = 0.985$$

Critical angle at core cladding interface

$$i_c = \sin^{-1}\left(\frac{\mu_2}{\mu_1}\right)$$

$$= \sin^{-1}(0.985)$$

$= 80.06°$ **Ans.**

Velocity of light in the core $= \dfrac{c}{\mu_1}$

$$= \frac{3 \times 10^8}{1.8} = 1.67 \times 10^8 \text{ m/s}$$ **Ans.**

Velocity of light in the cladding $= \dfrac{c}{\mu_2}$

$$= \frac{3 \times 10^8}{0.985\ \mu_1}$$

$$= \frac{3 \times 10^8}{0.985 \times 1.8}$$

$$= 1.69 \times 10^8 \text{ m/s}$$ **Ans.**

Example 3.27: Calculate the maximum value of Δ in the case of a single mode fibre of diameter 10 μm and core refractive index 1.5. Fibre is coupled to a light source of wavelength 1.3 μm. Also calculate the cladding refractive index and the acceptance angle. Given $V_{cut-off}$ for a single mode propagation = 2.405.

Solution: Given $d = 10\ \mu m$

$\mu_1 = 1.5$

$\lambda = 1.3\ \mu m = 1.3 \times 10^{-6}$ m

$V_{cut\text{-}off} = 2.405$

The normalized frequency (V - number) is

$$V = \frac{\pi d}{\lambda_0}\sqrt{\mu_1^2 - \mu_2^2}$$

For maximum value

$$V_{cut\text{-}off} = \frac{\pi d}{\lambda} NA_{cut\text{-}off}$$

$$NA_{cut\text{-}off} = \frac{V_{cut\text{-}off} \times \lambda}{\pi d}$$

$$= \frac{2.405 \times 1.3 \times 10^{-6}}{3.14 \times 10^{-5}}$$

$$= \frac{3.12}{3.14 \times 10^{-5}}$$

$$= 0.09957$$

∴ Acceptance angle = $\sin^{-1}(NA)$

$= 5.71°$ **Ans.**

Also numerical aperture $NA = \mu_1\sqrt{2\Delta}$

$$\therefore \quad (NA)^2 = \mu_1^2 \times 2\Delta$$

$$\therefore \quad \Delta = \frac{(NA)^2}{\mu_1^2} \times \frac{1}{2}$$

$$= \left(\frac{0.09957}{1.5}\right)^2 \times \frac{1}{2}$$

$$\Delta = 2.2 \times 10^{-3}$$ **Ans.**

Also $$\Delta = \frac{\mu_1 - \mu_2}{\mu_1}$$

$$\therefore \quad \mu_2 = \mu_1(1 - \Delta)$$

$$= 1.5\ (1 - 2.2 \times 10^{-3})$$

$$\mu_2 = 1.496$$ **Ans.**

Example 3.28: Consider a fibre with $\mu_1 = 1.48$, $\mu_2 = 1.46$ and core radius $a = 30\ \mu\text{m}$. Show that all rays making an angle $\theta < 9.43°$ with axis will be guided through the fibre.

Solution: Numerical aperture is given by

$$NA = \sqrt{\mu_1^2 - \mu_2^2}$$

$$= \sqrt{(1.48)^2 - (1.46)^2}$$

$$= \sqrt{2.1904 - 2.1316}$$

$$= \sqrt{0.0588}$$

$$NA = 0.2425$$ **Ans.**

The acceptance angle $\theta_m = \sin^{-1}(NA) = \sin^{-1}(0.2425)$

$$= 14.03°$$

i.e., $\theta_m > 9.43°$, therefore all rays making an angle $\theta < 9.43°$ with axis will be guided through the fibre. **Ans.**

3.8 APPLICATIONS OF OPTICAL FIBRE

Due to large bandwidth, low attenuation loss, very high tensile strength (flexibility) and low cost, the optical fibre system is widely used in the field of communication, medical, industry and defence applications.

3.8.1 Optical Fibre Communication

The block diagram of a fibre optic communication is shown in Fig. 3.17.

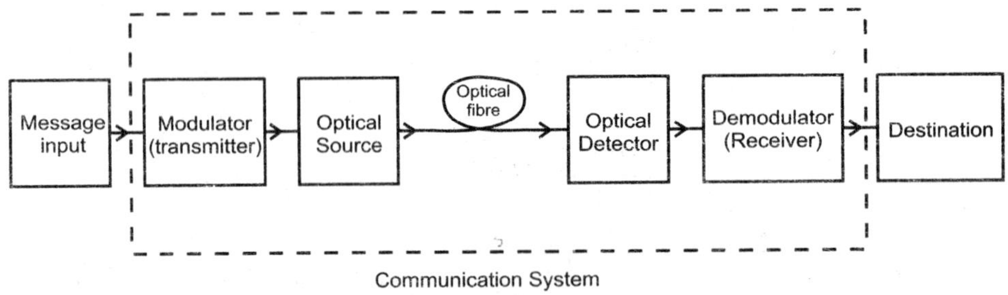

Fig. 3.17 Optical fibre communication system

In this system, a transducer changes a non-electrical message to an electrical signal, then transmitter converts the signal into a suitable form for propagation over the transmission medium by a technique is called modulation. In fibre communication, light waves are used as carrier waves. These waves are generally generated by a laser diode or a light emitting diode (LED). These carrier waves are modulated using various techniques *viz.* frequency modulation, amplitude modulation and digital modulation*. These modulated signals suffer attenuation losses due to distortion and dispersion in the optical fibre. Therefore for long distance applications repeators are used at several positions of the fibre to remove signal distortion. In this system, the receiver consists of an optical detector, which is used to convert the received optical signal into an electrical signal. The examples of optical detector are photodiodes, phototransistors and photoconductors. The output of the detector then passes through the signal processor to capture the original electrical signal from the carrier, *i.e.,* demodulation process is used. The signal output is finally put into a transducer to set it either in video or audio form depending upon the nature of the original message.

3.8.2 Fibre Optics in Medical Science

The fibre technology is commonly used in the field of medical science. The fabrication of flexible fibrescope in endoscopy for visualisation of interior parts of the human body is based on the optical fibre technology. In a fibrescope, two bundles of optical fibres are used. One for guiding light from the source onto the organ to be viewed. The other set of bundle gather light from the object and guide it to viewing screen as shown in Fig. 3.18.

Fibrescope not only permits the visualization of interior parts of the human body but also the selective cauterization of tissues using a laser beam.

* The digital modulation is preferred in optical communication over the analogue modulation as it can pass through a large distance with the same power.

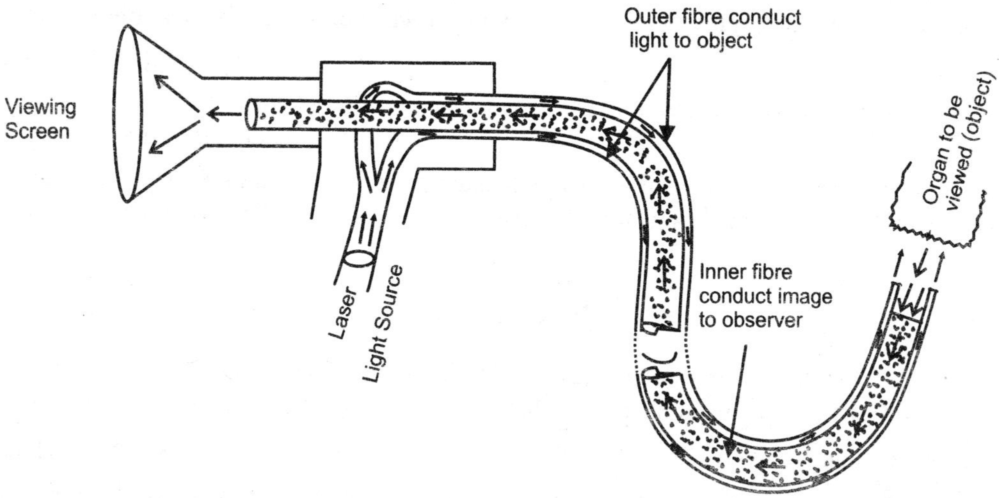

Fig. 3.18 Flexible fibrescope

QUESTIONS AND PROBLEMS

QUESTIONS

1. Classify coherence as temporal and spatial coherence. Describe how size of source is related to spatial coherence and monochromaticity to temporal coherence.

   [RTU B. Tech. Back 2012]

2. Define spatial and temporal coherence with their examples.

   [RTU B. Tech. Back 2008 & 2011]

3. Show that visibility is a measure of degree of coherence.

   [RTU B. Tech. II Sem. 2008, 2010]

4. Explain coherence length and coherence time.

   [RTU B. Tech. II Sem. 2008, , Back 2010]

5. Write two prominant applications of optical fibres.

   [RTU B. Tech. I Sem. 2008]

6. What do you mean by spatial and temporal coherence for propagating waves? Show the visibility is measure of degree of coherence. **[RTU B. Tech. II Sem. 2010 & 2011]**

7. What is an optical fibre? What do you mean by numerical aperture of an optical fibre? Find an expression for the numerical aperture of a step index optical fibre.

   [RTU. B. Tech. Back 2010]

8. How does an optical fibre function in transporting electromagnetic energy?

   [Raj. B.E. I 2003]

9. Write short note on optical fibre. **[Raj. B.E. I 2004]**

10. Explain clearly, the propagation of an electromagnetic wave inside an optical fiber. Use proper diagram. Derive the expression for the maximum acceptance angle of optical fibre in air.

[Raj. B.Tech. I 2007]

11. How optical fibres can be used in medical and communication fields?

[RTU B. Tech. II Sem. 2010]

12. Explain how light is propagated in a variable index fibre. Define numerical aperture and acceptance angle. Derive expressions for the same. **[RTU B. Tech. II Sem. 2012]**

PROBLEMS

1. The coherence length for sodium light is 2.945×10^{-2} m. The wavelength of sodium light is 5890 Å. Calculate (*i*) the number of oscillations corresponding to coherence length and (*ii*) the coherence time. **[5×10^4; 9.82×10^{-11}]**

2. Compute the numerical aperture, acceptance angle and the critical angle of the fibre having $\mu_{core} = 1.50$ and $\mu_{clad} = 1.45$. **[0.387; 22.78°; 75.2°]**

3. Calculate the refractive indices of the core and the cladding material of a fibre from the following data. $NA = 0.22$, $\Delta = 0.012$ **[1.42, 1.403]**

[RTU B. Tech. Back 2011]

4. A step index fibre has a normalized frequency $V = 2.66$ at a wavelength of 1300 nm. If the core radius is 25 μm, calculate the numerical aperture. **[0.22]**

5. A step index fibre is made with a core of index 1.52, a diameter of 29 μm and a fractional difference index of 0.0007. It is operated at a wavelength of 1.3 μm. Calculate the V-number and the number of modes. **[3.957, 8]**

6. The core diameter of multimode step index fibre is 60 μm. The difference in refractive indices is 0.0013. The core refractive index is 1.46. Determine the number of guided modes when the operating wavelength is 0.75 μm. **[1739.32]**

7. An optical fibre has an NA of 0.20 and a cladding refractive index of 1.59. Determine the acceptance angle for the fibre in water has a refractive index of 1.33. **[8.62°]**

[RTU B. Tech. II Sem. 2010]

8. For a step index fibre having core index 1.43 and cladding index 1.40. Calculate (*i*) Critical angle (*ii*) Critical propagation angle and (*iii*) Numerical aperture.

[(*i*) 78.23° (*ii*) 16.85° and (*iii*) 0.29]

[RTU B. Tech. II Sem. 2012]

REFERENCES

1. Edward A. Laey, Fibre Optics, Prentice Hall Inc. New Jersey, 1982.
2. John M. Senior, Optical Fibre Communication, Prentice-Hall of India Private Ltd. New Delhi, 1994.

3. G. Keisier, Optical Fibre Communications, McGraw Hill N.Y. (1991).
4. N.S. Kapany, Fibre Optics, Principles and Applications, Academic Press, New York, 1967.
5. A. Ghatak and K. Thyagarajan, Introduction to Fibre Optics, Cambridge Universtiy Press, Cambridge, 2004.
6. H. Kolimbiris, Fibre Optics Communications, Pearson Education Asia, New Delhi, 2004.
7. J.C. Palais, Fibre Optics Communications, Pearson Education Asia, New Delhi, 1998.

4

LASER AND HOLOGRAPHY

4.1 LASER

LASER stands for Light Amplification by Stimulated Emission of Radiation. The principle of laser is based on the stimulated emission. The idea of stimulated emission was first proposed by Einstein in 1917 but in 1954 Townes, Gordon and Zeiger were the first to use the principle of laser in the amplification of microwaves. That device was called Maser. In 1958, Townes and Arthur L. Schawlow used this principle in light amplification. The first successful Laser device was built by Maiman in 1960. It is ruby laser which consists of a three level Laser system.

4.1.1 Comparison of Laser Source with Ordinary Source

The light obtained from usual sources is spatially incoherent, it is temporarily incohernt also due to finite spectral bandwidth, *i. e.,* finite mean life time (spectral bandwidth $\Delta v = \frac{1}{\tau_c}$).

A laser source generates a highly spatially as well as temporarily coherent beam.With an ordinary source the resultant intensity at a point is the sum of intensities due to the constituent atomic sources, *i.e.,*

$$I = I_1 + I_2 + I_3 + ... + I_n$$

$$= nI_1 \text{ (if all the sources are alike)} \quad \textbf{(4.1)}$$

In a laser source, all the emissions are in step with each other, *i.e.,* they are spatially coherent so that all the disturbances reaching a point will be in phase and the resultant amplitude will be

$$A = A_1 + A_2 + A_3 + ... + A_n$$

$$= nA_1 \text{ (if all the sources are alike)}$$

$$\therefore \quad I = n^2 A_1^2 = n(nI_1) \quad \textbf{(4.2)}$$

Thus the laser source will be n times more intense (where n is the number of atoms taking part in the emission) than an ordinary source. Thus the laser produces a highly intense (due to spatial coherence) and truly monochromatic (due to temporal coherence) beam.

4.2. ABSORPTION, SPONTANEOUS EMISSION AND INDUCED EMISSION OF RADIATION

Albert Einstein(14 March 1879 – 18 April 1955) was a German-born theoretical physicist. He is best known for his theory of relativity. Einstein received the 1921 Nobel Prize in Physics for his discovery of the laws of Photoelectric Effect. In 1917 he gave the idea of stimulated emission of radiation.

Let us consider two energy levels in a particular atom. The lower level is designated as i and upper level as j (Fig. 4.1). If the atom is initially in state i, it can be raised to state j by absorbing a photon of light whose frequency is

$$\nu = \frac{E_j - E_i}{h} \tag{4.3}$$

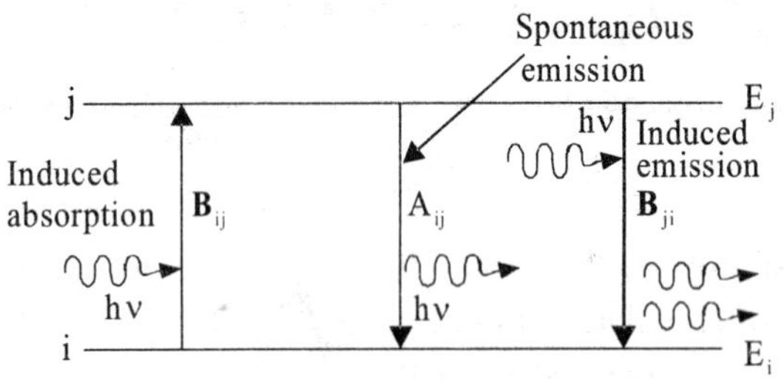

Fig. 4.1 Transitions between two energy levels in an atom can occur by induced absorption, spontaneous emission and induced emission

The possibility that the atom will actually undergo this transition is proportional to the rate at which the photons of frequency ν fall on it and therefore to the spectral energy density $u(\nu)$. The transition probability also depends on the properties of states i and j, but we can include this dependence in some constant of proportionality B_{ij}. Hence if we shine light of frequency ν and energy density $u(\nu)$ on the atom when it is in the lower state i, the probability for it to go to higher state j is

$$P_{i\to j} = B_{ij}u(\nu) \text{ (Absorption)} \tag{4.4}$$

If the atom is initially in the upper state j, two processes can occur:

(1) The atom has a small life time (mean lifetime $\approx 10^{-8}$ sec) in the excited state so that it can spontaneously drop to the state i by emitting a photon of frequency ν. This is called spontaneous emission (Fig. 4.1). The probability of such a process will, of course, depend on the two energy levels but it will be independant of the presence of any photon incident on the atom. Let us express this probability by the coefficient A_{ji} then

$$P_{j\to i} = A_{ji} \text{ Spontaneous emission} \tag{4.5}$$

(2) While discussing the interaction of radiation with matter, Einstein proposed that if a photon of right frequencey is present, it is not essential that it may always be absorbed, it can also interact with an atom in the excited state and induce that atom to emit a new photon (Fig. 4.1). Such a process of emission is known as *induced* or *stimulated emission*. The probability of induced emission will thus be proportional to the spectral energy density $u(\nu)$ and we can write

$$P_{j\to i} = B_{ji}u(\nu) \text{ Stimulated emission} \tag{4.6}$$

The total probability for an atom in state j to fall to the lower state i is therefore

$$P_{j\to i} = A_{ji} + B_{ji}u(\nu) \text{ Emission} \tag{4.7}$$

Now we consider an assembly of N_i atoms in the state i and N_j atoms in the state j, all in thermal equilibrium at the temperature T with light of frequency ν and energy density $u(\nu)$. The number of atoms in state i that absorb a photon and go the state j per second, *i. e.*, the rate of transition is given by

$$R_{ij} = N_i P_{i\to j} = N_i B_{ij} u(\nu) \tag{4.8}$$

The number of atoms that go from j to i per second will have two components

(a) due to spontaneous emission

$$(R_{ji})_{\text{spontaneous}} = N_j P_{j\to i} = N_j A_{ji}$$

(b) due to stimulated emission

$$(R_{ji})_{\text{stimulated}} = N_j P_{j\to i} = N_j B_{ji} u(\nu)$$

Thus the total rate of transition from state j to state i will be

$$R_{ji} = N_j P_{j\to i} = N_j \left[A_{ji} + B_{ji}u(\nu)\right] \tag{4.9}$$

In thermal equilibrium the number of atoms going from state i to state j per second must be equal to the number going from state j to i. This is in accordance with statistical principle of detailed balancing. Hence from Eqs. (4.8) and (4.9)

$$N_i P_{i\to j} = N_j P_{j\to i}$$

$$\Rightarrow \quad N_i B_{ij} u(\nu) = N_j \left[A_{ji} + B_{ji} u(\nu)\right] \tag{4.10}$$

Dividing both sides of this equation by $N_j B_{ji}$, we get

$$\left(\frac{N_i}{N_j}\right)\left(\frac{B_{ij}}{B_{ji}}\right) u(\nu) = \frac{A_{ji}}{B_{ji}} + u(\nu)$$

Solving for $u(\nu)$, we get

$$u(\nu) = \frac{A_{ji}/B_{ji}}{\left(\frac{N_i}{N_j}\right)\left(\frac{B_{ij}}{B_{ji}}\right) - 1} \tag{4.11}$$

Now, in thermal equilibrium, by Maxwell-Boltzmann distribution law*, the ratio between the populations of states i and j is given by

$$\frac{N_i}{N_j} = e^{\frac{(E_j - E_i)}{kT}} = e^{h\nu/kT} \tag{4.12}$$

With this result we have

* The Maxwell-Boltzmann distribution law is given by $N_i = N_0 e^{-E_i/k_BT}$

$$u(\nu) = \frac{A_{ji}/B_{ji}}{\left(\dfrac{B_{ij}}{B_{ji}}\right) e^{h\nu/kT} - 1} \qquad (4.13)$$

Eq. (4.13) is a formula for the energy density of photons of frequency v in equilibrium at the temperature T with atoms whose possible energies are E_i and E_j. This formula can be compared with the Planck's radiation law, according to which

$$u(\nu)d\nu = \frac{8\pi h\nu^3 / c^3}{e^{h\nu/kT} - 1} d\nu \qquad (4.14)$$

The comparison yields

$$\frac{A_{ji}}{B_{ji}} = \frac{8\pi h\nu^3}{c^3} \qquad (4.15)$$

and $$B_{ij} = B_{ji} \qquad (4.16)$$

It is inferred from Eq. (4.16) that the probability of induced emission per incident photon is equal to the probability of absorption per incident photon, *i.e.*, in the presence of an incident photon absorption and induced emission are equally probable. However the net absorption rate R_{ij} is usually much more than the net stimulated emission rate $(R_{ji})_{\text{stimulated}}$ because of large difference in the population in the two states. Let us now compare the probability of spontaneous emission to that of stimulated emission.

$$\frac{\left(P_{j\to i}\right)_{spontaneous}}{\left(P_{j\to i}\right)_{stimulated}} = \frac{A_{ji}}{B_{ji}u(\nu)}$$

$$= \frac{u(\nu)\left[\left(\dfrac{B_{ij}}{B_{ji}}\right) e^{h\nu/kT} - 1\right]}{u(\nu)}$$

$$= (e^{h\nu/kT} - 1) \qquad (4.17)$$

(1) If $h\nu >> kT$, the spontaneous emission will be much more probable than induced (stimulated) emission (which can even be considered negligible). This is ordinarily true for electronic transition in atoms or molecules and in radiative transitions in nuclei.

(2) If $h\nu << kT$, then

$$(e^{h\nu/kT} - 1) \approx \frac{h\nu}{kT}$$

so that

$$\frac{(P_{ji})_{spontaneous}}{(P_{ji})_{stimulated}} = \frac{h\nu}{kT} \tag{4.18}$$

In such a case the stimulated emission is no more negligible and it even predominates over the spontaneous emission. Such a situation can be achieved for radiation in microwave region.

4.3 PRINCIPLE OF LASER

The population of atoms in different energy states is governed by Maxwell-Boltzmann distribution law. Further, according to the process of detailed balancing the total emission and absorption rates, in thermal equilibrium, remain equal. In the presence of incident radiation the equilibrium is disturbed and the ratio of emission rates is given by

$$\frac{\text{Rate of emission}}{\text{Rate of absorption}} = \frac{R_{ji}}{R_{ij}} = \frac{A_{ji} + B_{ji}u(\nu)}{B_{ij}u(\nu)} \frac{N_j}{N_i}$$

$$= \left[1 + \frac{A_{ji}}{B_{ij}u(\nu)}\right] \frac{N_j}{N_i} \tag{4.19}$$

For laser action

(*a*) the emisssion rate must be greater than the absorption rate and

(*b*) the probability of spontaneous emission (which produces incoherent radiations) must be much smaller than the probability of stimulated emission.

Thus with the second condition, *i.e.*, $A_{ji} << B_{ji}u(\nu)$, we have

$$\frac{\text{Rate of emission}}{\text{Rate of absorption}} = \frac{N_j}{N_i} \tag{4.20}$$

Under normal conditions the number N_j in the upper energy state j is always smaller than the number N_i in the lower energy state. If, by some means, N_j is made greater than N_i the emission rate will become larger than the absorption rate and consequently amplification will take place.

To achieve the basic requirement, *i.e.*, $A_{ji} << B_{ji}u(\nu)$ the following possibilities can be used.

(1) $h\nu << kT$ such a situation exists at microwave frequencies. Ammonia maser is based on this principle. **MASER** stands for *Microwave Amplification by Stimulated Emission of Radiation.*

(2) **Metastable States:** There are certain energy levels in an atom from which transitions to the ground state are forbidden by the selections rules. Such states are called '*Metastable States*'. The atom can however, go from the metastable state to the ground state either by giving up the appropriate amount of energy to another atom during a collision process or it may absorb radiation and go to a higher energy state (to which transitions are not forbidden) and from there it may return to normal state by emission of radiation. Such processes make the mean life time of a metastable state ($\approx 10^{-3}$ sec) to be much longer than the mean life time of other excited

states ($\approx 10^{-8}$ sec). Thus if certain atoms are excited to the metastabe state, the probability of spontaneous emission will be quite negligible. However, induced transition, being a sort of forced oscillations can take place.

There are several mechanisms which can be used to invert the normal population density of the energy levels. This process of population inversion is called '*pumping*'. The commonly used methods are

(1) Optical pumping

(2) Electron impact

(3) Inelastic atom-atom collision

(4) Chemical pumping

(5) Gas dynamic pumping

4.3.1 Pumping and Population Inversion

A system in which population inversion is to be achieved is called an '*active system*' (for laser). Raising the molecules to higher energy state is called '*pumping*' and this can be done in many ways.

The principle of laser using optical pumping and metastable states is illustrated in Fig. 4.2.

0 represents the ground state, 1 represents a metastable state and 2 an excited state. Transition between 0 and 2 and 1 and 2 are allowed. If an intense beam of light of frequency $\nu' = \dfrac{E_2 - E_0}{h}$ is incident on the system, a number of atoms in the state 0 will be excited to state 2. From this level 2 the atoms can trickle down to state 1 by spontaneous emission of photons of frequency.

$$\nu'' = \frac{E_2 - E_1}{h}$$

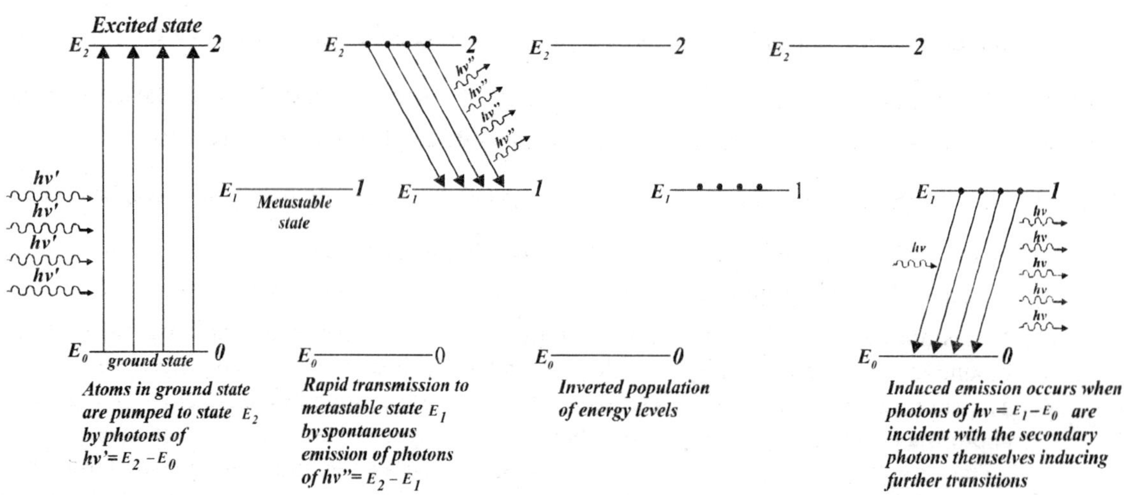

Fig. 4.2 Principle of masers and lasers

This state 1 being metastable, they will tend to remain in this state for a comparatively longer period so that the population in this state can be considerably increased. If now an input singal of frequency $\nu = \frac{E_1 - E_0}{h}$ is introduced, it will stimulate the emission from state 1 to state 0. Thus where we introduced one photon we will now get two.

These two can then trigger other atoms to radiate and this process builds up to give very *intense coherent beam*. The above scheme is termed as three level laser. It may be mentioned that though Einstein had visualized the possibility of amplification by stimulated emission as early as 1917 the first practical device, an Ammonia Maser, was developed by Gordon, Zeiger and Townes in 1954. The first laser device appeared in 1960 when Maiman constructed a ruby laser. Since then various types of lasers viz. gas lasers, chemical lasers, solid state lasers, organic lasers *etc*. have been developed.

4.3.2 Cavity Resonators and Shaping of a Beam (Operation of Laser)

In a laser, the active system is enclosed in an optical cavity, usually made in the form of a rod or a tube. The ends of the laser material are closed by accurately parallel reflectors, one of which is perfect reflector and the other has 90% reflectivity (or 10% transmission). The cavity makes the energy density large, whence stimulated emission dominates over the spontaneous emission. The multiple reflection also makes the stimulated emission more coherent. This process is called *shapping* of the beam.

Fig. 4.3 schematically shows how the beam shapes, with time. The hollow circles here represent the unexcited atoms and black dots represent the excited atoms. The left end mirror represents the full reflector and the right end mirror the partial reflector. The description stagewise is as follows:

(1) Almost all atoms are initially in the unexcited state. An outside agent (such as an electrical discharge for example) excites atoms inside the laser [Fig. 4.3 (*a*)].

(2) When the pumping radiation falls on the active material the atoms start reaching the excited state. (Population inversion not yet reached, black dots less than hollow circles). [Fig. 4.3 (*b*)].

(3) Pumping radiation continues. The majority of certain atoms in the active material area excited to the metastable states in which the excited atoms remain for an appreciable time. Population inversion is reached at this stage. The dominant radiations initially are the spontaneous ones. Two cases of coherent (stimulated) emission are shown (multiple-line arrows) one along the axis, another at a slant angle. [Fig. 4.3 (*c*)].

(4) One excited atom falls to a lower state and emits a photon. This photon wave travels past the other excited atoms and triggers them to fall to the lower state. In doing so, they emit photons with waves in step with the original photon wave. The beam travels back and forth again and again due to multiple reflection on the end faces. On each trip, it causes more excited atoms to emit waves in phase with it. Spontaneous component of emissions is still not neglible. [Fig. 4.3 (*d*)].

(5) Only those waves that travel accurately along the axis of the tube will remain in the tube after many trips up and down it, therefore, the beam is almost entirely composed of waves going in exactly the same direction. Its rays are accurately parallel. Thus radiations that are mutually coherent go on becoming dominat at successive stages. Spontaneous component of emisssion is now very weak [Fig. 4.3 (*e*)] .

(6) Almost the entire emission becomes coherent. The wave shapes shown at the exit on the right show phase agreement of waves. This is called the *laser beam* [Fig. 4.3 (*f*)].

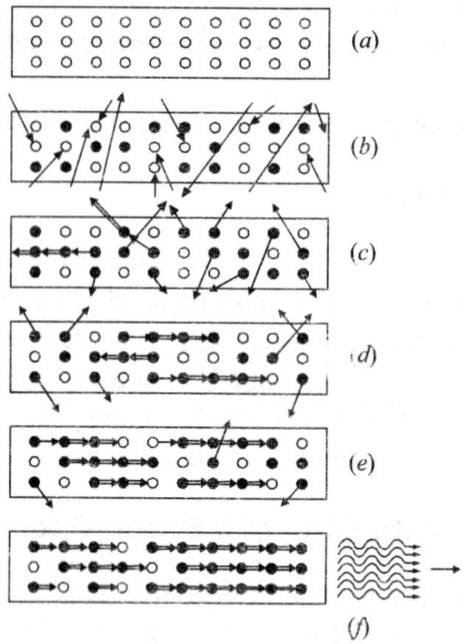

Fig. 4.3 Shaping of a Laser beam

4.4 THE HELIUM-NEON LASER

The first gas laser to be operated successfully was the He-Ne laser. This laser was first fabricated and was put into operation by Ali Javan, Bennett and Harriott in 1961 at Bell Telephone Laboratories in U.S.A.

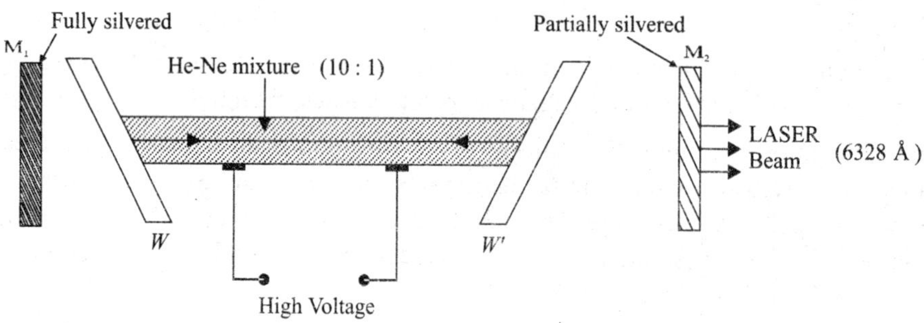

Fig. 4.4 He-Ne gas laser

This laser contains a mixture of helium and neon gas as an active medium. A schematic diagram of He-Ne is shown in Fig. 4.4. It is composed of narrow glass tube nearly 1m long and contains helium at a pressure of approximately 1 torr (1 mm of Hg) and neon at a presssure of about 0.1 torr in the ratio 10 : 1 of their weights. The gas system is enclosed between a pair of plane or concave mirrors so that a cavity resonator system is formed. One of the mirrors is a perfect reflector while the other is a partially transparent so that energy may be coupled out of the system. The spacing between the two mirrors is equal to an integral multiple of half wevelength of laser light.

The laser output is made linearly polarized by the inclusion of windows WW' at Brewster angle at the end of discharge tube. At Brewster angle (Polarization angle) the windows presumably have 100% transmission for light whose electric field component is parallel to the plane of incidence.

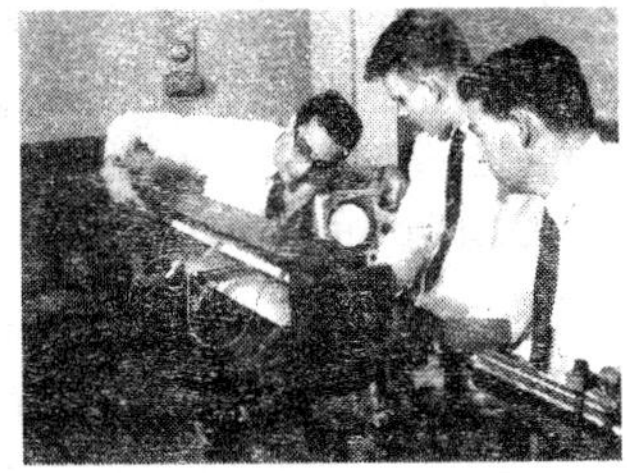

Ali Javan and his associates William Bennett Jr. and Donald Herriott at Bell Labs were first to successfully demonstrate a continuous wave (CW) helium-neon laser operation (1960–1962).

The first few energy levels of He and Ne atoms are shown in Fig. 4.5. When an electric discharge is passed through the gas, the electrons travelling down the tube collide with helium atoms and excite them to E_2^{He} and E_3^{He} levels. These levels are metastable, *i.e.*, He atoms excited to these states stay in these levels for sufficiently long time before losing energy through collision. Through these collisions, the Ne atoms are excited to E_4^{Ne} and E_6^{Ne} levels which have nearly the same energy as the levels E_2^{He} and E_3^{He}. Thus when He atoms in levels E_2^{He} and E_3^{He} collide with unexcited Ne atoms, they raise them to the levels E_4^{Ne} and E_6^{Ne} respectively. As a consequence the levels E_4^{Ne} and E_6^{Ne} are highly populated.

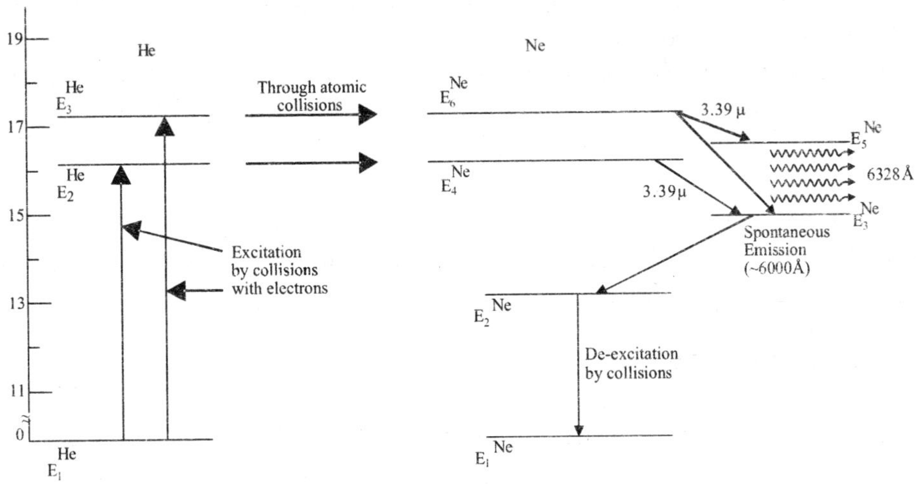

Fig. 4.5 Energy levels in Helium-Neon laser

The population in these levels happens to be much more than those in the lower levels E_3^{Ne} and E_5^{Ne}. Thus a state of population inversion is achieved and any spontaneously emitted photon can trigger laser action in any of the three transitions shown in Fig. 4.5. The Ne atoms then drop down from the lower laser levels to the level E_2^{Ne} through spontaneous emission. From the level E_2^{Ne} the Ne atoms drop back to the ground state through collisions with walls . The transitions $E_6^{Ne} \rightarrow E_5^{Ne}$, $E_4^{Ne} \rightarrow E_3^{Ne}$ result in the emission of radiations having wavelengths 3.39 μm and 1.15 μm respectively. But these radiations do not fall in the visible region. The transition $E_6^{Ne} \rightarrow E_3^{Ne}$ results in the emission of radiation having wavelength $6328\mathring{A}$ in the visible region which corresponds to the well known red light of He-Ne laser.. By proper design of the resonator, laser action in Ne is obtained at large number of transitions in the visible region. The conditions of the laser (frequencies of pumping radiation, pressure of gas mixture *etc.*) must be such that there is and efficient transfer of energy from He to Ne atoms. The level marked E_2^{Ne} is metastable, electrons colliding with atoms in level E_2^{Ne} may excite them to level E_3, thus decreasing population inversion. The tube containing the gaseous mixture is also made narrow so that Ne atoms in level E_2^{Ne} can get de-excited by collisions with the walls of the tube.

The gas lasers are found to emit light which is more directional and more monochromatic. Gas lasers are capable of operating continuously without need for cooling.

4.5 SEMICONDUCTOR LASER

4.5.1 Introduction

Semiconductor laser (alternatively known as the diode laser) is a heavily doped *p-n* junction device, which emits coherent light when it is forward biased. Highly doped *p-n* junction is made from direct bandgap semiconductors (GaAs, InP, GaSb *etc.*) and the two ends are cleaved and polished to form an optical resonator. In 1962, Robert N. Hall and his team made the first semiconductor laser using gallium arsenide (GaAs), which when operated at low temperature (known as Homojunction Laser) emitted light in the near IR region. The first semiconductor laser capable of operating continuously at the room temperature (known as Heterojunction Laser) was produced by 1970 after the invention of liquid phase epitaxy technology. Now diode lasers are made to emit light almost everywhere in the spectrum from UV to IR.

Today semiconductor laser plays a central role in electro-optics, primarily because of its spectral purity, high efficiency ($\eta = 40\%$) * ruggedness, ability to be modulated at extremely rapid rates, long lifetimes, moderate power (200 mW) despite its very small size.

4.5.2 Theory of Semiconductor

Semiconductors are materials that have electrical conductivity intermediate between the high conductivity of metals and the low conductivity of insulators. In a semiconductor, the outer electrons are usually bound to their parent atoms, but a small fraction of them can migrate through the material, so that there is a small amount of electrical conductivity. The allowed energy values of the valence electrons in semiconductors occur within two well defined energy bands separated by an energy gap known as band gap. A pure semiconductor crystal has exactly enough electrons to fill all the states in the lower band namely valence band. However, when covalent bond is just broken, an electron just becomes free. Then we say that the electron jumped into the upper band, namely conduction band leaving behind a vacancy in the valence band. The vacancy is called a hole and is assigned a positive charge and a mass equivalent to that of an electron.

In a pure semiconductor, for each covalent bond broken an electron and a hole are generated. Therefore the number of electrons in the conduction band and the number of holes in the valence band are equal. Pure semiconductors have a high electrical resistance at low temperature and lower resistance at higher temperature. By adding small percentages of impurity elements into the semiconductor, the

* Because of small size the population inversion density becomes quite large making these lasers highly efficient.

electrical conductivity of a semiconductor can be increased. The presence of the small traces of impurity elements can yield extra charge carriers, which are free to move through the material.

The most common semiconductor material that has been used in lasers is gallium-arsenide (GaAs) which is a compound of chemical element gallium-31, and chemical element arsenic -33. In the compound gallium arsenide, each gallium atom has three electrons in its outermost shell of electrons and each arsenic atom has five. This gives an average of four electrons per atom in the compound. When a trace of an impurity element with two outer electrons such as zinc, beryllium and cadmium is added to the crystal, the result is the shortage of one electron from one of the pairs. This shortage sets up an imbalance in which there is a place in the crystal for an electron but there is no electron available. This is commonly called a "hole". This forms the so called *p*-type semiconductor in which the conduction of electricity is by motion of the hole from one atom to another. When a trace of an impurity elemernt with six outer electrons, such as selenium, sulfur or tellurium is added to a crystal of GaAs, it provides an additional electron that is not needed for the bonding. This electron can be free to move through the crystal. Thus, it provides a mechanism for electrical conductivity. This type is called an *n*-type semiconductor.

4.5.3 Homojunction Laser

Homojunction means the laser device uses the same band gap materials for both the *p*-and *n*-sides of the junction. These lasers are operated at very low temperature (77 K). In these lasers, the active region is not a good dielectric waveguide due to small difference (0.02) in refractive indices of active and passive regions. Therefore the beam can be spread out to the surrounding region. Hence vigorous pumping is needed to enhance laser action. This requires high threshold current density (400 A /mm ^{2}) and it increases with temperature. Therefore the junction region leads to a high ohmic dissipation of heat, which can rapidly lead to damage of the material. Thus the homojunction laser operates at very low temperature (77 K) and produces the output in pulse form.

4.5.4 Construction

In homojunction semiconductor laser, a heavily doped *p-n* junction is an active medium which consists of a layer of about 1 μm thickness. It is made from gallium arsenide. Starting with a heavily doped *n*-type GaAs material and *p*- region is formed on its top by diffusing zinc atoms into it as shown in Fig. 4.6. The diode is quite small in size (0.2-1 mm long). The top and bottom faces are metallized and metal contacts are provided to pass current through diode.

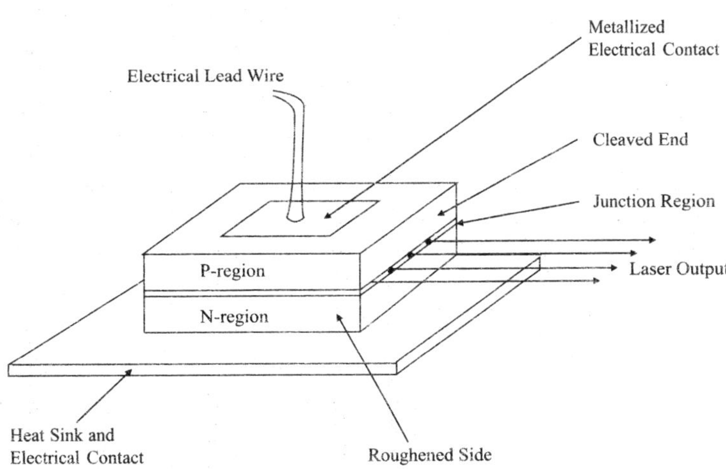

Fig. 4.6 Gallium arsenide laser diode

In laser diode, the two ends of the crystal are cleaved (110 planes of GaAs)* to form a

* The (110) crystallographic plane in III- V semiconductor crystals (GaAs, InP, GaSb *etc*) is weak compared to other planes.

Fabry-Perot resonator or an optical resonant cavity. Because the material has high refractive index, the reflectivity at the interface between gallium arsenide and air is approximately 36%. Coating the surfaces with metal film may increase the reflectivity and decrease the threshold current for laser operation. The two remaining sides are purposely roughened to reduce reflection and prevent lasing action in that direction.

4.5.5 Principle of Operation

Semiconductor lasers are excited by electric current flowing through *p-n* junction. To start laser action there must be population inversion, *i.e.*, there should be many excited electrons and holes present together in the active region. This is achieved by forward biased heavily doped *p-n* junction. The energy band diagram of a heavily doped *p-n* junction without bias is shown in Fig. 4.7(*a*).

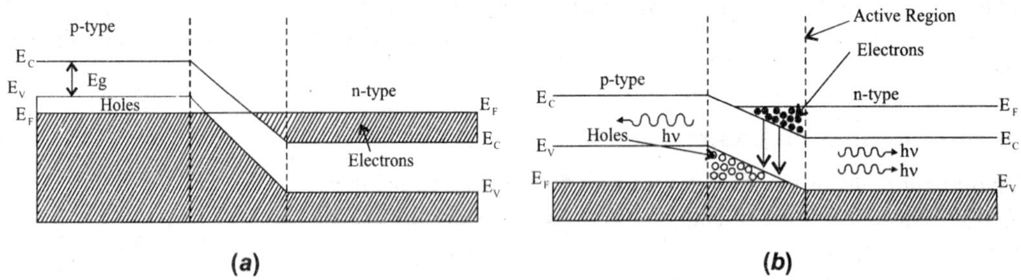

Fig. 4.7(a) Heavily doped *p-n* junction without bias **(b)** Heavily doped *p-n* junction with forward bias

Because of very high doping on *n*-side the Fermi level is pushed into the conduction band and electrons occupy the portion of the conduction band lying below the Fermi level. Similarly on the heavily doped *p*-side the Fermi level lies within the valence band and holes occupy the portion of the valence band that lies above the Fermi level [as shown in Fig. 4.7 (*b*)]. At thermal equilibrium, the Fermi level is uniform across the junction.

When the junction is forward biased, holes and electrons are injected from opposite sides of the *p-n* junction into the depletion region. Holes are injected from the *p*- doped, and electrons from the *n*-doped semiconductor. When forward current is low the recombination of electron-hole at the junction emits the spontaneous photons and the junction acts as a LED (Light Emitting Diode). This spontaneous emission increases linearly with current until threshold is reached where the carrier concentration in the junction region will rise to very high value. As a result, the junction region contains a large concentration of electrons within the conduction band and simultaneously a large number of holes within the valence band. Thus the upper energy levels in the narrow region are having a high electron population while the lower energy levels in the same region are vacant. Therefore, the condition of population inversion is achieved in the narrrow junction region.

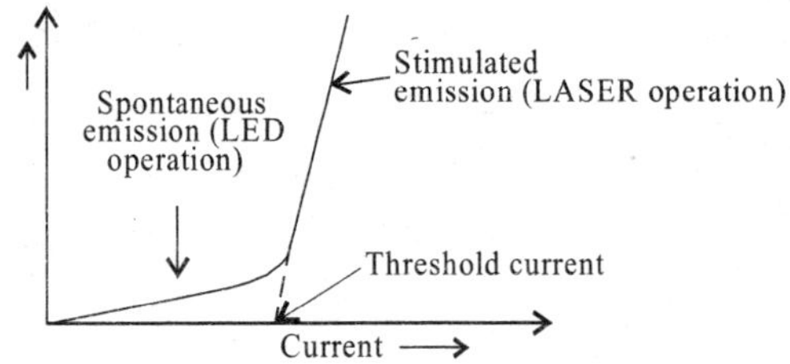

Fig. 4.8 Output light from a semiconductor laser versus laser current

Now the spontaneous photon propagating in the junction region stimulates the conduction electrons to jump into vacant states of valence band. This generates another photon of the same frequency travelling in the same direction with the same polarization and phase as the first photon. These photons will travel along the waveguide and be reflected several times from each end face before they are emitted. As a light wave passes through the cavity, it is amplified by stimulated emission. This means that stimulated emission provides coherent radiations. GaAs laser emits light at a wavelength of 9000 Å in *IR* region. The output power available from this laser is limited by the loop gain available within the laser cavity. The amplifier gain of the active medium is dependent on the current density through the junction. Higher currents produce greater power but higher currents also increase heating effects that can damage the device.

4.5.6 Heterojunction Laser

To operate semiconductor larers at room temperature, the threshold current density should be low. To get low current density, the carriers should be confined in a narrow region and build a dielectric waveguide around the optical gain region. It can be achieved by heterostructured laser diodes and quantum–well lasers*.

The first heterojunction diode lasers were single heterojunction lasers. These lasers utilized aluminum gallium arsenide (AlGaAs) p-type injectors situated over n-type gallium arsenide layers grown on the substrate by liquid phase epitaxy (LPE) technology. An admixture of aluminium replaced gallium in the semiconductor crystal and raised the bandgap of the p-type injector over that of the n-type layers beneath. In this heterojunction laser, the charge carriers are confined to a much smaller region than in homojunction laser. The heat dissipation is much lower while the threshold current density (the recombination radiation) are still at a sufficiently high level for laser output. Hence the single-heterostructure diode lasers did not function continuously at room temperature. Therefore, the concept of double heterostructure lasers was introduced to operate the semiconductor lasers continuously at room temperature with low threshold current density.

4.5.7 Double Heterostructure Lasers

All practical semiconductor laser diodes are double heterostructures (DH). Heterostructures consist of layers of semiconductor crystal having varying bandgap and refractive index. In these devices, a layer of low bandgap material is sandwiched betweeen two high bandgap layers. One commonly used pair of materials is gallium arsenide (GaAs) with aluminium gallium arsenide ($Al_xGa_{(1-x)}As$). Each of the junction between different bandgap materials is called a heterostructure, hence the name "double heterostructure laser".

The advantage of a DH laser is that the active region (region where free electrons and holes exist simultaneously) is confined to the thin middle layer. This means that many more of the electron hole pairs can contribute to amplification. In addition the light is confined to the active region due to the reflection of light from the heterojunction. The heat produced is also dissipated away efficiently by cementing the diode to a tin plate which acts as a heat sink. The first laser diode to achieve continuous wave operation was a double heterostructure demonstrated in 1970 by Zhores Alferov and his team.

---

* The structure of quantum well laser is similar to the double heterostructure laser except the thickness of active layer, which is very small (10-20 nm) in case of QW laser.

4.6 APPLICATIONS OF LASERS

Lasers are used in variety of applications. These applications are based on four important properties of the laser beam viz; high intensity, high coherence, high monochromaticity and high directionality. The important applications are listed below–

1. Lasers can be used to accurately focus energy on small area of the order of a thousand of a millimeter in diameter. This gives very high intensity of the order 10^{12} watt/cm^2 as compared to intensity of sun light which is 0.1 watt/cm^2. This energy when applied on an opaque object, gets converted into heat energy.
2. The intense concentration of energy can be used for drilling fine holes and for precision welding.
3. Lasers can be used for accurate measurements of length ranging from a millionth of a centimetre to the distance nearer planets and satellites.
4. Lasers are widely used in the field of medicine. It can be used to destroy certain kinds of tumours.
5. Laser beams have also been used in the intertial confinement of plasma.
6. Laser finds applications in communication engineering, *e.g.*, radar, radio transmission etc.
7. During war time lasers are used to detect and destroy enemy missiles.
8. In space science lasers are used to control rockets and satellites in directional radio communication like fibre optics telephony.
9. Laser is very useful in science and research. Laser torch is used to see objects at long distances.
10. Laser is used in holography and non-linear optics.
11. Semiconductor lasers find wide use in optical fiber communication, in CD players, CD-ROM and DVD technology (IR Lasers), laser pointers, optical reading, high speed laser printing etc. High power laser diodes are used in industrial appplications such as heat treating, cladding, seam welding and for pumping other lasers, such as diode pumped solid state lasers.

The disadvantages of semiconductor lasers are their large spatial divergence, less monochromatic and highly temperature sensitive.

4.7 CHARACTERISTICS OF THE LASER LIGHT

The most outstanding characteristics are the following

(1) **Directionality:** The laser beam is highly directional. The laser rays are almost perfectly parallel, that is why it can travel long distances without spreading.

(2) **Monochromaticity:** The laser beam is almost perfectly monochromatic. A laser produces light in more or less single wavelength, *i.e.*, the line width associated with laser beams are extremely narrow.

(3) **Coherence:** The laser beam is completely coherent spatially as well as temporarily. It is possible to observe interference effects from two independent laser beams.

(4) **High Intensity:** The laser beam is highly intense as compared to ordinary sources of light. It is because of its high energy density and directional property a laser beam can produce temperatures of the order of 104°C at a focussed point.

Example 4.1: A laser beam has a wavelength of 8000 Å and aperture 5×10^{-3} m. The laser beam is sent to moon. The distance of moon is 4×10^{5} km from the earth. Calculate (*i*) the angular spread of the beam and (*ii*) the areal spread when it reaches the moon. **[MREC, B.E.I 2002]**

Solution: (*i*) Let $d\theta$ is the angular spread, then

$$d\theta = \frac{\lambda}{d} \qquad \text{(for rectangular aperture)}$$

Here $\lambda = 8\times10^{-7}$ m and $d = 5\times10^{-3}$ m

$$\therefore \quad d\theta = \frac{8\times10^{-7}}{5\times10^{-3}} = 1.6\times10^{-4} \text{ radian}$$ **Ans.**

(*ii*) Areal spread $= (D\cdot d\theta)^2$

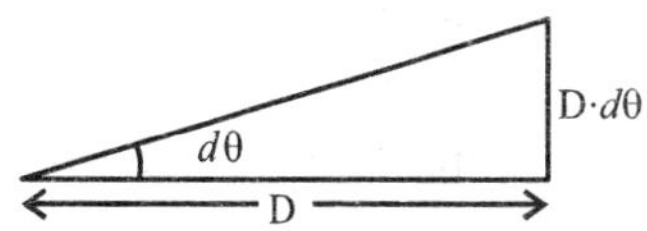

Here $D = 4\times10^{5}$ km $= 4\times10^{8}$ m

$$\therefore \text{ Areal spread} = \left(4\times10^{8}\times1.6\times10^{-4}\right)^2$$

$$= 4.096\times10^{9}\,\text{m}^2$$ **Ans.**

Example 4.2: A laser source of wavelength $\lambda = 6000$Å, coherence width 4 mm and power 10 mW shines on a surface 100 m away. Deduce the (*i*) angular spread (*ii*) areal spread and (*iii*) intensity of illumination.

Solution: (*i*) The angular spread for circular aperture $= d\theta = 1.22\dfrac{\lambda}{a}$

Here $\lambda = 6\times10^{-7}$ m and $a = 4$ mm $= 4\times10^{-3}$ m

$$d\theta = \frac{1.22\times6\times10^{-7}}{4\times10^{-3}} = 1.83\times10^{-4}\,\text{radian}$$ **Ans.**

(*ii*) The areal spread is given by

$$A = \pi r^2 = \pi(D\cdot d\theta)^2$$

Here $A = 3.14\times\left(10^{4}\times1.83\times10^{-4}\right)^2 = 10.51\text{ cm}^2$ **Ans.**

(*iii*) Intensity of illumination $= \dfrac{\text{Power}}{\text{Area}} = \dfrac{P}{A}$

Here $P = 10$ mW and $A = 10.51\text{cm}^2$

$$\therefore \quad \text{Intensity of illumination} = \frac{10}{10.51} = 0.95\text{ mW/cm}^2$$ **Ans.**

Example 4.3: A laser beam has a power of 50 mW. It has an aperture of 5×10^{-3} m and it emits light of wavelength 7200Å. The beam is focussed with a lens of focal length 0.1 m. Calculate the area and the intensity of the image. **[RTU B. Tech. II Sem. 2012]**

Solution: The angular spread of the beam is given by

$$d\theta = 1.22\frac{\lambda}{a}$$

Area of the image is given by

$$A = \pi r^2 = \pi\left(fd\theta\right)^2$$

Here $f = 0.1$ m, $\lambda = 7.2\times10^{-7}\,\text{m}$, $a = 5\times10^{-3}\,\text{m}$

$$\therefore \qquad A = 3.14\times\left(\frac{0.1\times1.22\times7.2\times10^{-7}}{5\times10^{-3}}\right)^2 = 9.69\times10^{-10}\,\text{m}^2 \qquad \textbf{Ans.}$$

Intensity of image is given by $I = \dfrac{P}{A}$

Here $P = 50\times10^{-3}\,\text{W}$ and $A = 9.69\times10^{-10}\,\text{m}^2$

$$\therefore \qquad I = \frac{5\times10^{-3}}{9.69\times10^{-10}} = 5.15\times10^{8}\ \text{W/m}^2 \qquad \textbf{Ans.}$$

Example 4.4: Find the intensity of laser beam of 20 mW and diameter 1.5 mm. Assume the intensity to be uniform throughout the beam. **[RTU B. Tech. Sem. II 2010]**

Solution: Given $P = 20\ \text{mW} = 20\times10^{-3}\ \text{Watt}$

and $\Delta = 1.5\ \text{mm}$

$\therefore$ $r = 0.75\times10^{-3}\,\text{m}$

$r = 7.5\times10^{-4}\,\text{m}$

$$\text{Intensity } I = \frac{P}{A}$$

$$= \frac{20\times10^{-3}}{3.14\times(7.5\times10^{-4})^2}$$

$$= \frac{20\times10^{-3}}{1.77\times10^{-6}}$$

$$= 11.29\times10^{3}\ \text{W/m}^2$$

$$I = 11.29\ \text{kW/m}^2 \qquad \textbf{Ans.}$$

4.8 Q-SWITCHING

It was first discovered and demonstrated in 1962 by R. W. Hellwarth and F.J. Meclung using electrically switched Kerr cell shutters in a ruby laser.

Q-switching is a technique to obtain energetic short pulses (10^{-9} seconds) from a Laser by modulating the intracavity losses. Since Laser is an oscillator, its resonator cavity is characterised by the quality factor (Q), which is defined as 2p times the ratio of the energy stored in the system to the energy losses per cycle, *i.e.*

$$Q = 2\pi \frac{\text{Energy stored in the resonator}}{\text{Energy loss per cycle}}$$

Therefore a high Q factor corresponds to low resonate losses and vice versa.

4.8.1 Principle of Q-Switching

Consider a laser cavity in which a shutter is introduced in front of one of the mirrors. If the active medium is continuously pumped keeping the shutter closed (Q is low) the laser action is prevented and the population inversion will go on increasing and far exceeds the threshold population holding when the shutter is absent.

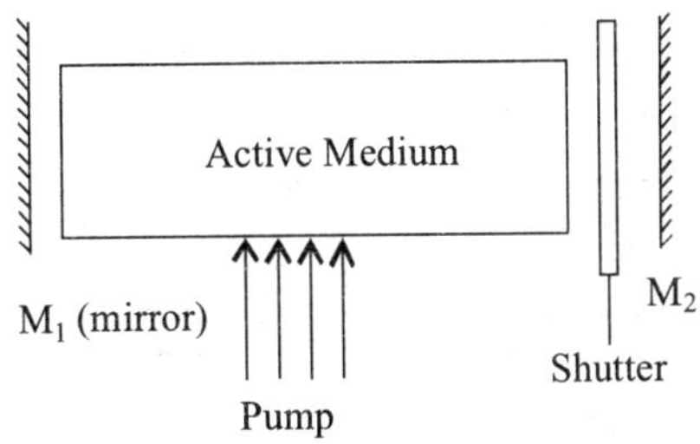

Fig. 4.9 Principle of Q-Switching

If the shutter is now suddenly opened (Q is changed from low to high), then the existing population inversion will correspond to a value much above the threshold value for oscillations and the laser will exhibit a gain that greatly exceeds the losses. Therefore the radiation in the cavity mode will build up very rapidly. This rapid increase in the intensity will deplate the population inversion which will go below threshold. The stored energy may then be released in the form of a short and intense light pulse*. Since this operation involves switching the cavity Q factor, value changes from low to high. Therefore this technique is known as Q-switching.**

4.8.2 Evolution of a Q-Switched Laser Pulse

The following four requirements must be satisfied to produce the necessary inversion density required for Q-switching.

(1) The life time of the upper energy level must be longer than the cavity build up time (when gain equals losses), so that the upper level can store the extra energy pumped into it over the extended pumping time.

(2) The pumping flux duration must be longer than cavity build up time and preferably at least as long as the upper level life time.

(3) The initial cavity losses must be large during the pumping duration to prevent beam growth and oscillations.

(4) After Q-switching, the cavity losses must be reduced almost instantaneously, so that the beam could evolve and extract the extra energy from the upper level of the gain medium.

---

* If the shutter is opened in a time much shorter than the time required for the building of laser oscillation, the output would be a series of pulses having smaller peak power.

** LASER to which the Q-switching technique is applied is called Q-switched laser.

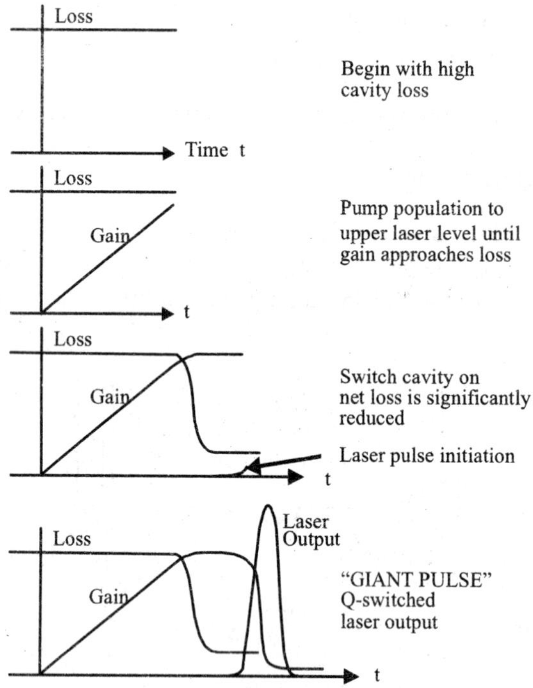

Fig. 4.10 Evolution of a Q-Switched laser pulse

Satisfying these requirements would produce a giant pulse laser output. Fig. (4.10) shows a high initial cavity loss during the intense pumping and consequent gain build up over the life time of upper level. The cavity loss is rapidly reduced, producing a pulse output within a time of the order of cavity build up time after the cavity Q is changed.

4.8.3 Types of Q-Switching

There are two types of Q-switching

(A) Active Q-switching

(B) Passive Q-switching

(A) Active Q-switching

In this switching, the external devices are used to control Q-factor of cavity. The rotating mirror, electro-optic shutter (Kerr and Pockels effects) and acousto optic shutter are the examples of active Q-switching. To understand the concept of Q-switching, the working of electro-optic shutter is discussed in detail.

Electro-optic shutter: In this method, a shutter is typically an electro-optic crystal that becomes birefringent when an electrical field is applied across the crystal. Since these shutters operate by rotating the polarization of the beam, the laser must have a polarizing element within the cavity like Brewster angle windows. In this case, Brewster angle windows allow a low-loss beam to transit the cavity polarized in the plane of the paper. When the voltage is on, the shutter rotates the plane of polarization of the beam by 45° as it passes through the cell. The beam reflects from the mirror and returns through the cavity, travelling in the opposite direction whereafter its plane is rotated by an additional 45° by the shutter. It thus arrives back at the laser amplifier with a polarization of 90° with respect to that when it left the amplifier. This 90° polarized beam is rejected by the Brewster window and directed out of the cavity. In such a position, the losses in the cavity would be large. When the voltage is turned off, the cell loses its birefringence and the cell does not rotate the plane of polarization. In this position, the losses are small and correspond to an open shutter. Two types of electro-optic shutters that can be used for Q-switching are the Kerr cell and the Pockels cell*. The Pockels cell is generally preferred over the Kerr cell because of the lower voltage needed to produce the desired effect.

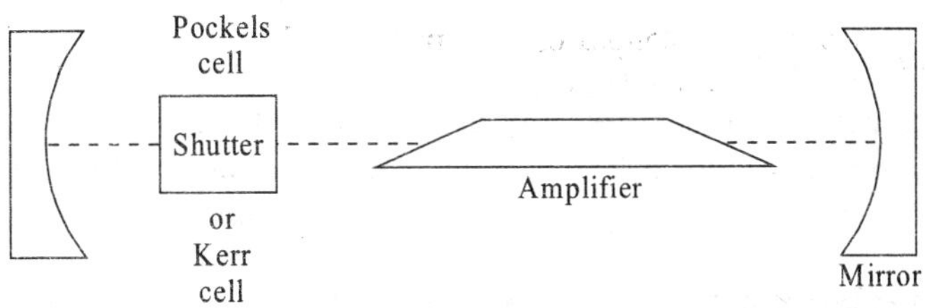

Fig. 4.11 Q-switching with an electro-optic shutter

(B) Passive Q-Switching

In this switching, the losses inside the cavity are automatically modulated with a saturable absorber. Therefore, it is also known as self Q-switching. Saturable absorber is a material whose transmission increases when the intensity of light exceeds some threshold. The material may be an ion-doped crystal, a bleachable dye or a passive semiconductor device.

Saturable Absorber: It can be used as a passive Q-switching** element by placing it within the laser cavity. The operation of such a device may be understood as follows—

Because of the small transmittance of the saturable absorber, laser oscillations cannot start until the attainment of a large population invension. As the power level inside the cavity goes on increasing the dye begins to be bleached. This bleaching results in a larger transmittance which in turn increases the power level inside the cavity. The increased power becomes almost transparent. Since in this condition the inversion is much more than the threshold inversion, the gain is much more than the losses and thus a giant pulse is produced.

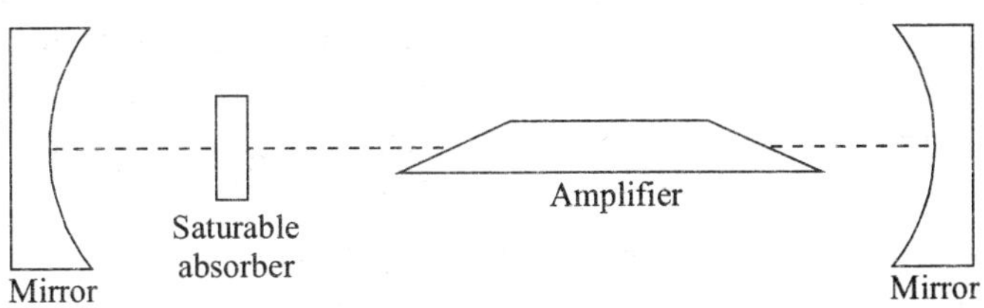

Fig. 4.12 Q-switching with saturable absorber

4.8.4 Applications

The Q-switched lasers are used in following applications.

(1) Applications which demand high laser intensities in nano second pulses such as dentistry, metal cutting or pulsed holography.

* When the index of refraction of an electro-optic material changes linearly with the applied electric field, it is termed as Pockels effect, while a quadratic dependence is termed the Kerr effect.

** Passive Q-switching is simple and cost effective and is suitable for very high pulse repetition rates. However, the pulse energies are typically lower and external triggering of the pulses is not possible.

(2) 3D optical data storage and 3D microfabrication where the high peak power of these lasers is useful.

(3) Also used for range finding (distance measurements).

(4) Used to remove tattooes.

4.9 MODE LOCKING

Mode locking is a technique for producing periodic, high power and extremely short duration* laser pulses. The duration order of pulses is of picoseconds ($10^{-12} s$) to femtoseconds ($10^{-15} s$). The basis of this technique is to induce a fixed phase relationship between the modes of the laser's resonant cavity. Then the laser is said to be phase-locked or mode locked. Interference between these modes causes the laser light to be produced as a train of pulses of very short duration.

4.9.1 Longitudinal Modes of the Laser Cavity

Laser light is not a single pure frequency light but produces light over some natural bandwidth or range of frequencies. Therefore, the range of frequencies that a laser may operate over is known as the gain bandwidth.

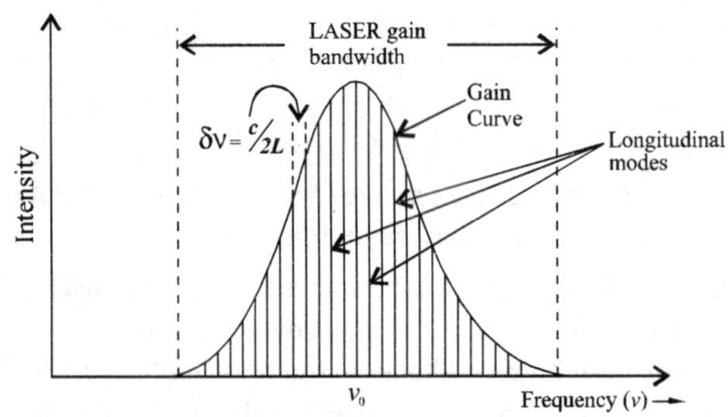

Fig. 4.13 Laser output spectrum

We know that the resonant cavity of the laser consists of two plane mirrors facing each other and active medium. Since light is a wave, when bouncing between the mirrors of the cavity the light will constructively and destructively interfere with itself leading to the formation of standing waves between the mirrors. These standing waves form a discrete set of frequencies, known as the longitudinal modes of the cavity. These modes are due to constructive interference.

For simple plane mirror cavity, the allowed modes are those for which the separation distance of the mirrors L is an exact multiple of half the wavelength of light λ * *i.e.*

---

* Q-switched pulses are limited to minimum pulse duration of few ns.

** If η be the refractive index of laser medium then the length of optical path is given by $\eta L = \frac{n\lambda}{2}$

$\therefore$ Frequency separation between any two adjacent modes $\delta v = \frac{c}{2L\eta}$ and pulse separation time $= \frac{2L\eta}{c}$

$$L = \frac{n\lambda}{2} \tag{4.21}$$

where n is an integer known as the mode order. From Eq. (4.21)

$$\lambda = \frac{2L}{n} = \frac{c}{v}$$

$$\Rightarrow \qquad v = \frac{cn}{2L} \tag{4.22}$$

Therefore, frequency separation between any two adjacent modes n and $n+1$ is given by

$$\delta\nu = \nu_{n+1} - \nu_n = \frac{c(n+1)}{2L} - \frac{cn}{2L}$$

$$\Rightarrow \qquad \delta\nu = \frac{c}{2L} \tag{4.23}$$

4.9.2 Mode Locking Theory

To achieve the mode locking, a number of distinct longitudinal modes of a laser having slightly different frequencies are combined in phase. The comparison between non-mode locked and mode locked laser output for three different frequencies ω, 2ω and 3ω is shown in Fig. 4.14.

In a simple laser, amplitudes and phases of various longitudinal modes (frequencies) are randomly distributed and these produce randomly varying total field amplitude and intensity (Fig. 4.14 (*a*)). When these three frequencies (modes) are added in phase, then they combine to produce a total field amplitude and intensity output in the form of repetitive pulses of high intensity (Fig. 4.14(*b*)). These pulses are called mode locked pulses.

In mode locked laser, the intensity is periodic, *i.e.*, in the time interval $\tau = \frac{1}{\delta\nu} = \frac{2L}{c}$, which is time taken for the light to make a complete round trip of the cavity, *i.e.*, the pulse separation time or pulse repetition rate is

$$\tau = \frac{1}{\delta\nu} = \frac{2L}{c} \tag{4.24}$$

The duration or width of each pulse of light is determined by the number of modes which are oscillating in phase. If there are N modes locked with a frequency separation δv, the overall mode locked bandwidth is $N\delta\nu$. Therefore pulse duration or width is

$$t_p = \frac{1}{N\delta\nu} = \frac{1}{\Delta\nu} \tag{4.25}$$

$$t_p = \frac{2L}{cN} \tag{4.26}$$

From equations (4.24) and (4.26), it is clear that the ratio of the pulse spacing to the pulse width is approximately equal to the number of modes (N). Thus, to obtain high power short duration pulses there should be a large number of modes. This requires a broad laser transition and a long laser cavity.

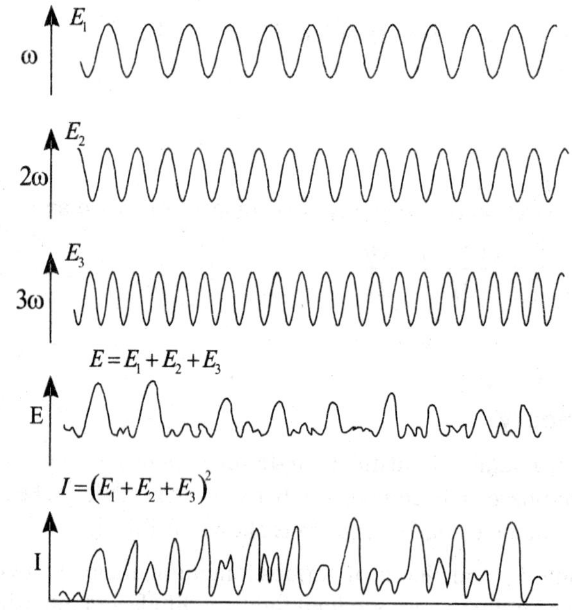

(*a*) Non-mode Locked (Simple laser) output

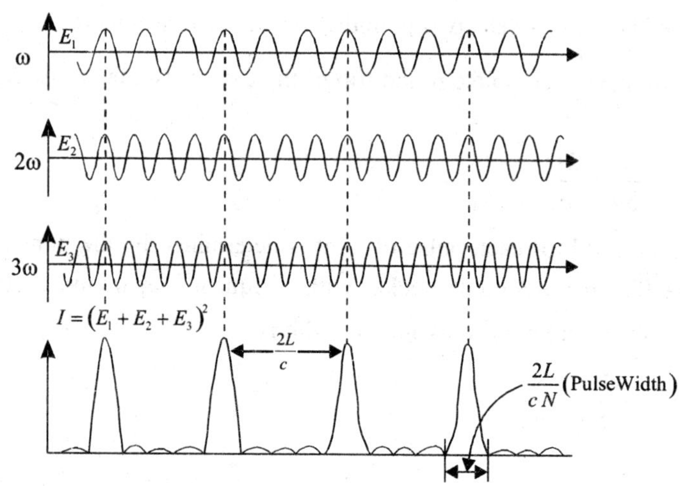

(*b*) Mode Locked output (Mode locked laser)

Fig. 4.14 Comparison of non-mode locked and mode locked laser output

4.9.3 Mode-Locking Methods

There are two methods for producing mode-locking in a laser.

(1) Active mode-locking.

(2) Passive mode-locking.

Active methods typically involve using an external signal to induce a modulation of the intra-cavity light.

Passive methods do not use an external signal but rely on placing some element into the laser cavity which causes self modulation of the light.

4.9.4 Application of Mode Locked Laser Output

There are many uses of very short duration laser pulses in the fields of digital communications, diagnostics of ultrafast processes and ablation of materials without causing significant heating of the material.

Example 4.5: Calculate the number of modes, pulse separation and pulse duration in a mode locked Nd : YAG laser where the fluoroscent linewidth is 1.1×10^{11} Hz and the laser rod is 0.1 m long.

Solution: Given that the linewidth $\left(\frac{1}{t_p}\right) = \frac{cN}{2L} = 1.1\times10^{11}\,\text{Hz}$, and $L = 0.1\,\text{m}$

As we know that the pulse duration is

$$t_p = \frac{2L}{cN} = \frac{1}{1.1\times10^{11}}\ \text{sec.}$$

The pulse separation time is

$$\tau = \frac{2L}{c} = \frac{2\times0.1}{3\times10^8} = 0.66\times10^{-9}\,\text{sec} = 0.66\ \text{ns}$$ **Ans.**

$$\text{No. of modes } (N) = \frac{\text{Pulse spacing}}{\text{Pulse duration}}$$

$$= \frac{0.66\times10^{-9}}{9.09\times10^{-12}}$$

$$= 0.7333\times10^{2}$$

$$N = 73.33$$ **Ans.**

Example 4.6: A continuous laser of wavelength $\lambda = 6000$ Å is Q-switched into 0.5 ns pulses. Compute its coherent length, bandwidth and linewidth.

Solution: Given that $\Delta\tau = 0.5\ \text{ns} = 5\times10^{-10}\,\text{s}$

and $\lambda = 6000\text{Å} = 6 \times 10^{-7}\,\text{m}$

The coherence length $L = c\Delta\tau = 3 \times 10^8 \times 5 \times 10^{-10} = 0.15\text{m}$

Bandwidth $\Delta\nu = \dfrac{1}{\Delta\tau} = \dfrac{1}{5 \times 10^{-10}} = 2 \times 10^9\,\text{Hz}$

Linewidth $\Delta\lambda = \dfrac{\lambda^2}{c}\Delta\nu$

$$= \frac{\left(6 \times 10^{-7}\right)^2 \times 2 \times 10^9}{3 \times 10^8}$$

$$= \frac{36 \times 2 \times 10^{-5}}{3} \times 10^8$$

$$= 24 \times 10^{-13}$$

$$= 0.024 \times 10^{-10}\,\text{m}$$

$\Delta\,\lambda = 0.024\text{Å}$ **Ans.**

Example 4.7: Calculate the pulse width produced by a mode locked Nd : glass laser assuming that the laser cavity is 0.2 m long and there are 3000 participating longitudinal modes. What is the time separation of the pulse?

Solution: Given L = 0.2 m, N = 3000 modes

Pulse width $= \dfrac{2L}{c \times N} = \dfrac{2 \times 0.2}{3 \times 10^8 \times 3000}$

$$= \frac{0.4}{9} \times 10^{-11} = 0.44 \times 10^{-12}\,\text{s}$$

$$= 0.44\ \text{ps}$$

Pulse separation time $\tau = \dfrac{2L}{c}$

$$= \frac{2 \times 0.2}{3 \times 10^8} = \frac{0.4}{3} \times 10^{-8}$$

$$= \frac{4}{3} \times 10^{-9} = 1.33 \times 10^{-9}\,\text{s}$$

$\tau = 1.33\ \text{ns}$ **Ans.**

HOLOGRAPHY

4.10 INTRODUCTION

Holography is a two step process by which (1) an object illuminated by coherent light is made to produce interference fringes in a photosensitive medium, and (2) reillumination of the developed interference pattern by light of the same wavelength produces a three dimensional image of the original object.

Dennis Gabor (June 5, 1900, Budapest – February 9, 1979, London) was a Hungarian electrical engineer and inventor, most remembered for inventing holography in 1947, for which he received the Nobel Prize in Physics in 1971.

Holography was invented in 1947 by Dennis Gabor of U.K., but the lack of strong coherent-light sources limited the field of holography until the invention of laser in 1960. The first practical hologram was made using laser beam by two Americans E.N. Leith and Y.U. Patneiks in 1964.

In conventional photography, a lens system (camera) focuses the light reflected from an illuminated three dimensional object on a photographic plate upon which a two dimensional image of the object is formed. This image can be seen by illuminating the developed photographic plate by ordinary (incoherent) light.

In holography, one does not record the image of the object being photographed, rather records the light waves reflected from the object. The photographic record is called a 'hologram'.The hologram has no resemblance with the object in a sort of 'optical code'. When it is illuminated by the same coherent light source (laser) a three dimensional image of the object is formed. The process of formation of an observable image from the hologram is known as the 'reconstruction process'.

Thus, holography is a two step process; the construction of the object into a hologram and reconstruction of the hologram into the image of the object. No lens is needed in either process.

4.11. THE BASIC PRINCIPLE OF HOLOGRAPHY

The basic principles of 'holography' (Greek meaning 'whole writing') were first presented by Dennis Gabor, for which he was awarded Nobel Prize in 1971. Consider Fig. 4.15 P is a point object. A parallel beam of monochromatic laser light is incident towards right. As the plane wavefront strikes the point object, it gets scattered. The scattered wavefronts are characterised by their amplitude and phase. Thus two sets of monochromatic coherent waves superimpose on a photographic film. One set the 'reference beam' consists of plane waves.

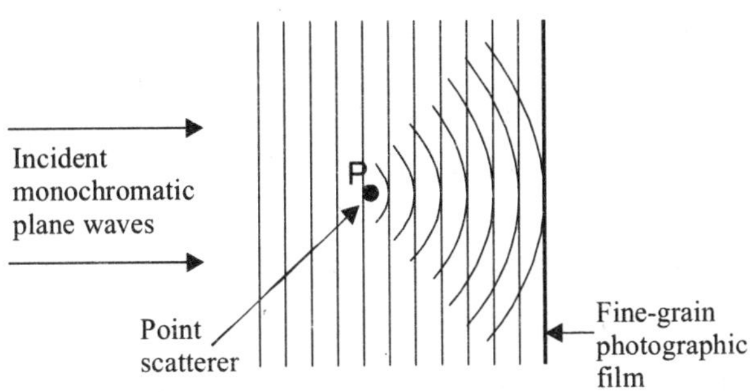

Fig. 4.15. Incident and scattered waves produce interference pattern on the film

The other set is the light scattered from the point object (spherical waves). At the film, the interference of these two sets of coherent waves produces a pattern of bright and dark rings. Upon development, the film will have opaque and clear regions, forming a 'Gabor Zone Plate', similar to a Fresnel zone plate. The developed film is known as 'hologram'. The bright fringe areas on the hologram have the greatest transparency.

Now we illuminate the hologram, as obtained above, with monochromatic plane-wave light of the same wavelength λ that we used initially. In Fig. 4.16 a strong diffraction maximum occurs at P'. That is light after diffraction from the hologram converges to P' and, therefore P' is a real image of point object P. This is not the entire diffraction pattern, however, the interference of the wavelets that spread out from all the transparent areas forms a second spherical wave that is diverging rather than converging. When this wave is traced back behind the film (Fig. 4.16.), it appears to be spreading out from point P. Thus the total diffraction pattern from the hologram consists of the superposition of spherical waves converging to form a real image at P' and a spherical wave that diverges as though it had come from virtual image point P.

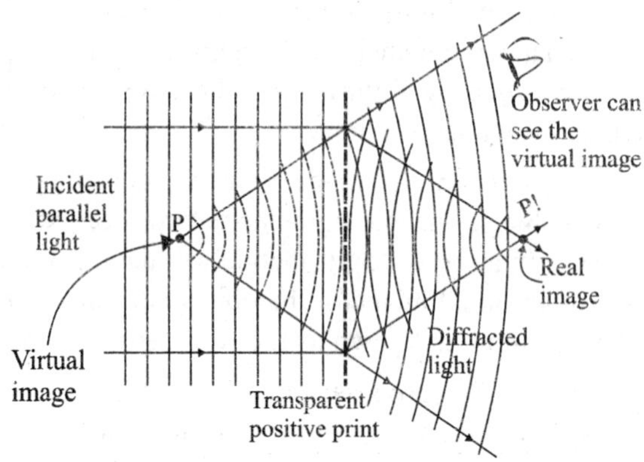

Fig. 4.16. Formation of real and virtual images by a hologram

If a small extended object is used instead of a point object, each point of the object forms its own Gabor zone pattern, which superimposes with patterns for all the other points. The resulting hologram is a very complicated array of fringes that contains the full information about the pattern for each point on the object.

4.11.1. Obtaining a Hologram (Construction of Hologram)

In order to obtain a hologram of an object, the photographic film is to be illuminated simultaneously by

(*i*) a plane carrier beam (reference beam) and

(*ii*) an object bearing beam. Since both belong to the same laser wavefront, the beam interfere on the plate.

Let the carrier wave (reference wave) is given by the equation

$$y_1 = a_0 \cos(\omega t - kx) \quad \textbf{(4.27)}$$

where a_0 is the amplitude of the carrier wave and its phase is taken to vary linearly with x. Let the object bearing wave is given by

$$y_2 = a(x) \cos\{\omega t - \phi(x)\} \quad \textbf{(4.28)}$$

It has amplitude $a(x)$ and phase $\phi(x)$ coming from the sum up of all the secondary wavelets from all the points of the object. According to the principle of superposition the resultant disturbance is given by

$$y = y_1 + y_2 = a_0 \cos(\omega t - kx) + a(x)\cos\{\omega t - \phi(x)\} \quad \textbf{(4.29)}$$

Since the incident beam is spatially coherent, the relative phase of the wavelets which add up to the object bearing wave depend upon the angle of viewing and the depth differences only. Hence $a(x)$ and $\phi(x)$ carry this information.

Since photographic plate records only the intensities, we get from equation (4.29)

$$I = \langle y^2 \rangle = \langle [a_0 \cos(\omega t - kx) + a(x)\cos\{\omega t - \phi(x)\}]^2 \rangle$$

$$\text{or } I = \langle a_0^2 \cos^2(\omega t - kx) \rangle + \langle [a^2(x)\cos^2\{\omega t - \phi(x)\}]^2 \rangle$$

$$+ \langle [2a_0 a(x)\cos(\omega t - kx)\cos\{\omega t - \phi(x)\}] \rangle$$

$$\because \quad \langle \cos^2(\omega t - kx) \rangle = \langle \cos^2\{\omega t - \phi(x)\} \rangle = \frac{1}{2}$$

$$\therefore \quad I = \frac{a_0^2}{2} + \frac{a^2(x)}{2} + a_0 a(x) \langle [\cos(2\omega t - kx - \phi(x) + \cos\{kx - \phi(x)\} \rangle$$

Again

$$\because \quad \langle \cos\{2\omega t - kx - \phi(x)\} \rangle = 0$$

$$\therefore \quad I = \frac{a_0^2}{2} + \frac{a^2(x)}{2} + a_0 a(x)\cos\{kx - \phi(x)\} \quad \textbf{(4.30)}$$

The significant point is that the intensity on the photographic plate has modulation involving both $a(x)$ and $\phi(x)$. The plate has only interference fringes which in their complexity store all the information needed to reconstruct in three dimensions the object holographed.

The experimental arrangement for obtaining a hologram is shown in Fig. 4.17. The object is illuminated by a laser light beam. Each point on the object reflects light to the entire photographic plate.

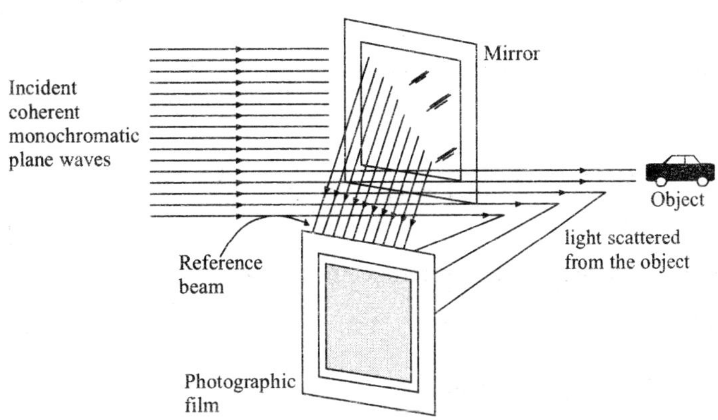

Fig. 4.17 Obtaining a hologram

Apart of the incident laser beam is made to fall on a plane mirror which reflects it towards the photographic plate. This is the '*reference (carrier) beam*'. It interferes with the signal beam reflected by the object and produces an interference pattern on the photographic plate (hologram). The hologram when saw under magnification appears to consist of highly irregular fringes having no apparent relationship with the object. The information on the hologram regarding the object is embodied in the fringe-contrast and the fringe-shaping.

4.11.2 Viewing the Object (Reconstruction of an Image from Hologram)

The holograms recorded above acts like a diffraction grating with grooves in complex geometry and so light is diffracted from it. Reconstruction wave is taken the same reference wave as was taken for recording the hologram. The transmitted wave is, therefore, given by

$$y_t = y_1 I$$

Using equations (4.27) and (4.30), we get

$$y_t = a_0 \cos(\omega t - kx)\left[\frac{a_0^2}{2} + \frac{\{a(x)\}^2}{2} + a_0 a(x)\cos\{kx - \phi(x)\}\right]$$

or

$$y_t = \frac{a_0^3}{2}\cos(\omega t - kx) + \frac{a_0\{a(x)\}^2}{2}\cos(\omega t - kx)$$

$$+ a_0^2 a(x)[\cos(\omega t - kx)\cos\{kx - \phi(x)\}]$$

or

$$y_t = \left[\frac{a_0^3}{2} + \frac{a_0\{a(x)\}^2}{2}\right]\cos(\omega t - kx) + \frac{a_0^2 a(x)}{2}[\cos\{\omega t - \phi(x)\}$$

$$+ \cos\{\omega t - 2kx + \phi(x)\}]$$

or

$$y_t = \left[\frac{a_0^3}{2} + \frac{a_0\{a(x)\}^2}{2}\right]\cos(\omega t - kx) + \frac{a_0^2 a(x)}{2}\cos\{\omega t - \phi(x)\}$$

$$+ \frac{a_0^2 a(x)}{2}\cos\{\omega t - 2kx + \phi(x)\}$$

or

$$y_t = T_1 + T_2 + T_3 \quad \textbf{(4.31)}$$

There are three terms on the R.H.S. of equation (4.31)

The first term

$$T_1 = \left[\frac{a_0^2}{2} + \frac{\{a(x)\}^2}{2}\right]a_0 \cos(\omega t - kx) \quad \textbf{(4.32)}$$

represents the attenuated incident reference beam.

The second term

$$T_2 = \frac{a_0^2 a(x)}{2}\cos\{\omega t - \phi(x)\} \quad \textbf{(4.33)}$$

represents a beam indentical to the beam scattered by the object which forms a virtual image at the original place of the object.

The third term

$$T_3 = \frac{a_0^2 a(x)}{2} \cos\{\omega t - 2kx + \phi(x)\} \quad (4.34)$$

represents a beam having phase ϕ opposite to the object beam and forms a real image which can be recorded on a photographic plate.

In the reconstruction process, the hologram is illuminated by the same collimated laser beam which undergoes diffraction, (Fig. 4.18). One of the diffracted beams emerging from the hologram, when traced back, appears to diverge from an apparent object located in the position of the original object. This beam forms a virtual image which is seen by looking through the hologram as if it were a window. The other beam produces a real image which can be recorded on a photographic plate.

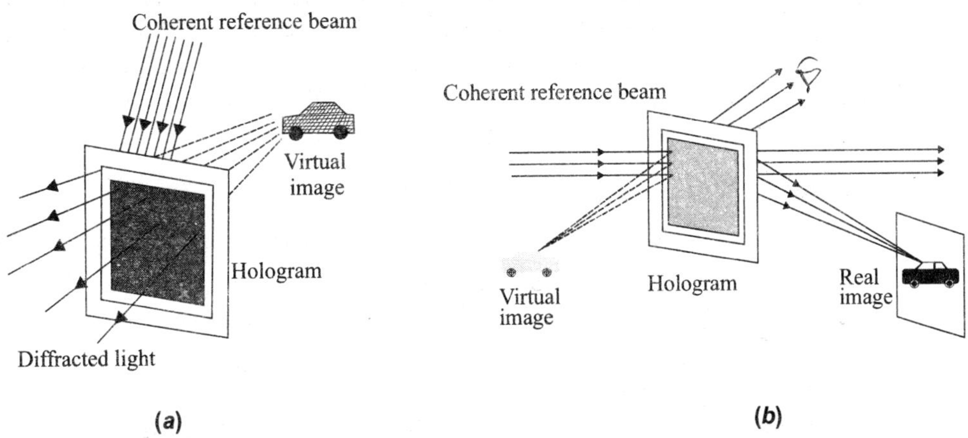

Fig. 2.18 Viewing the object by a hologram

4.12. TYPES OF HOLOGRAMS

The commonly used holograms are listed below.

(*a*) Transmission Hologram

In this hologram, both the reference and the object waves are introduced from the same side of the recording medium. The incident reconstruction wave is reflected towards the other side of the hologram so that the reconstructed wave appears as the wave transmitted through the hologram. In transmission hologram, greater depth of the scene is possible.

(*b*) Reflection Hologram

In this hologram, the reconstructed wave is on the same side of the hologram and thus appears as a wave reflected from the hologram. Reflection holograms are useful since laser light is not necessary for reconstruction, *i.e.*, these holograms are viewed with white light.

(c) Rainbow Hologram or White Light Hologram

This is a new type of transmission hologram, which can be viewed in white light. A double holographic process is involved in making a rainbow hologram. First, a conventional transmission hologram of the object is made which is used as the object and a second hologram is made through a slit. A horizontal slit limits the vertical perspective of the first image so that there is no vertical parallex. In this hologram, the coherence requirement can be removed by using slit process.

(d) Embossed Hologram

In this hologram, the original hologram is recorded in a photosensitive material called photo resist. These holograms offer an effective method to protect against any forms of manipulation. Therefore, these are widely used in most of the security applications like credit cards, passports *etc.*

(e) Volume Hologram

Volume holograms are the transmission type holograms, which are produced when the thickness of the recording material is much larger than the light wavelength used for recording. These are also known as thick holograms, which are mainly considered as a high density data storage technology. One important feature of volume hologram is the interdependence of the wavelength and the scattering angle according to the Bragg's Law.

4.13 HOLOGRAPHY VERSUS PHOTOGRAPHY

There are following interesting properties of holograms which make it preferable to a photograph.

(*i*) Holography produces a three dimensional (3-D) image of the original object (3-D) whereas photography produces a two dimensional (2-D) image of a three dimensional (3-D) object.

(*ii*) Hologram records intensity as well as the phases of the waves coming from various parts of the object and hence the idea of depth is inbuilt in hologram whereas photograph records simply the intensity variation only.

(*iii*) The image reconstructed from hologram is always positive whereas photography produces negative.

(*iv*) In holography, no lens is needed in either process (recording and reconstruction) while in photography a lens system (camera) is used to produce an image of the object.

(*v*) In holography, a highly coherent source (LASER) is used while in photography, an ordinary light source is used.

(*vi*) In hologram, the size of image can be altered by converging or diverging the reconstruction beams, while in photography, a photo once recorded its size cannot be changed.

(*vii*) Several images can be recorded in a hologram simultaneously but it is not possible in photography.

(*viii*) The information holding capacity of a hologram is very high while in photography it is low.

(*ix*) If the hologram is broken into pieces, each piece still would show the whole object while in photography it is not possible.

4.14 BASIC REQUIREMENTS OF A HOLOGRAPHIC LABORATORY

(*i*) Since, holography is an interference phenomenon, therefore coherence is an important requirement for holography. The coherence length of light should be greater than the maximum path difference between the object wave and reference wave. To meet this requirement the amplitude of reference wave should be quite large as compared to the amplitude of object wave. Further, the spatial coherence is important so that the wave scattered from different regions of the object can interfere with reference beam.

(*ii*) The reconstruction source and reference source should be identical (in term of wavelength and relative position), so that the reconstructed image does not suffer from any aberrations.

(*iii*) Another critical requirement in making hologram is stability of the recording arrangement. Thus entire set up (photographic plate, object and mirrors) must be motionless during the exposure.

(*iv*) The resolution of photographic plate used to record a hologram should be high.

(*v*) To get good resolution in the reconstructed image, the reconstructing source must be a narrow band of wavelengths.

The experimental arrangement of the basic requirements for holographic laboratory involving double illumination is shown in Fig. 4.19.

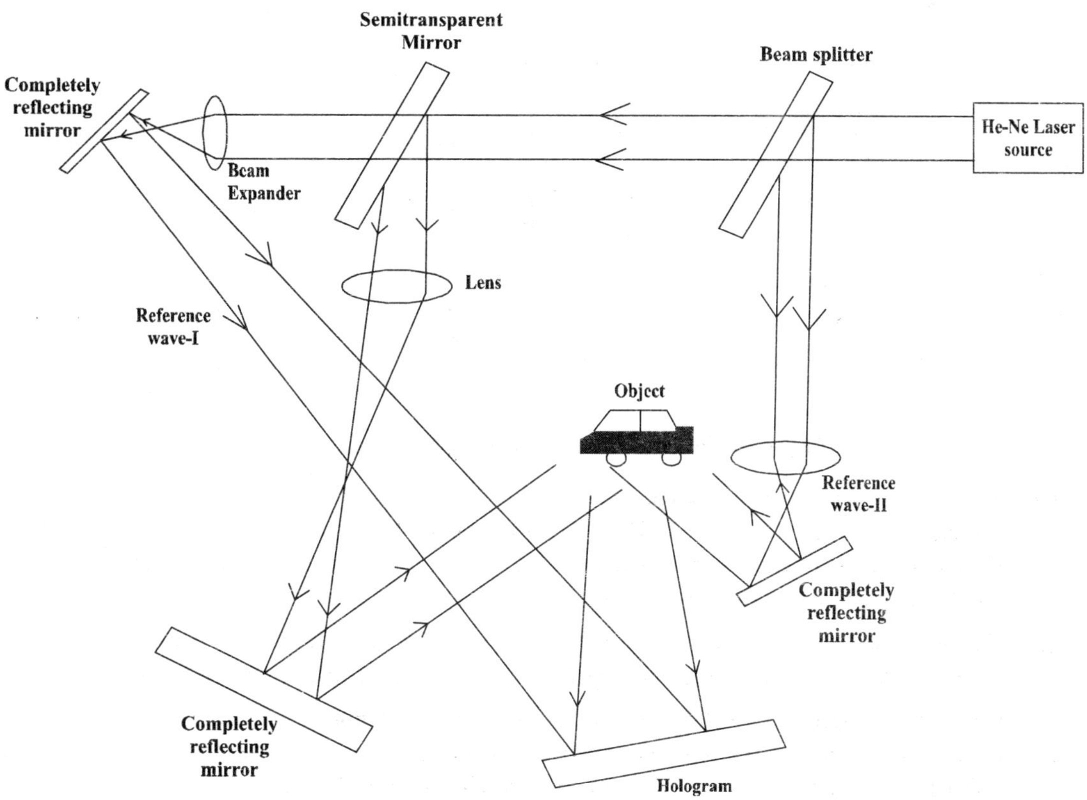

Fig. 4.19 Experimental arrangement of holographic laboratory.

In above experimental set up, a laser beam from He-Ne laser falls on semitransparent mirrors to split into two parts. One part of this beam produces the object beam and another part produces the reference beam. These beams produce interference pattern on the photographic plate (hologram). As we know that no lenses are required in the process of holography, in our experimental arrangement, these play auxiliary role of beam expander only.

4.15 SOME SPECIAL FEATURES OF A HOLOGRAM

There are following interesting features of hologram which make it preferable to a photograph.

(*i*) One special feature of a hologram is that if the plate is broken into pieces; each piece still would show the whole object. Only the quality of the image becomes poorer.

(*ii*) Several images can be recorded in a hologram either by placing several objects in differently oriented locations and using a single reference beam for simultaneous recording, or by exposing the plate in succession, each time turning the hologram plate by an angle. Fig. 4.20 shows how from different eye positions E_1 and E_2 on the same part of the hologram shows the images of different objects earlier holographed.

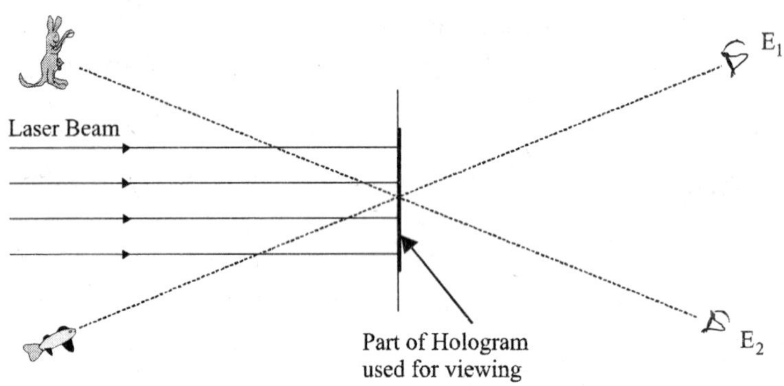

Fig. 4.20 Viewing different objects in a single hologram

(*iii*) The reconstruction of the image of the hologram can be done with reference beam of any wavelength if it is coherent and identical to the original reference beam. If the wavelength λ of the reconstructing beam is greater than that λ_0 of the reference beam, the reconstructed image will be proportional to the ratio of the two wavelengths.

(*iv*) The depth and sideways width of a scene which can be recorded on a holograph usefully are limited by the coherence length and spatial coherence of the laser beam respectively.

(*v*) Colour holography has been developed : In a thick photographic emulsion, waves travelling into the emulsion can interfere with those reflected from successive layers in the emulsion, forming a three-dimensional standing wave pattern. Such volume holograms, when viewed in white light, give reconstruction in full colour.

(*vi*) Interestingly, it has been stated that the use of holography allows the storage of information in a three-dimensional medium with such compactness that a 1 cm^3 crystal may accommodate the information contained in a library of a million books.

(*vii*) Reconstruction always gives a photographic positive. If the hologram was copied by contact printing, the opaque parts in the hologram would become transparent in the copy and vice-versa. The image reconstructed from the copy would still be a positive, being indistinguishable from the image reconstructed from the original hologram.

4.16 APPLICATIONS OF HOLOGRAPHY

4.16.1 Holographic Interferometry

One of most promising application of holography lies in the field of interferometry. This technique can be used for testing stresses, strains and surface deformations in objects. For this purpose double exposure holographic ineterferometry technique is used. In this technique, the original object and the object after deformation are recorded in two successive holograms on the same photographic plate. When the images are reconstructed, interference effects will occur if the object has changed in any way. Thus, any distortions of the object will appear in the form of fringes. It is also known as real time holographic interferometry.

4.16.2 Holographic Microscopy

In this technique, one laser beam (object wave) after passing through the object and microscope interfere with reference beam and produce the hologram on photofilm as shown in Fig. 4.21. The reconstructed image may be seen at any cross-section and at any depth. Therefore, the point to point observation is possible in holographic microscopy. Thus the holographic microscopy has a high depth of field and this ability finds application in studying transient microscope events. If one has to study some transient phenomenon which occurs in a certain volume, then using ordinary microscope techniques it becomes difficult to first locate the position and make observation. If a hologram is recorded of the scene, then the event gets frozen into the hologram. Now we can focus through the depth of the reconstructed image and study the recorded scene in detail.

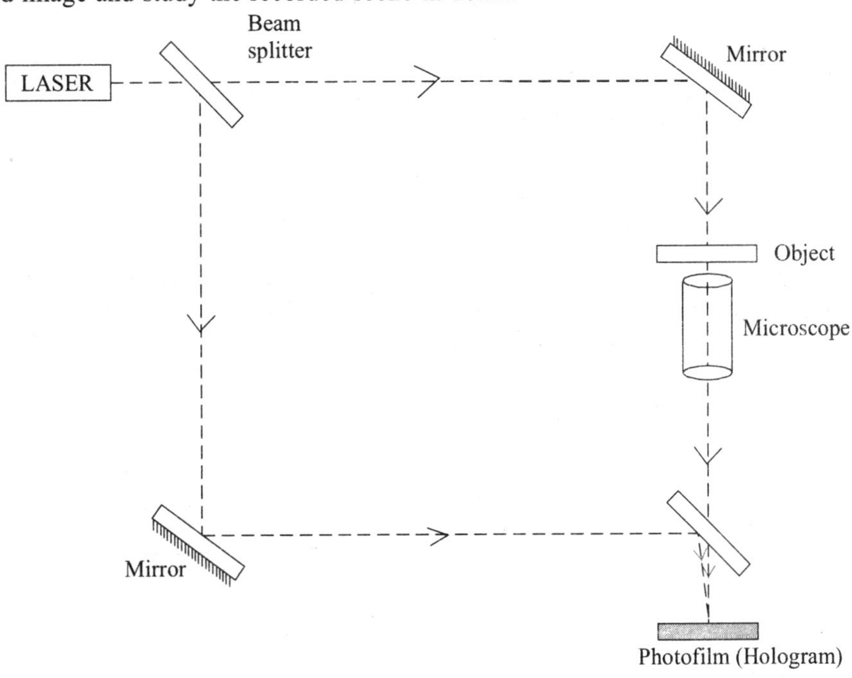

Fig. 4.21 Holographic Microscopy

4.16.3 Acoustic Holography

Acoustic holography uses sound waves instead of light waves to create a hologram. This holography requires coherent sound sources. It is also known as liquid surface acoustic holography. In this holography, an ultrasonic generator and an object, both are kept inside the water as shown in Fig. 4.22. Ultrasonic generator generates reference beam and a beam that is directed towards the underwater object. The ultrasonic beam reflected form the object forms the object beam. The object beam and the reference beam interfere and produce ripples on the surface of water. The amplitude of the ripples depends on the intensity of the coherent sound beams used. The ripples pattern forms the hologram and can be photographed by a laser. This laser is located above the surface and laser light directed onto the surface ripples. The reflected light is photographed. The plate becomes a recording of the hologram and can be illuminated with laser light to reconstruct the image of the object.

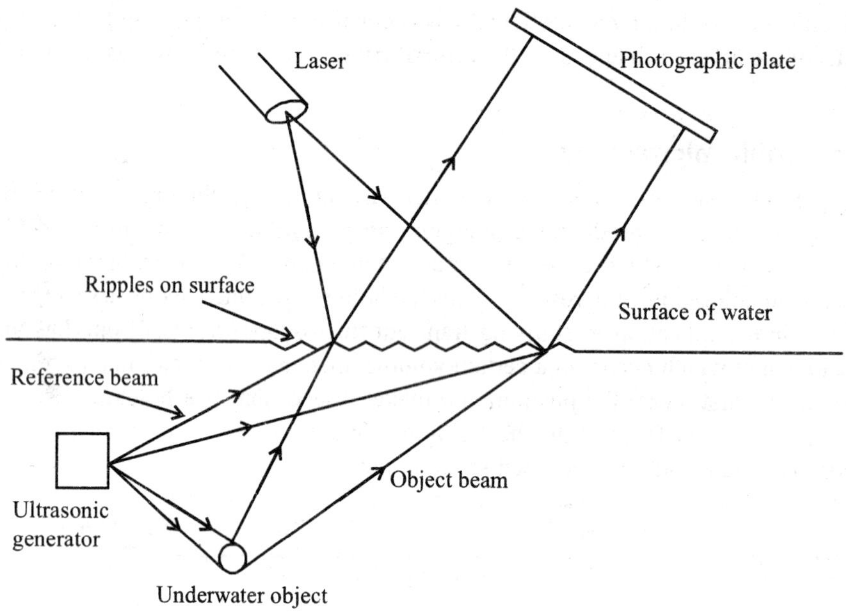

Fig. 4.22 Acoustic Holography

Example 4.8: A hologram is recorded by using He-Ne laser of wavelength 6328 Å. The object and the reference source are equidistant 1.5 m from holographic plate and these are separated by a distance of 0.5 m. Calculate the fringe spacing in hologram.

Solution: The fringe spacing is given by $\beta = \dfrac{D\lambda}{d}$

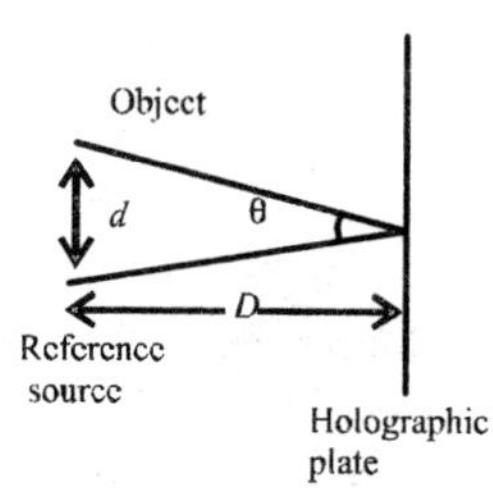

Given $\lambda = 6328 \times 10^{-10}$ m

$D = 1.5$ m

$d = 0.5$ m

$$\therefore \quad \beta = \frac{1.5 \times 6328 \times 10^{-10}}{0.5} = 18984 \times 10^{-10} \text{ m} = 1.89 \ \mu\text{m}$$ **Ans.**

QUESTIONS AND PROBLEMS

QUESTIONS

1. Explain the term absorption, spontaneous emission and stimulated emission. Derive a relation between Einstein's coefficients. **[RTU B. Tech. Back 2012]**
2. What is a 'laser' ? How is the output of a laser different from that of an ordinary source of light?
3. Explain the difference between spontaneous and stimulated emission. How can we achieve a higher probability of stimulated emission as compared to that of spontaneous emission?
4. Explain the principle of 'laser'. What are the essential requirements of a laser? Explain how these requirements are achieved. **[RTU B. Tech. II Sem. 2012]**
5. Describe the laser action of He-Ne laser with energy level diagram. Describe various applications of lasers. **[Raj. B.E. I 1999]**
6. What are the essential requirements to produce a laser? How are these fulfilled in a He-Ne laser ? **[RTU B. Tech. II Sem. 2011]**
7. Obtain threshold condition for sustained laser action. **[RTU B. Tech. Back, 2010]**
8. Describe the construction and working of He-Ne laser. How would you use it to produce a hologram?
9. (*a*) In He-Ne laser, what is the function of He atoms? Explain the answer with the help of energy level diagram for He-Ne. Describe with a neat sketch the working of He-Ne laser. **[RTU B.Tech. I 2007]**

   (*b*) Derive the relation between Einstein's coefficients and discuss the result. **[Raj. B.E. I 2002]**
10. Describe the construction and working of He-Ne Laser. How is population inversion achieved in such a laser? **[MERC B.E. I 2002, RTU B. Tech. Back, 2010 & 2011]**
11. Distinguish between spontaneous and induced emission. How does induced emission dominate in He-Ne laser?
12. What are spontaneous and induced emissions? How could probability of stimulated emission be increased? **[Raj. B.E. I 2004]**
13. What are active medium, population inversion and optical pumping in reference to laser action? **[Raj. B.E. I 2005]**
14. (*a*) Outline various processes that occur when atoms interact with radiation field. Find analytically the condition when stimulated emission dominates other process. **[RTU B. Tech. I Sem. 2008]**

    (*b*) Give the reasons for the following basic properties of a laser (i) High intensity (ii) High directionality. **[RTU B.Tech. I 2007]**

15. What do you understand by spontaneous emission and stimulated emission? Obtain a relation for Einstein's coefficients A and B for spontaneous and stimulated emission. What are essential requirements for laser action? **[RTU B. Tech. II Sem. 2008]**

16. How do pumping and population inversion are achieved in a semiconductor laser? What are advantages of using heterojunction over homojunction in semiconductor lasers? **[RTU B. Tech. II Sem. 2008]**

17. What is basic requirement of semiconductor laser? Draw its labelled diagram and explain its working with necessary theory. Write down applications of semiconductor laser. **[RTU B. Tech. II Sem. 2010]**

18. Explain the essential requirements for producing leser action. Describe the construction and working of semiconductor lasers. **[RTU B. Tech. Back, 2010]**

19. Describe the method of recording the hologram and reconstruction of image from it. **[Raj. B.E. I 2001, RTU B.Tech. II Sem. 2008]**

20. Explain the basic principle of holography. **[Raj. B.E. I 2000, RTU B.Tech. II Sem. 2008]**

21. Write short note on holography. **[Raj. B.E. I, MREC. B.E. I 2001, Raj. B.E. I 2003]**

22. Discuss construction and reproduction of a hologram. In brief discuss application of a hologram. **[Raj. B.E. I 2005, Raj. Back 2007]**

23. Explain the construction of a hologram and reconstruction of image by hologram using suitable diagram. **[RTU B.Tech. I Sem. 2007, II Sem. 2010]**

24. What is holographic microscopy? With illustrative diagram show outlay of a holographic microscope and explain its working. **[RTU B.Tech. II Sem. 2008 & 2010, Back 2010]**

25. What is holography? Describe the construction of a hologram and reconstruction of an image by hologram. **[RTU B.Tech. I Back 2008]**

26. Give two fundamental differences between a hologram and photograph. **[RTU B.Tech. I Sem. 2008]**

27. Outline basic theory of holography. With suitable diagram explain basic requirements of a holographic laboratory. **[RTU B.Tech. I Sem. 2008]**

28. Describe the method of recording the hologram and reconstruction of image from it. Can holography be studied with ordinary light? **[RTU B. Tech. Back, 2010]**

29. Write short notes on the following:

    (*i*) Q-switching (*ii*) Mode-locking (*iii*) Optical resonator and (*iv*) Spontaneous emission **[RTU B.Tech. II Sem. 2010]**

30. Write short notes on the following:

    (*i*) Q-switching (*ii*) Mode-locking (*iii*) Holographic Microscopy and (*iv*) Semiconductor laser. **[RTU B.Tech. II Sem. 2011]**

31. Discuss a technique for obtaining powerful pulses of short duration from laser source of low power. **[RTU B.Tech. Back 2012]**

PROBLEMS

1. A laser beam $\lambda = 6000Å$ on earth is focussed by a lens of diameter 2 m on to a crater on the moon. The distance of the moon is 4×10^8 m. How big is the spot on the moon? Neglect the effect of earth's atmosphere. **[1.44×10^4 m^2]**

2. A three level laser emits a light of wavelength of 5500 Å.

   (*a*) What will be the ratio of population of upper level (E_2) to that of the lower energy level (E_1) if the optical pumping mechanism is shut off? Assume $T = 300$ K **[1.3×10^{-38}]**

   (*b*) At what temperature for the conditions of (*a*) would the ratio of populations be ½ ? **[37,800 K]**

3. Calculate the power per unit area (intensity) delivered by a laser pulse of energy 4×10^{-3} Joule and the pulse length in time as 10^{-9} sec, when the pulse is focussed on target to a very small spot of radius 1.5×10^{-5} m. **[5.7×10^{-15} W/m^2]**

4. Calculate the population density of the excited state of active medium that produces light of wavelength 6328 Å at 3000 K. The population density of the lower state is 1020 atoms per unit volume.

   [Hint: $\dfrac{N_i}{N} = e^{\frac{(E_j - E_i)}{k_B T}} = e^{\frac{h\nu}{k_B T}}$]

   [5.1 × 1016]

5. A semiconductor heterojunction laser is grown with GaAsP active region layer with lattice constant 0.56 nm. Assume that laser transition occurs from the bottom of conduction band involving energy band gap 1.80 eV. Determine wavelength of laser emitted out. **[6901 Å]**

6. Laser action occurs by stimulated emission from an excited state to a state to a energy 30.5 eV. If the wavelength of laser light emitted is 690 nm, what is the energy of the excited one? **[32.29 eV]**

   [Hint: $E_n - E_m = h\nu = \dfrac{hc}{\lambda}$]

   [RTU. B. Tech. II Sem. 2012]

7. Determine number of longitudinal modes for gas laser operating in blue region (= 400 nm) and the mirror separation is 0.40 m. **[2×106]**

8. A cavity is 10 cm long. Calculate the number of modes present for light possessing a 4 nm wavelength spread about central wavelength of 500 nm. **[3200]**

9. If the degree of coherence between two interfering beams is 0.5 and on superposition the maximum receives 15 units of light intensity. What is the intensity at minimum? **[5 Units]**

10. The pulse width of laser of wavelength 1064 nm is 25 ms. If the average power output per pulse is 0.8 W, then **[RTU. B. Tech. II Sem. 2012]**

    (*i*) What is the energy released per pulse? **[0.02 J]**

    (*ii*) How many photons does each pulse contain? **[10.70×10^{16}]**

11. The ratio of population of two energy levels at 300 K is 10 : 30. Find the wavelength of the radiation emitted. **[48.6 μm]**

12. Calculate the number of photons emitted by ruby laser of output power 100 W. The lasing wavelength of the ruby laser is 6943 Å. **[3.49 ×1018]**

13. A laser beam of intensity 50 W/m^2 falls on a perfectly reflection plane mirror for an hour, the area of the mirror is 0.5 m^2. Calculate
    (*i*) Momentum imparted to the mirror in this time. **[6×10^{-4} kgms^{-1}]**
    (*ii*) Average force acting on this mirror. **[1.67×10^{-7} N]**

14. The output of a laser has a pulse of 40 ms and average output power of 0.5 W per pulse. If the wavelength of the laser light of 640 nm
    (*i*) How much energy is deposited per pulse and (*ii*) How many photons does each pulse contain? **[0.02 J, 6.44×10^{16}]**

15. Calculate the intensity of laser beam of 12 mW power and having a diameter of 1.5 mm. Assume that the intensity is uniform across the beam. **[6.79 kW/m^2]**

REFERENCES

1. J.R. Meyor-Arendt, Introduction to Classical and Modern Optics, Prentice-Hall of India, New Delhi, 1984.
2. W.T. Silfvast, Laser Fundamentals, Cambridge University Press, (1998).
3. A.K. Ghatak and K. Thyagrajan, Lasers Theory and Applications, MacMillan India Ltd., New Delhi, 1981.
4. B.A. Lengyel, Introduction to Laser Physics, John Wiley and Sons, New York, 1967.
5. H.M. Smith, Principles of Holography, New York, Wiley-Interscience, 1969.
6. M.J. Beesley, Lasers and their Application, Tylor and Francis Ltd. London, 1976.
7. C.M. Vest, Holographic Interferometry, New York, John Wiley & Sons, 1979.
8. T.C. Lee and P.N. Tamura (Eds.), Proc. SPIE 215, Recent Advances in Holography, 1980.

5

NUCLEAR RADIATION DETECTORS

5.1 INTRODUCTION

The instruments which detect the nuclear particles or radiations are called nuclear radiation detectors or instruments. These are based on the principle of excitation or ionization of atoms of medium in which the incident charged particles pass through. In general, the radiations are categorised into three types viz.

(*a*) Charged particles like α -particles, protons, neutrons and fast moving electrons.

(*b*) Uncharged particles like neutrons.

(*c*) Electromagnetic radiations like X-rays and γ-rays.

When the charged particles enter the detector medium, these interact simultaneously with many electrons of detector atoms via Coulomb interaction. The interaction raises the electrons of the medium atoms to higher energy level (excitation) or completely remove the electron from the atom (ionization). The counters working on excitation principle are called scintillation counters. Many radiation detectors like ionization chamber, proportional counters, G.M. counters, semiconductor radiation detectors etc. are based on the ionization of atoms by the charged particles.

The uncharged particles carry no charge and therfore can't create direct ionization or excitation of the detector material. Therefore an indirect process is used to detect these particles, which is as follows:-An X-or γ-ray photon can transfer all or part of its energy to electrons within the medium via photoelectric absorption, compton scattering and pair production. The resulting secondary electrons bear a close similarity to the fast electron radiations. In contrast to X-rays or γ-rays, the secondary radiations resulting from neutron interactions are almost always heavy charged particles, which can be detected directly.

5.2 GAS FILLED DETECTORS

Several of the oldest and most widely used types of radiation detectors are gas filled detectors. Ionization chamber, proportional counter and Geiger-Müller counter are the oldest types of gas filled detectors. These are based on sensing the direct ionization created by the passage of the radiation. In each of the above type of gas filled detectors, an electric field is applied to a volume of gas enclosed in a chamber.

A schematic diagram of gas filled detector is shown in Fig. 5.1. External voltage V is applied between the wall of the gas-filled chamber (cathode) and the central wire (which acts as anode) through an external resistance R. The capacity of the electrode and the counting system is C_0. In this way, an electric field is set up in the volume of the gas.

Exposed to the nuclear radiation, charged particles-either primary α-rays or β-rays or secondary electrons formed in the gas during the ionisation by γ-rays will travel through the gas and produce positive and negative ions by inelastic collisions with atoms or molecules. In the absence of the electric field, the ion pairs thus created will just recombine. However, in the presence of the applied electric field the positive and the negative ions will move along the radial electric lines of force towards the outer wall (cathode) and the central wire (anode) respectively. Normally the negative ions (usually electrons) move with much higher drift velocities (10^4 m/s) as compared to the positive ions. The net result is that a charge Q collects on the anode, thus changing its potential by Q / C_0. The changes in the potential drop across R will give rise to an electrical pulse signal. Thus, the passage of a nuclear radiation through the detector will give rise to a pulse signal which can be processed by a preamplifier etc. for counting.

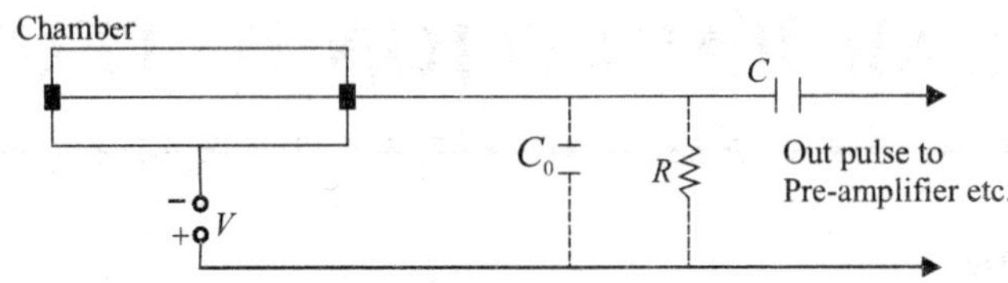

Fig. 5.1 Gas filled detector

A plot of output pulse height at the anode (or relative number of negative ions collected by the anode) will depend on (*i*) external voltage applied, (*ii*) initial ionising event-that is, whether the pulse has been initiated by the passage of α-particle or β-particle. This behaviour is shown is Fig. 5.2.

The variation of the number of ions collected with the applied voltage is usually described by dividing the graph of Fig. 5.2 into six regions as shown.

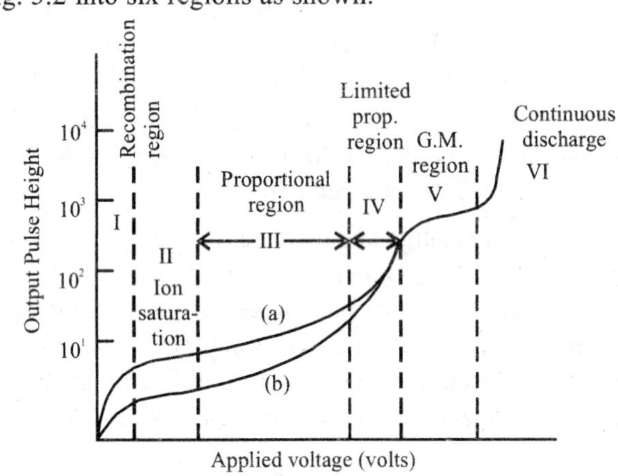

Fig. 5.2. Output pulse height from a gas filled detector versus applied voltage

(*a*) **Recombination region:** Initially, when the applied voltage is low, the electric field is not so effective in removing the ions for collection at the electrodes. In region I, the ions face a competition between two processes (*i*) loss of ion pairs through recombination and (*ii*) removal to the electrodes by the electric field. As the electric field is increased, the ions move faster leaving less time for possible recombination and thus a relatively larger number of ions reach the electrodes. This region I is called *recombination region.*

(*b*) **Ionization or saturation region:** As the voltage is increased further we reach region II where the loss of ions through recombination is almost negligible. The ions collected at the electrodes give rise to pulse signal. The region is called the *ionization chamber or saturation region.*

(*c*) **Proportionality region:** In region III, the phenomenon of gas multiplication sets in. Because of increased voltage, the electrons which are liberated by the primary ionising event get sufficient kinetic energy to cause secondary ionization. This secondary ionization increases the amount of collected charge. In the first part of region III, the gas multiplication factor M is strongly dependent on the particle energy for a given applied voltage. In other words, in this region, the detector will give rise to pulses of different heights depending on whether the initial ionization is caused by α-particle or β-particle. This proportionality between the pulse height and the initial ionization allows us to use the detector to distinguish between particles of different energies and ionizing powers.

(*d*) **Limited proportionality region:** As the applied voltage is increased the proportionality breaks down. Increasing the applied voltage still further can introduce nonlinear effect. These are mainly related to the positive ions, which are also created in each secondary ionization process. The free electrons are quickly collected, the positive ions move much more slowly. Therefore each pulse within the counter creates a cloud of positive ions, which drifts slowly towards the cathode. If the concentration of these ions is sufficiently high, they represent a space charge that can significantly alter the shape of the electric field within the detector. Because further gas multiplication is dependent on the magnitude of the electric field, some nonlinearities will begin to be observed. Thus region (IV) is called the region of limited proportionality in which the pulse amplitude still increases with increasing of initial ion pairs, but not in a linear fashion.

(*e*) **Geiger-Müller region:** Beyond the proportionality region, the pulse size is completely independent of the initial ionization and all particles produce pulses of same height irrespective of their energy and primary ionization. Since at sufficiently high voltage the space charge created by the positive ions can become completely dominant in determining the subsequent history of the pulse. Under these conditions, the avalanche proceeds until a sufficient number of positive ions have been created to reduce the electric field below the point at which additional gas multiplication can take place. The process is then self-limiting and will terminate when the same total number of positive ions have been formed regardless of the number of initial ion pairs created by the incident radiation. Then each output pulse from the detector is of the same amplitude and no longer reflects any properties of the incident radiation. This is the Geiger-Müller (V) region of operation, and detectors operated in this region are called Geiger-Müller (GM) counters.

(*f*) **Continuous discharge region:** If the voltage is increased beyond the GM region, there will be an onset of continuous electrical discharge.

5.3 IONIZATION CHAMBER

An ionization chamber can have various geometries, the basic system being similar to one shown in Fig. 5.1. The external voltage applied between the two electrodes is properly adjusted for operation in region II of Fig. 5.2. The gas used in the chamber is either dry air at normal pressure or some other suitable gas.

When ionization chamber is to be used for intensity measurements, one usually measures its ionization current with a system shown in Fig. 5.3. The ionization current of a chamber exposed to nuclear radiation first increases with applied voltage but soon saturates to a saturation current value Fig. 5.4, because at this voltage all the primary ions are collected before they can recombine.

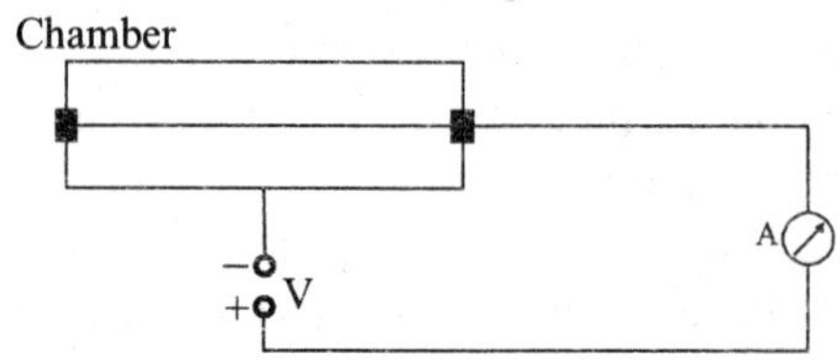

Fig. 5.3. Ionization Chamber

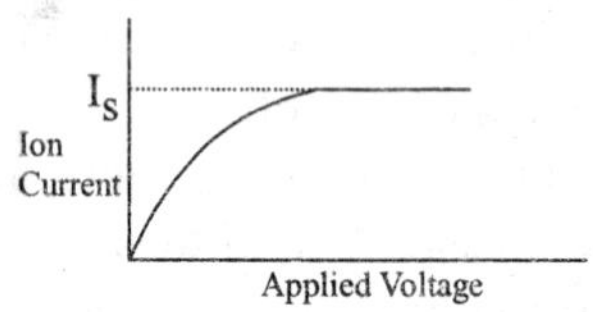

Fig. 5.4 Current versus voltage characteristics

The voltage at which saturation sets is determined by the intensity of the incident nuclear radiation. If the number of ion pairs produced per second is N, the average ionization current I_S at saturation is given by $I_S = Ne$ where e is the electronic charge. Thus measurement of I_S can give us the integrated effect of the total ionizing radiation. The ionization current is measured either by micro ammeter or by some other sensitive method.

When an individual particle or a group of simultaneous particles is to be measured one uses the system as shown in Fig. 5.1. The pulse coming out of the ionization chamber is taken through an amplifier for further counting. The voltage-time dependence of the pulse shows characteristics as in Fig. 5.5.

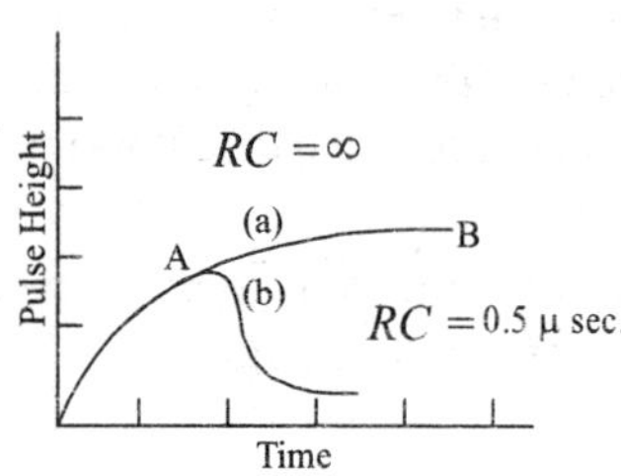

Fig. 5.5. Voltage time characteristics of the pulse

The upper curve ($RC = \infty$) shows the behaviour for large values of the RC constant. Here, the initial fast rise of the pulse (upto point A) is due to the arrival of fast moving electrons. This is followed by a slow rise (upto point B) which is attributed to the arrival of slow moving positive ions. For the case of a finite time constant $RC = 0.5$ μ sec the pulse has a shape as shown by (*b*).

5.3.1 Advantages

(*i*) The operation of this counter is not affected by voltage fluctuations.

(*ii*) It is also used to distinguish the ionization effects produced by α-particles and β-particles.

(*iii*) It is most suitable to detect heavy particles α, *p*, *d etc.*

5.3.2 Disadvantages

(*i*) One drawback of this type of detector is that after the detection of one particle the chamber has to wait for several milliseconds before it is ready to detect the next particle. This is a serious limitation if high counting rates are involved.

(*ii*) The ionization current is quite small and hence external electronic amplifier is a must with an ionization chamber.

(*iii*) It is not able to detect low energy ionizing particles.

(*iv*) It is not suitable for γ-ray detection.

Example 5.1: If a charged particle when completely stopped in the gas of an ionization chamber produces 3×10^5 ion pairs per 10 MeV of energy, calculate the pulse height if total capacitance of the detector is 50 pF.

Solution: The pulse height is given by

$$V=\frac{Q}{C}=\frac{ne}{C}$$

Here $n=3\times10^5$, $e=1.6\times10^{-19}$ Coulomb

$$C=50\times10^{-12}F$$

$$\therefore \quad V=\frac{3\times10^5\times1.6\times10^{-19}}{50\times10^{-12}}=10^{-3}\text{Volt}$$ **Ans.**

Example 5.2: An α-particle is stopped in an ionization chamber in which it produces 15×10^4 ion pairs. Each time the α-particle produces an ion pair, it losses 35 eV of energy. What is the kinetic energy of α-particle? Calculate the amount of charge collected by each plate.

Solution: Kinetic energy of the α-particle

$$E = NE_1$$

Here N = number of ion pairs produced by an α-particle

$$= 15\times10^4$$

E_1 = energy required to produce an ion-pair

$$= 35 \text{ eV}$$

$$\therefore \quad E=15\times10^4\times35=5.25\times10^6\text{eV}=5.25 \text{ MeV}$$ **Ans.**

Charge collected by each plate

$$Q=Ne=15\times10^4\times1.6\times10^{-19}$$

$$=2.4\times10^{-14}\text{Coulomb}$$ **Ans.**

Example 5.3: Fifty percent of β-particles emitted at a rate of 10^3 disintegrations per minute enter into an air filled ionization chamber and spend an average 8 cm path in the chamber. Calculate the current and corresponding potential drop detected by the instrument. Given $R=10^{11}$ ohms and β-particle produces 10 ion pairs per mm of path.

Solution: The total number of ion pairs produced per second

$$N = n_\beta nl$$

where n_β is the number of β-particles entering the chamber per sec.

n = number of ion pairs produced per mm by each β-particle

l = path length traversed by β-particles

Current due to charge flowing per second

$$\therefore \quad I = Ne = n_\beta nle$$

Here $n_\beta = \dfrac{1000 \times 0.5}{60}$, $n = 10, l = 80$ mm

$$e = 1.6 \times 10^{-19}\text{Coulomb}$$

$$\therefore \quad I = \frac{1000 \times 10 \times 80 \times 1.6 \times 10^{-19} \times 0.5}{60}$$

$$= 1.1 \times 10^{-15}\text{ amp}$$ **Ans.**

Potential drop $V = IR$

$$= 1.1 \times 10^{-15} \times 10^{11} = 1.1 \times 10^{-4}\text{Volt}$$ **Ans.**

Example 5.4: A γ-ray of energy 4 MeV produces an electron-positron pair. The two particles move in opposite directions with equal speeds. The electron enters a helium field detector and looses all its kinetic energy in producing ion pairs. Calculate the number of ion pairs produced and the pulse height, assuming that the average ionization energy per ion pair in helium is 42.6 eV and the capacity of the system is $10\,pf$.

Solution: The interaction equation of the reaction is

$$E_\gamma = 2m_0c^2 + 2E$$

where E is the kinetic energy of each particle which is given by

$$E = \frac{E_\gamma - 2m_0c^2}{2}$$

Here $E_\gamma = 4$ MeV, $m_0c^2 = 0.51$ MeV

$$\therefore \quad E = \frac{4 - 2 \times 0.51}{2} = 1.49\text{ MeV}$$

The number of ion pairs produced in He is

$$n = \frac{E}{E_1} = \frac{1.49 \times 10^6}{42.6} = 3.5 \times 10^4$$ **Ans.**

The pulse height is given by

$$V = \frac{Q}{C} = \frac{ne}{C}$$

$$= \frac{3.5 \times 10^4 \times 1.6 \times 10^{-19}}{10^{-11}}$$

$$= 56 \times 10^{-5} \text{ Volt}$$ **Ans.**

5.4 PROPORTIONAL COUNTER

When the low energy ionizing particles produce a pulse height less than 30 μV, in an ionization chamber the instrument does not respond properly and detection of such particles becomes difficult. It is seen that if voltage is increased to the region III, negative ions produced acquire sufficient energy so as to produce secondary ionization. Thus the successive build up of ions called *avalanche,* takes place in the chamber and considerable multiplication ($10 - 10^4$) of initial ionization takes place. This effect is called *gas amplification.* It is further observed that gas multiplication remains the same whatever be the initial ionization and the total ionization produced is proportional to the primary ionization. Hence the gas chamber operating in the region III is known as *proportional counter.*

5.4.1 Construction

A proportional counter is usually built in a cylindrical geometry (Fig. 5.6), with a hollow metal cylinder forming the outer electrode (cathode) while a fine tungsten wire (diameter about 0.1 mm) running along the axis forming the central electrode (anode). Proportional counters are usually filled with gases like methane, methane-argon mixtures, etc. It should be pointed out that a proportional counter filled with a pure noble gas is not a satisfactory system. The problems arise because a positive ion of argon can neutralize itself at the cathode by pulling out an electron from the surface of the cathode. Some energy is released in this process in the form of recombination radiation.

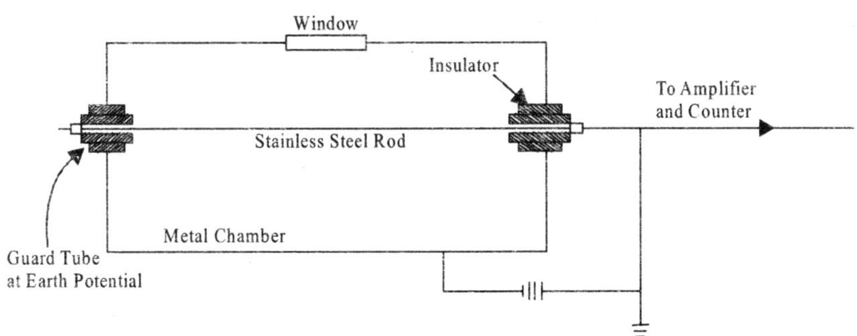

Fig. 5.6 Proportional Counter

This radiation, when incident on the cathode, can liberate electrons and an unavoidable electron avalanche can start in the counter. To avoid this one usually uses a mixture of a noble gas (*e.g.* argon 90%) and some polyatomic gas (*e.g.* 10% of methane or ethyl alcohol vapour). In such a mixture the positive argon ion has to undergo many collisions with the neutral polyatomic molecules before it can reach the cathode. In this way the argon ion exchanges its charge with the neutral molecules and has little or no chance to start an avalanche.

A proportional counter has a thin window on the side wall through which the ionizing particles can enter. The end effects in the geometry of the counters are reduced by using guard tube at earth potential. As the pulse heights in proportional counters are usually of the order of millivolts, they are amplified before counting. If the ionizing particles enter at a rapid rate in the chamber then a scaling circuit is used before mechanical counter.

5.4.2 Operation

The voltage applied to the counter is adjusted so as to be in the proportionality region III (Fig. 5.2). The proportional counter functions satisfactorily at voltages 500 to 800 volts or above at a pressure of about one atmosphere. For a cylindrical geometry, the strength of the electric field at a radial distance r from the central wire is given by

$$E = \frac{V_0}{r \ln\left(\frac{b}{a}\right)} \propto \frac{1}{r} \qquad \textbf{(5.1)}$$

where V_0 is the applied voltage a and b are the radii of the central wire and the cylindrical electrode respectively. The electric field in the neighbourhood of the central wire is very high; for example, for typical values of $a = 0.1$ mm, $b = 1.0$ cm, $r = 0.1$ mm and $V_0 = 1000$ volts, $E = 2.17 \times 10^4$ Volts/cm. Such high electric field causes gas multiplication and one usually gets multiplication factor M in the range of $10^2 - 10^4$.

The electrons and positive ions produced by the incoming nuclear particle move towards anode and cathode respectively. Since the electric field is very high near the anode, as a result avalanche develops near the anode due to the formation of secondary ions by collisions with electrons.

As avalanche takes place near anode, the all secondary electrons are quickly (10^{-9} sec) collected at anode and produces a very small voltage pulse. The total time taken by the positive ions to move towards cathode is about 500 μs. Therefore main output (pulse) comes from the motion of positive ions towards the cathode, electron motion simply delays it. Initially positive ions are in high field region, therefore pulse rises quickly initially and thereafter slowly.

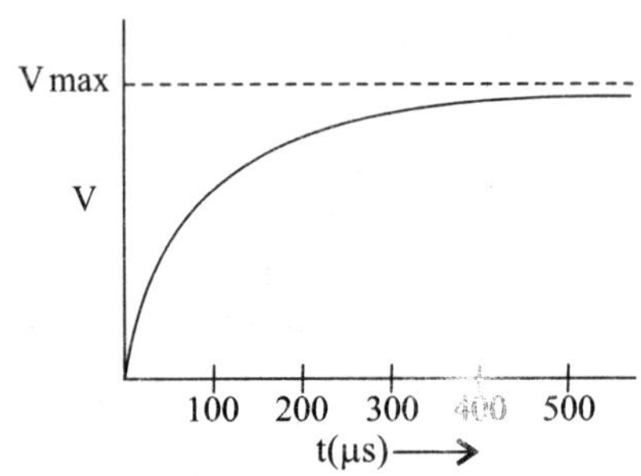

Fig. 5.7 Output pulse in a proportional counter

5.4.3 Gas Multiplication Factor

Let us now understand the gas multiplication factor more quantitatively.

Let us assume that every electron in the primary ionization gives rise to a total of n secondary electrons by collisions. The production of secondary electrons during the collisions in the gas will also give rise to photons which in turn can produce photoelectrons in the volume of the counter. Let p be the probability that each secondary electron will give rise to a photoelectron. Thus there will be np photoelectrons in all. If each photoelectron, in turn, produces n electrons by further collisions, we shall have a second generation avalanche of n^2p electrons, which in turn will give rise to more photoelectrons and so on. Finally, the total number of electrons will be

$$M = n + n^2 p + n^3 p^2 + ... = n\left(1 + np + n^2 p^2 + ...\right) \tag{5.2}$$

where M is a gas multiplication factor. For practical proportional counters $np < 1$ and the series in Eq. (5.2) converges so M can be written as a sum of geometric series, *i.e.,*

$$M = \frac{n}{1 - np} \tag{5.3}$$

It is thus seen that in a proportional counter the total number of secondary electrons is proportional to the number of initial ion pairs.

The voltage pulse shapes from a proportional counter are similar to those shown in Fig. 5.5. In practice, suitable clipping time constants ($RC \approx 0.1$ μ sec) are used in the output circuit so that better time resolutions are available.

The pulse height in the proportional counter is given by

$$V = \frac{Mne}{C} \tag{5.4}$$

5.4.4 Advantages of Proportional Counter

(*i*) The proportional counter has got an added advantage over grided ionization chamber that it is used to detect the particles of low energy.

(*ii*) The counter is successfully employed to detect α-particles in the presence of β-particles or vice-versa just by adjusting the operating voltage.

(*iii*) Large-sized proportional counters are used to study samples of low specific radioactivity. In such applications the weak samples are mounted inside the counter and the β-particles (emitted in the decay) are constrained to move within the counter by an externally applied magnetic field.

(*iv*) The proportional counter has excellent energy resolution at low energies and hence it is widely used for energy analysis of *X*-rays and low energy γ-rays (*i.e.,* in Mössbauer spectroscopy).

(*v*) It is also used for detection of neutrons, fission fragments *etc.*

5.4.5 Disadvantage of Proportional Counter

The main disadvantage of proportional counter is that the amplification of output pulse in it depends upon the applied voltage, therefore the applied voltage must be maintained constant.

Example 5.5: Calculate the electric field at the surface of the wire of a proportional counter with a wire radius 0.1 mm and a cylinder radius 1 cm when 1500 Volt is applied between the two.

Solution: The electric field strength at radius r is given by

$$E = \frac{V_0}{r \log_e\left(\frac{b}{a}\right)}$$

Here $V_0 = 1500$ Volt, $r = 0.01$ cm, $b = 1$ cm and $a = 0.01$ cm

$$\therefore \qquad E = \frac{1500}{0.01 \times 2.3027 \log_{10}\left(\frac{1.00}{0.01}\right)} = 32500 \text{ Volt/cm} \qquad \textbf{Ans.}$$

Example 5.6: Estimate the number of ion pairs produced in a proportional counter by a 10 MeV proton if the counter size and pressure are large enough for all the proton's energy to be absorbed. If the gas multiplication factor is 10^3, how many coulombs flow in the counter when this proton is absorbed ? If the pulse of current flows for about 10^{-3} sec and if the resistance is 10^4 ohms, estimate the height of the voltage of pulse. Given that the energy required for production of one ion pair is 35 eV.

Solution: Number of ion-pairs produced by a proton of energy 10 MeV

$$n = \frac{10 \times 10^6}{35} = 2.86 \times 10^5$$

After amplification the total number of ion pairs

$$= nM = 2.86 \times 10^5 \times 10^3 = 2.86 \times 10^8$$

Charge carried by these ions

$$\Delta Q = 2.86 \times 10^8 \times 1.6 \times 10^{-19} = 4.57 \times 10^{-11} \text{ Coul.} \qquad \textbf{Ans.}$$

The current passed through the resistor

$$\Delta I = \frac{\Delta Q}{T} = \frac{4.57 \times 10^{-11}}{10^{-3}} = 4.57 \times 10^{-8} \text{ amp} \qquad \textbf{Ans.}$$

The height of voltage pulse

$$\Delta V = \Delta I \times R$$

$$= 4.57 \times 10^{-8} \times 10^4$$

$$= 4.57 \times 10^{-4} \text{ Volt} \qquad \textbf{Ans.}$$

Example 5.7: In a proportional counter an 11 MeV α-particle destroys all its energy in producing ion pairs. 35 eV energy is required to produce an ion pair. If the multiplication factor for the proportional counter is 1000 and the capacitance between its central wire and the ground is $30\,pf$, then find the height of the voltage pulse.

Solution: Number of initial ion-pairs produced by α-particle

$$n = \frac{11 \times 10^6}{35}$$

Number of total ion pairs produced by the α-particle = nM

$\therefore$ Charge collected on the electrodes

$$q = \frac{11 \times 10^6 \times 10^3 \times 1.6 \times 10^{-19}}{35}$$

$$= 5.04 \times 10^{-11} \text{ Coulomb}$$

$$\text{Height of voltage pulse} = \frac{q}{C}$$

$$= \frac{5.04 \times 10^{-11}}{30 \times 10^{-12}}$$

$$= 1.68 \text{ Volt}$$ **Ans.**

5.5 GEIGER-MÜLLER COUNTER

Geiger-Müller (G.M.) counters are gas-filled detectors operating in the region V of Fig. 5.2. It has been one of the most widely used nuclear detectors. The instrument is essentially an ionization chamber. In this the gas amplification is so large that the total ionization recorded is independent of primary ionization and hence the instrument does not distinguish the type of radiations that enter the chamber.

5.5.1 Construction

G.M. counters can be built in various geometries, the most common are shown in Fig. 5.8. The end window type G.M. counter has a thin mica window at one end so that α-or β-particles can be detected.

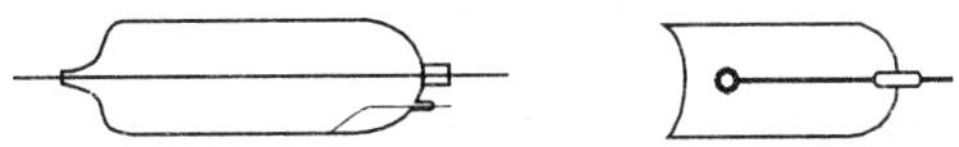
(a) (b)

Fig. 5.8 Geiger-Müller Tubes

The G.M. counter tube consists of a central electrode (anode) of tungsten wire of about 0.01 cm diameter. The outer electrode (cathode) is a metal tube which can be contained inside a glass envelop Fig. 5.9. The metal tube is well insulated from the central electrode. The tube is filled with a gas mixture of about 90% argon and 10% alcohol at a pressure of 10 cm of Hg. The applied voltage is in the neighbourhood of 1000 volts. The external circuit contains a high leak resistance R.

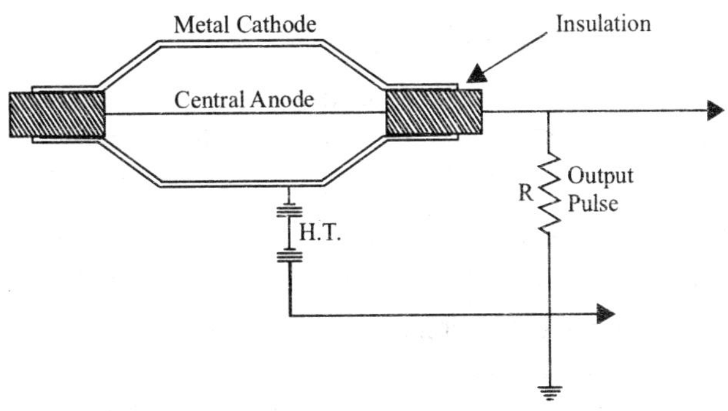

Fig. 5.9. G.M. Counter

5.5.2 Operation

The operation of G.M. tube can be understood if fresh review of G.M. region of Fig. 5.2. is made. The situation is illustrated in Fig. 5.10, when G.M. counter is connected with electronic circuits which record pulse height (≈ 0.25 Volt) proper to this region and not the small pulses.

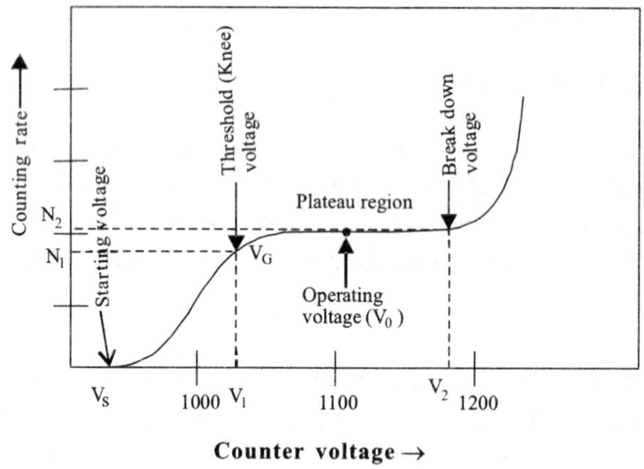

Fig. 5.10 Variation of counting rate with voltage in G.M. Region

It is seen that until the voltage reaches the starting voltage V_S the pulses are too small to be detected. As voltage increases above this limit the count rate increases until the threshold voltage V_G of G.M. region is reached. Above V_G for about 200–300 volts counting rate almost remains constant. The range of potential over which this happens is called *Geiger plateau*. Beyond the plateau a continuous discharge takes place and counting is not possible. The slope of the plateau region is not greater than

$$= \frac{N_2 - N_1}{N_{av}} \times \frac{100}{V_2 - V_1} \times 100 \text{ \% per 100 Volt}$$

$$= 5 \text{ \% per 100 Volt} \qquad \textbf{(5.5)}$$

In the G.M. region, the ionization pulse depends on the general characteristics and construction of the tube such as physical dimensions, voltage, type of gas employed and not on the initial ionization. The initial ionization produced by entering particle simply initiates the avalanche and single particle is sufficient to start it. The primary ionization produces a gas amplification of the order of 10^8 which results in a pulse height upto 50 volts. The pulse can be detected without preamplification.

The operating voltage in the G.M. region is high enough for the electrons to raise some atoms to excited states followed by ultraviolet radiation, the absorption of which gives rise to new avalanches*. Thus in a short time avalanche spreads over whole length of the tube. The electrons produced in the series of avalanches reach the anode in very short time (10^{-9} sec) but positive ions move slowly and they form a sheath around the anode for a short while. The ion sheath depresses electric field near the anode

* Each electron produced by primary radiation get enough energy to produce the secondary ion pairs and these secondary electrons are also capable to repeat the process. This chain reaction leads to avalanche multiplication of electrons, called Townsend avalanche. Thus, each electron of gas multiplication induces many avalanches elsewhere.

below V_G and no further pulse can be detected, *i.e.*, for the time this sheath is removed from anode, the instrument becomes inoperative.

When positive ions reach the cathode, they detach secondary electrons from the cathode which neutralize the positive charge leaving some atoms in the excited states. These atoms radiate high energy photons and on striking the cathode eject photoelectrons which cause fresh avalanches. The counter would therefore, be kept in state of continuous avalanching.

5.5.3 Quenching

If in a state of continuous avalanching, measurements are made, the counter would confuse the two pulses; one due to continuous avalanching and other due to fresh event. Hence to remove the state of confusion, the continuous avalanching is suppressed. The method of suppression is called *quenching*. The suppression by adding a quenching agent (alcohol, or some poly atomic gas) in counter gas is known as *self quenching*. The quenching done by employing a suitable electronic circuit is known as *external quenching.*

Self quenching

This type of quenching is achieved internally by adding a polyatomic gas, like ethyl alcohol vapour, to the argon gas (ratio argon 90% by weight, ethyl alcohol 10%). The quenching action of alcohol is as follows:

The ionization potential of alcohol (11.3 eV) is lower than that of argon (15.7 eV). As a result the ions moving out towards the outer cathode consist mostly of alcohol ions. These alcohol ions, however, do not give rise to secondary avalanche when they are neutralized at the cathode. Thus, there is no multiple pulsing and the discharge is quenched soon (fraction of a millisecond) after the initial ionization. The alcohol ions, however, dissociate after being neutralized and thus after about 10^8 counts alcohol amount in a G.M. counter is exhausted, resulting in poor characteristics of the G.M. tube.

5.5.4 Dead Time and Recovery Time

The presence of positive ion-sheath around the anode makes the instrument inoperative for a period of time which is known as dead time (t_d) which is nearly the time taken by positive ions to reach the cathode. After a dead period there is a time during which a small pulse is produced. This is known as recovery time (t_r). These are about 10^{-4} sec (100 μs). The sum of these two times $(\tau = t_d + t_r)$ is called paralysis time (τ). Therefore the counter will count another ionizing particle after the paralysis time from the time of entrance of earlier ionising particle.

If n_0 is the observed counting rate, n is number of nuclear particles entering the counter and τ be the paralysis time, then the counter will remain dead for $n_0\tau$ time. Thus the number of counts not observed by the counter will be $(n_0\tau)n$

* The quenching action with halogen gas (Cl_2 or Br_2) seems to be similar to that of the organic agents. However, the dissociated molecules of halogen combine at the end of the discharge, so the tube has unlimited life.

$$\therefore \qquad n - n_0 = (n_0\tau)n$$

$$\Rightarrow \qquad n(1 - n_0\tau) = n_0$$

The true counting rate is

$$n = \frac{n_0}{(1 - n_0\tau)} \tag{5.6}$$

Therefore, the efficiency of the counter is

$$\eta = \frac{n_0}{n} = 1 - n_0\tau \tag{5.7}$$

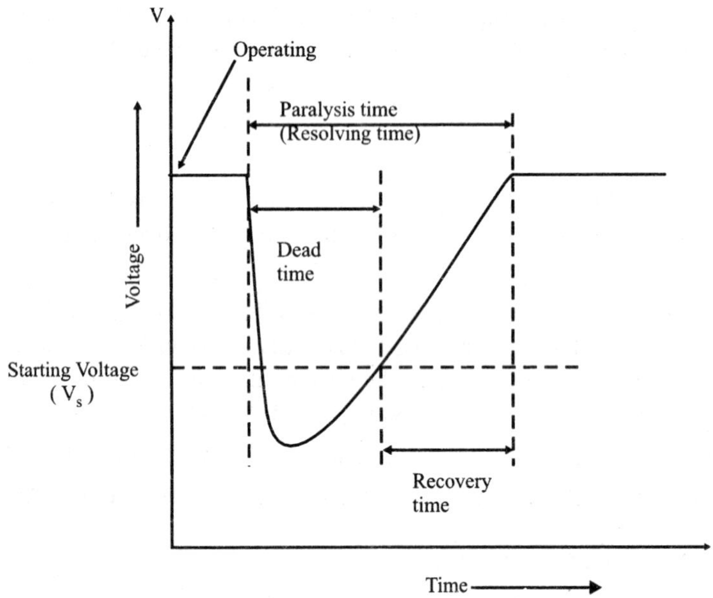

Fig. 5.11 Dead time, recovery time and paralysis time of pulse

5.5.5 Main Features of a G.M. Counter

(*i*) G.M. counter gives constant output pulse size independent of initial ionization.
(*ii*) It is sensitive to the production of even a single ion pair.
(*iii*) There is a long insensitive time to allow entry of each particle.
(*iv*) Its life time is infinite specially with halogen quenching*.
(*v*) It can detect α- and β- particles and γ- and cosmic rays.

5.5.6 Precautions in using G.M. Tube

(*i*) The tube should never operate at a voltage higher than its normal voltage.
(*ii*) In case a continuous discharge is produced, the voltage should be decreased.

(*iii*) The tube must be disconnected when measurements are over.

(*iv*) The high voltage must be relatively stabilized.

(*v*) Introduction of light must be prevented in order to avoid photoelectric effects.

(*vi*) The background counting rate must be subtracted.

5.5.7 Comparison between Proportional Counter and G.M. Counter

(*i*) The operating voltage of proportional counter is lower (500 V) than that of G.M. counter (1000 V).

(*ii*) Proportional counter works on the principle of gas multiplication whereas G.M. counter works on the principle of avalanche multiplication.

(*iii*) The output pulse in proportional counter is proportional to the energy of incident ionising particles whereas the output pulse in G.M. counter is independent of the energy of incident ionizing radiations.

(*iv*) Proportional counter can detect low energy radiations like *X*-rays, γ-rays and β-particles, whereas G.M. counter cannot detect low energy radiations.

(*v*) Proportional counter can also detect uncharged particles like neutrons, whereas G.M. counter is not able to detect uncharged particles.

Example 5.8: A halogen quenched G.M. counter works at 1 kV. anode wire has radius 0.2 mm and radius of cathode is 20 mm. The guarantee period for the counter is 10^9 counts. Find maximum radial field. If the counter is operated 30 hrs a week and count rate is 3000 counts/min, find the life of the counter (take 1 year = 50 weeks).

Solution: The field at a radial distance *r* is given by

$$E_r = \frac{V_0}{r \log_e \left(\frac{b}{a}\right)}$$

Here $V_0 = 10^3$ Volt. $r = 0.2$ mm, $b = 20$ mm and $a = 0.2$ mm

$$\therefore \quad E_{\max} = \frac{10^3}{0.2 \ \log_e \left(\frac{20.0}{0.2}\right)}$$

$$= \frac{10^3}{0.2 \times 2.3027 \times 2} = 1.087 \times 10^3 \text{ V/mm}$$ **Ans.**

If its life is *T*, then

$$T \times 50 \times 30 \times 60 \times 3000 = 10^9$$

$$\therefore \quad T = \frac{10^9}{50 \times 30 \times 60 \times 3000}$$

$$= 3.7 \text{ years}$$ **Ans.**

Example 5.9: A Geiger-Müller counter operates at 1500 volts and the central wire has radius of 0.1 mm. The inner radius of the outer cylinder is 2 cm and the tube has a guarantee life time of 10^9 counts. Calculate the maximum radial field and the life time of the counter assuming that the counter works on an average for 50 hours per week at 3000 counts per minute.

Solution: The maximum radial field at the surface of the central wire is given by

$$E_{max} = \frac{V}{2.303\ \log_{10}\left(\frac{b}{a}\right)}$$

Here $V = 1500$ Volt, $r = 0.01$ cm, $b = 2$ cm, $a = 0.01$ cm

$$E_{max} = \frac{1500}{2.303 \times 0.01 \times \log_{10}\left(\frac{2}{0.01}\right)} = 2.83 \times 10^4 \text{ V/cm}$$ **Ans.**

If the life time of the tube is T years, the total number of counts recorded will be equal to guaranted count. Hence

$$T \times 0.5 \times 60 \times 50 \times 3000 = 10^9$$

$$\Rightarrow \quad T = 2.22 \text{ years}$$ **Ans.**

Example 5.10: A GM tube with a cathode 5.0 cm in diameter and a wire diameter of 0.012 cm is filled with argon to a pressure such that the mean free path is 7.8×10^{-4} cm. Calculate the value of the voltage that must be applied to just produce an avalanche. Given the ionization potential of argon is $I = 15.7$ Volt.

Solution: The radial electric field is given by

$$E_r = \frac{V}{r \log_e\left(\frac{b}{a}\right)}$$

Avalanche will just start when the integral of this equation, taken over one mean free path is equal to the ionization potential of the filling gas, *i.e.,*

$$E_a \lambda = \frac{V\lambda}{a \log_e\left(\frac{b}{a}\right)} = I$$

$$\Rightarrow \quad V = \frac{I\ a\ \log_e\left(\frac{b}{a}\right)}{\lambda}$$

$$= \frac{15.7 \times 0.006 \times 10^{-2} \times 2.3026 \log(0.025/0.00006)}{7.8 \times 10^{-6}}$$

$$= 730 \text{ volts}$$ **Ans.**

Example 5.11: A GM counter wire collects 10^8 electrons per discharge. When the counting rate is 500 counts/minute, what will be the average current in the circuit?

Solution: Counting rate = 500 counts/min

Total number of electrons collected per minute

$$= 500 \times 10^8 = 5 \times 10^{10}$$

Charge on each electron $= 1.6 \times 10^{-19}$ Coul

$$\text{Charge per second} = \frac{5 \times 10^{10} \times 1.6 \times 10^{-19}}{60} = 1.33 \times 10^{-10} \text{Amp}$$ **Ans.**

Example 5.12: A GM counter operating at 1000 volt has a plateau region slope equivalent to 5 % per 100 Volt. If the counter is restricted to be 0.1%, what is the maximum voltage fluctuation in its applied voltage ?

Solution: Given slope (S) = 5% per 100 Volt and $\frac{N_2 - N_1}{N_{av}} = \frac{0.1}{100} = 10^{-3}$ the slope of the curve in the plateau region is

$$S = \frac{N_2 - N_1}{N_{av}} \times \frac{100}{V_2 - V_1} \times 100\% \text{ per 100 Volt}$$

$$\therefore \quad 5 = \frac{10^{-3} \times 100 \times 100}{V_2 - V_1}$$

$$\Rightarrow \quad V_2 - V_1 = \frac{10}{5} = 2 \text{ Volt}$$ **Ans.**

Example 5.13: A GM counter counts to a maximum of 5000 counts per minute. If its efficiency is 90%, calculate the dead time of the counter.

Solution: Given $n_0 = 5000$ per minute $= \frac{5000}{60}$ and $\eta = 90\% = 0.9$

The efficiency of the counter $\eta = 1 - n_0\tau$

$$\Rightarrow \quad \tau = \frac{1-\eta}{n_0} = \frac{(1-0.9)60}{5000}$$

$$\tau = 1.2 \times 10^{-3} \text{ sec or } 1.2 \text{ ms}$$ **Ans.**

Example 5.14: A GM counter counts 6000 counts per minute. If the dead time of the counter is 600 μs, calculate actual count rate.

Solution: Given $n_0 = 6000$ counts per minute, $\tau = 600\ \mu s = \frac{600 \times 10^{-6}}{60} = 10^{-5}$ minute

The counting rate is $n = \dfrac{n_0}{1 - n_0\tau}$

$$= \frac{6000}{1 - 6000 \times 10^{-5}}$$

$$= \frac{6000}{0.94}$$

$$n = 6382.97 \approx 6383/\text{minute}$$

The actual count rate $= n \pm \sqrt{n}$

$= (6383 \pm 79.89)$/minute **Ans.**

Example 5.15: A GM counter with a dead time of 300 μs records 16000 counts/mim. What is the dead time loss in the counting rate ? **[RTU B. Tech. Sem. II 2010]**

Solution: Given $n = 16{,}000$ counts/min

$$\tau = 300\ \mu\text{s} = \frac{300 \times 10^{-6}}{60}$$

$$\tau = 5 \times 10^{-6}\ \text{min}.$$

The count rate of incident particles

$$N = \frac{n}{1 - n\tau} = \frac{16000}{1 - 16000 \times 5 \times 10^{-6}}$$

$$N = \frac{16000}{1 - 0.08}$$

$$N = \frac{16{,}000}{0.92} \approx 17391$$

The actual count rate $= N \pm \sqrt{N}$

$= 17391 \pm 131.87$

$\therefore$ Loss in counting rate $= \pm 131.87$ **Ans.**

5.6. SCINTILLATION DETECTOR

5.6.1 Introduction

There are several substances (certain organic and inorganic materials) which emit light flashes or scintillations when charged particles, X-rays or γ-rays strike them. These substances are called scintillators. In 1947, Marshall and Cottman in the U.S.A. and Kallman in Germany introduced a new technique in which the light flashes from a suitable scintillator were detected by a device known as a photomultiplier tube.

In a scintillation detector, these light flashes are allowed to fall on the photocathode of a photomultiplier tube and a pulse is extracted out to signal the passage of nuclear radiation. The light of this output pulse can be proportional to the energy dissipated by the ionizing radiation in the scintillator. Thus a scintillation detector can be used not only for counting but for energy analysis also. The schematic diagram of a scintillation detector is shown in Fig. 5.12.

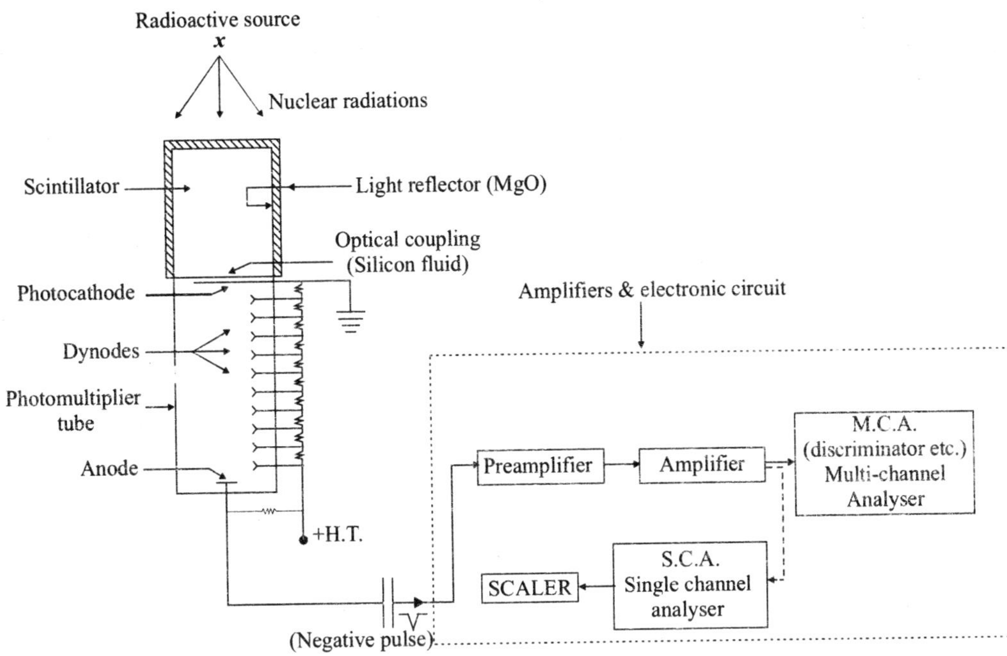

Fig. 5.12 Schematic representation of scintillation detector

5.6.2 Construction

The scintillation counter is one of the commonest devices for the detection of nuclear radiation. It consists of three basic parts.

(*i*) **Scintillating material (phosphor) and scintillation process:** There are certain organic and inorganic materials, which produce light flashes when nuclear radiations strike them. These materials are known as the scintillators. Thallium activated sodium iodide NaI (Tl) is the most common type of scintillator used for the detection of γ-rays. A less sensitive alternative is the thallium activated cesium iodide CsI (Tl), which is very popular scintillator for detection of protons and α-particles.

Organic scintillators like Anthracene and Stilbene are useful for counting the γ-rays in the presence of β-rays but show less efficiency for detection and counting of heavy particles. Zinc sulphide and europium activated lithium iodide LiI (Eu) are quite suitable for detection of α-rays and neutrons respectively.

The plastic scintillators are quite useful in counter telescopes used in high energy physics for defining the path of a particle.

The scintillation process

The scintillator absorbs energy when it undergoes excitation by the electrons. This absorbed energy appears either as heat energy (in more than 60% case) or as luminescence photons. In the latter process, the excited states of the scintillator material de-excite to lower states by light emission within 10^{-8} second or less. This emitted light is known as scintillation. The time dependence of the photon emission follows an exponential law

$$N_p = \text{constant } (1 - e^{-t/\tau})$$

where N_p = number of light photons emitted in the time t after the ionizing radiation has arrived

τ = decay time or time required for the emission of 63% of the photons.

Each scintillation material has a characteristic mean life τ. The properties of some commonly used scintillators are given in Table 5.1.

Table 5.1

Material	Density g/cm³	Wavelength of maximum emission Å	Relative beta ray pulse height	Decay time (seconds)
(1) Anthracene crystal	1.25	4,400	100	2.7×10^{-8}
(2) Trans-stilbene crystal	1.15	4,100	60	3 to 7×10^{-9}
(3) NaI (Tl)	3.67	4,100	210	2.5×10^{-7}
(4) Plastics				
(a) NE 102	1.03	4,250	65	2.2×10^{-9}
(b) NE 111	10.3	3,700	55	1.7×10^{-9}

(*ii*) **Photomultiplier tube:** It detects the light flashes and produces an electrical impulse. It consists of a photocathode to convert light energy into electrical energy. The photocathode in a photomultiplier tube is usually a semitransparent coating of some photosensitive material (*e.g.*, antimony-cesium) on the inner surface of the flat top of the tube envelope. The spectral sensitivity of such photocathode usually follows a response curve shown in Fig. 5.13.

The structure of a photomultiplier tube * is such that there are several dynodes and these are maintained at successively higher electrical potential (about 100 Volts per dynode stage as shown in Fig. 5.14). Further the dynode surfaces have a property of emitting more (say 3-5)

* The quality of a photomultiplier tube is mostly determined by two properties-

(A) The transit time or the time taken by the electrons to reach the anode through the dynode structure. This property is important for the timing characteristics of the photomultiplier tube.

(B) The stability of the electron multiplication factor. This property determines the energy resolution characteristics of the photomultiplier tube.

secondary electrons when one incident electron strikes them. Thus, a photoelectron emitted by a photocathode is accelerated by the electric field to the first dynode where it produces a bunch of secondary electrons. These electrons are now accelerated towards the next dynode where they, in turn, produce more electrons and this process is repeated at each dynode. In a photomultiplier tube there are usually 10 or more dynode stages and one can usually achieve a gain upto $10^7 - 10^8$ by the time electrons reach the last stage which is called the anode. In this way the initial ionization radiation gives rise to a burst of electrons at the anode where an electrical pulse is taken out for further analysis.

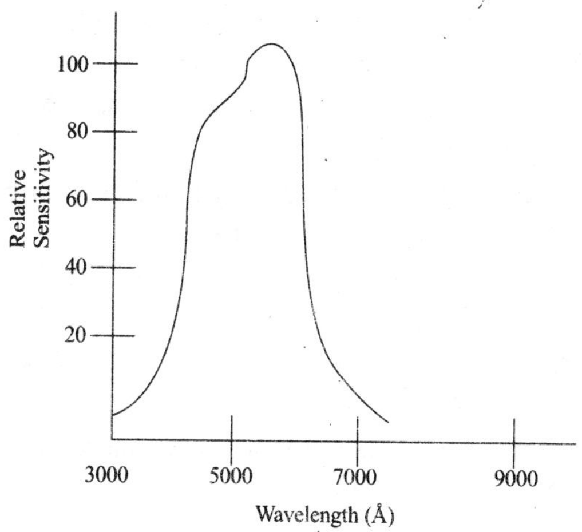

Fig. 5.13 Spectral sensitivity of photomultiplier tube

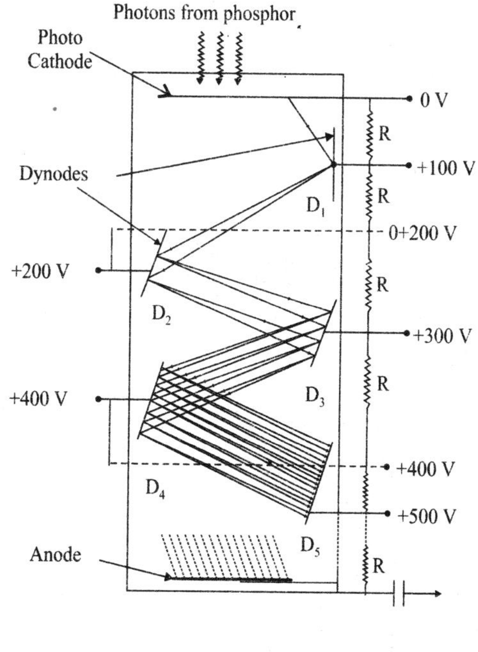

Fig. 5.14 Photomultiplier tube

(*iii*) **Amplifiers and electronic circuits:** These are used to record and count the electrical impulses from the photomultiplier tube. These circuits consist of preamplifier, amplifier, discriminators (to remove noise from pulse) and scaler to record negative output pulse (produced by a flash of light falling on the photocathode). The final pulse height in a scintillation detector is proportional to the initial energy of the incident particle and hence after amplification this pulse can be recorded by scaler.

5.6.3 Advantages and Disadvantages of Scintillation Detectors

Advantages

(*i*) Scintillation counter is capable for fast counting rate because the dead time and resolving time are of the order of 10^{-9} sec as against 10^{-5} sec in GM counter.

(*ii*) With large size and highly transparent phosphor, it displays very high efficiency.

(*iii*) It is most efficient for γ-rays counting.

(*iv*) As the pulse height is proportional to the energy of incident radiation, it is used for investigation of energy distribution of nuclear radiations.

Disadvantages

(*i*) The light output varies with type of radiation.

(*ii*) The overall efficiency is less due to the conversion of nuclear radiations into light flashes and reconversion of light flashes into electrons.

Example 5.16: Calculate the scintillation efficiency to anthracene if 2 MeV of particle energy creates 20,000 photons with the average 548 nm.

Solution: Given $E = 2\times10^6$ eV, $N = 20{,}000$ and $\lambda = 548\times10^{-9}$ m

$$\text{The efficiency } \eta = \frac{\text{Energy converted into light}}{\text{Energy observed}}$$

$$\eta = \frac{N \times E_p}{E}$$

$$\text{where } E_p = \text{Energy of emitted photon } = h\nu = \frac{hc}{\lambda}$$

$$= \frac{6.625\times10^{-34}\times3\times10^{8}}{548\times10^{-9}}$$

$$E_p = \frac{19.875}{548}\times10^{-17}$$

$$\frac{19.875\times10^{-17}}{548\times1.6\times10^{-19}}\text{ eV}$$

$$= \frac{1987.5}{876.8}$$

$$E_p = 2.26 \text{ eV}$$

$$\therefore \quad \eta = \frac{20000\times2.26}{2\times10^{6}}$$

$$= 0.0226$$

$$\eta = 2.26\ \%$$ **Ans.**

Example 5.17: An anthracene crystal and a ten stage photomultiplier tube are to be used as scintillation spectrometer. It is desired that a 10 kev β-particle incident on the scintillator produces a 2 mV pulse on the photomultiplier output circuit, which has a capacity of 120 *pF*. What average electron multiplication per stage is required in the photomultiplier ? Assuming a light collection efficiency unity and a photocathode efficiency 0.1. Anthracene is supposed to give highest yield of photons of 15 per kev energy dissipated in the crystal.

Solution: The number of photons emitted per β-particle

$$= 15 \times 10 = 150$$

Since the photocathode efficiency is 0.1, hence number of photoelectrons emitted $= 150 \times 0.1 = 15$. If each dynode produces n secondary electrons, after passing through 10 stages the electrons are multiplied to n^{10}.

Hence the output pulse

$$\Delta V = \frac{\Delta q}{C} = 2 \times 10^{-3}\,\text{Volt}$$

$$\Rightarrow \quad \frac{15 \times n^{10} \times 1.6 \times 10^{-19}}{120 \times 10^{-12}} = 2 \times 10^{-3}$$

$$\Rightarrow \quad n^{10} = 10^{5} = \left(\sqrt{10}\right)^{10}$$

$$\Rightarrow \quad n = \sqrt{10} = 3.162$$ **Ans.**

SUMMARY

(*i*) Classification of nuclear detectors

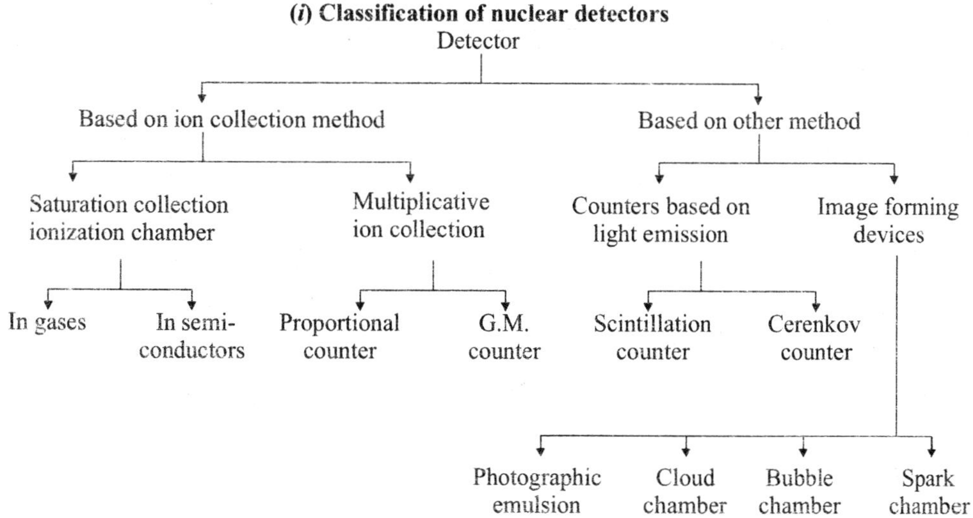

Detector	Detecting system	Particles detected	Application
Ionisation chamber	Air	α	Measurement of activity
Proportional counter	Argon + Cl_2	α, β, electrons	Energy measurement
G.M.counter	Argon + Cl_2	α, β, γ, Cosmic rays	Determination of intensity (Counts pulses)
Scintillation detector	NaI (Thallium activated)	γ, high energy particles	Energy measurement
Solid organic scintillators detector	Anthracene stilbene	β^-, p	Energy selection for electrons, detection of β-particles in the presence of γ-rays and protons
Cloud chamber	Alcohol + water	Charged particles	Detection of energy, particle detection
Bubble chamber	Propane, Xenon, liquid H_2	Charged particles	Detection of high energy particles
Spark chamber	Metallic plates	High energy charged particles	Localization of trajectories
Photographic emulsion	Emulsions	Charged particles	Determination of energy and charge, To study particle interaction.

QUESTIONS AND PROBLEMS

QUESTIONS

1. Compare the ionization chamber, proportional counter and Geiger-Müller counter. Draw a graph between the applied voltage and the pulse height. Depict the parts used for G.M. counter, Ionization chamber and proportional counter.

   [RTU B.Tech. II Sem. 2012]
2. Describe the construction and working of a Geiger-Müller counter. Give its important applications. **[RTU B.Tech. 2007 & Back 2011]**
3. Describe a scintillation counter and its mode of operation. What advantages has it over other types of counters? **[RTU B.Tech. I Sem., Back 2008 & 2011]**
4. Through a schematic diagram, explain operation of a scintillation counter. How will you select phosphor for protons, β-particles, slow neutrons and fast neutrons? Obtain an expression for output voltage pulse when radiation of energy E falls on the counting system.
5. (*a*) What do you mean by "dead time" in G.M. counter ? Mention some of its applications. Draw a neat diagram of the G.M. counter.

   [MREC Aut 2000, 2002, Raj. B.E. 1997, RTU B.Tech. II Sem 2010]

   (*b*) Describe the construction and working of scintillation counter.

   [Raj. B.E. I 2001, RTU B.Tech. II Sem 2010]
6. Describe the construction and working of Geiger-Müller counter. Explain the terms dead time and quenching. **[Raj. B.E. I 2000, RTU B.Tech. II Sem. 2010 & 2012]**
7. Draw curves to show qualitatively the variation of ionization current (assuming incident particle energy E in one case and $2E$ in another) versus voltage applied to an ionization chamber. Indicate operation regions of proportional counter and Geiger-Müller counter and give physical explanation to curves. What factors govern counting rate of a G.M. counter ?

   [Raj. B.E. I 2000]
8. Give a neat diagram of proportional counter and explain its working as a radiation detector.
9. Explain construction and working of Geiger-Müller counter, give its important applications. How quenching is achieved in this counter.

   [MREC B.E. I 2002, RTU B.Tech. II Sem. 2011]
10. Write short notes on the following

    (*a*) Photomultiplier tube **[Raj. B.E. I 2002]**

    (*b*) Scintillation counter **[MREC B.E. I 2002]**

    (*c*) G.M. counting system. **[Raj. B.E. I 2003]**

    (*d*) Proportional counter **[Raj. B.E. I 2004, 2006]**
11. With suitable diagram explain various components of a scintillation counting system.

    [RTU B.Tech. I Sem. 2008, RTU B. Tech. Back 2010]
12. What are advantages of proportional counter over G.M. counter?

    [RTU B.Tech. I Sem. 2008 & II Sem. 2011]
13. Give basic principle of Geiger-Müller counter. How quenching is achieved in Geiger-Müller counter? **[RTU B.Tech. II Sem. 2008]**

PROBLEMS

1. Calculate the pulse amplitude generated in the ionization chamber when 1 MeV charged particle loses all its energy. The circuit capacitance is 10^{-10} F. Assume that the energy required per ion pair is 35 eV. **[45.8×10^{-6} V]**
2. An ionisation chamber exposed to a beam of a-particle registers a current of 4.8×10^{-13} Ampere. On the average 20 α-particles enter the chamber per second. Assuming that in producing ion-pairs 35 eV per ion pair energy is needed, calculate the energy of the α-particle. **[5.25 MeV]**
[RTU B. Tech. II Sem. 2012]
3. In an air-filled ionization chamber, 15 α-particles enter the chamber from 5 MeV α-particle beam. If 35.2 eV energy is needed to produce an ion-pair in air. Calculate the ionization current. **[0.34×10^{-12} Amp.]**
4. An ionization chamber is charged to a potential 1000 volts. If its capacity be 50 *pf* by what percentage its charge would reduce in passing an α-particle producing 2×10^5 ion-pairs.
[RTU B. Tech. Back 2010]
5. β-particles emitted by a radioactive sample at the rate of 1000 disintegrations per minute enter into an air filled ionization chamber and travel a distance of 8 cm length. If in each cm it produces on average 900 ion pairs, calculate the charge collected on 5 *pf* condenser connected at the output. **[19.2×10^{-15} *C*]**
6. Determine the total number of ion-pair produced in a proportional counter when protons of 10 MeV energy enter into it. The gas multiplication factor in the proportional region is 1000.
If the current pulse time is 10 μs and the resistance between the electrodes is 10^4 Ω, calculate the height of the pulse. The energy required per ion pair is 34 eV. **[0.047 Volt]**
7. Calculate the electric field at the surface of the wire of a proportional counter with a wire radius 0.1 mm and cylinder radius 1 cm, when 2000 volt is applied between the two. **[4.34×10^4 Volt/cm]**
8. If a GM counter on an average 10^{10} electrons/counts are collected. If count rate is 10 per second, then find the ionisation current. **[1.6×10^{-4} Amp.]**
[RTU B. Tech. Back 2011]
9. A GM counter counts 944 counts per minute when 1000 charged particles are incident per minute on it. Determine the efficiency of GM counter. **[94.4 %]**
10. A GM counter reads 4000 counts per minute. If the dead time is 300 μs, calculate actual count rate. **[5200/minute]**
11. A GM tube collects 10^7 electrons in each discharge. What is the average current in the tube when it registers 1000 counts per minute? **[30 *p* Amp.]**
12. The efficiency of a GM counter is 90%. If it counts maximum 6000 counts/minute, then calculate the paralysis time of counter. **[1 ms]**
[RTU B. Tech. II Sem. 2012]
13. A GM counter works at 2000 V. The radii of wire and cylinder are 0.1 mm and 1.0 cm, respectively. The life of the tube is 10^{10} counts. Find the value of maximum radial electric field if the counter is operated 10 hours per day at the rate of 2000 counts per minute. What is the age of the counter? **[4.3×10^6 V/m, 2.28 Years]**

14. Calculate the scintillation efficiency of anthracene if 1 MeV of particle energy creates 20,300 photons with the average wavelength 447 nm. **[5.56 %]**

REFERENCES

1. J.Sharpe, Nuclear Radiation Detectors 2nd Ed. Methuen & Co. London, 1964.
2. G.F. Knoll, Radiation Detection and Measurement, New York, John Wiley & Sons 1979.
3. R.E. Lapp and H.L. Andrews, Nuclear Radiation Physics, 4th Ed. Englewood Cliffs, N.J. Prentice-Hall, 1972.
4. S.S. Kapoor and V.S. Ramamurthy, Nuclear Radiation Detectors, New Age International Publishers, New Delhi, 2001.
5. Singru, Experimental Nuclear Physics, Wiley Eastern Private Ltd., New Delhi, 1972.
6. W.H. Taif, Radiation Detection, Butter Worths, London, 1980.

EXPERIMENTS

0

GENERAL EQUIPMENTS

0.1 INTRODUCTION

The purpose of laboratory training is to develop a way of thinking that is somewhat different from the classroom one. It further develops a habit in the students to do things efficiently and regularly. The experiments in a course of engineering practical physics are designed to provide an opportunity to the students to familiarise themselves with various equipments which they read in theory and which find many applications in engineering , industry and everyday life.

0.2 HOW TO WRITE AN EXPERIMENT IN RECORD BOOK?

The record of each experiment should be made in a practical notebook according to a definite scheme as given below.

(*a*) Date on which the experiment is performed.

(*b*) The serial number of experiment.

(*c*) The object of experiment.

(*d*) The name of apparatus used in experiment.

(*e*) A brief theory of the experiment with formulae.

(*f*) Draw the block diagram or circuit diagram of the experiment.

(*g*) Observations.

(*h*) Calculation.

(*i*) Result.

(*j*) % error.

(*k*) Precautions.

(*l*) A brief discussion of the result with source of error, if any.

0.3 HOW TO WRITE THE TOTAL READING?

It can be written as

Total reading = Main scale reading + L.C. × Vernier or Circular scale reading (Coincide point of two scale)

Here L.C. is least count. It can be determined by

$$\text{L.C.} = \frac{\text{Value of smallest division on main scale}}{\text{Total number of divisions on vernier or circular scale}}$$

0.4 HOW TO CALCULATE THE % ERROR?

$$\% \text{ error} = \frac{\text{Standard value} \sim \text{Experimental value}}{\text{Standard value}} \times 100$$

0.5 VARIOUS IMPORTANT INSTRUMENTS TO MEASURE LENGTH, RADIUS & RADIUS OF CURVATURE

0.5.1 Vernier Callipers

A vernier callipers consist of a rectangular steel bar graduated in inches on one side and in centimetres on the other side as shown in Fig. 0.1. This is known as the main scale. Over this scale slides a small scale called the vernier scale.

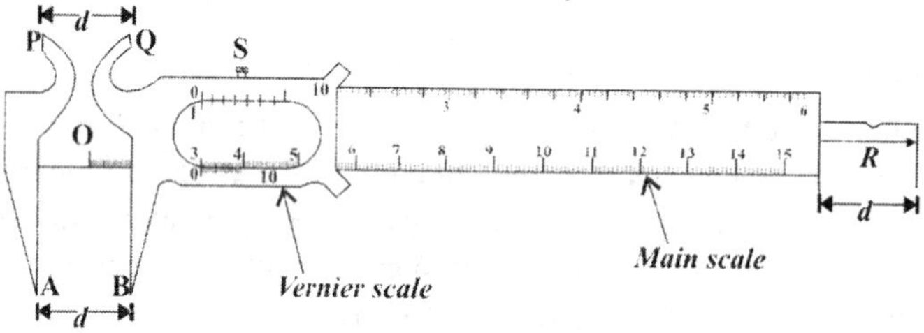

Fig. 0.1

The instrument has got two jaws, A and B. The jaw A is fixed at the end of the rectangular bar on the zero side, while the other jaw B is movable and can slide along the main scale. Each jaw is at right angles to the main scale and the movable jaw can be fixed at any position by the screw S. The two vernier scales are attached the movable jaw as shown. Usually when the two jaws are touching each other, the zero of the vernier scale coincides with zero of the main scale.

In some forms of the instrument the jaws project in the upper part as shown at P and Q. These projecting jaws are used to measure the internal diameters of the tubes. The movable jaw also carries a thin rectangular rod R which is used to measure the depth of vessels. As shown the distances AB, PQ and the length of projecting rod R are the same and can be measured on the scale.

To find the varnier constant (Least count)

(*i*) Fing the magnitude of the smallest division on the main scale.

(*ii*) Count the total number of divisions on the vernier scale.

(*iii*) Slide the movable jaw so that the zero mark (the first division) of the vernier scale coincides with any make of the main scale divisions.

(*iv*) Find the number of scale divisions, which coincide with the total number of vernier divisions. Thus least count or vernier constant

$$\text{L.C.} = \frac{\text{Value of smallest division on main scale}}{\text{Total number of divisions on vernier or circular scale}}$$

Therefore total reading can be written as

$$\text{T.R.} = \text{Main scale reading} + \text{least count} \times \text{vernier scale reading}$$

0.5.2 Screw Gauge

It is based upon the principle of a screw and consists of a U-shaped frame F having a fixed end at A. Through the other end B passes a fine and an accurately cut screw of a uniform pitch. A cap fits on to the screw and carries on its inner edge H, 50 or 100 equal division marks. This is called the head scale or circular scale and is used to measure the fraction of a revolution. The number of complete revolutions can be read on the main scale S which is graduated on the nut parallel to the axis of the screw. The reading of the circular scale is taken against a line known as the reference line. A screw gauge is used to measure accurately small thicknesses or diameters of the wires.

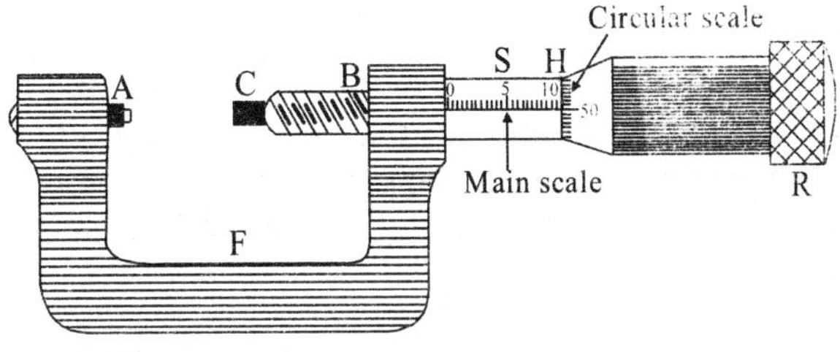

Fig. 0.2

In some screw gauges the screw head is provided with a ratchet arrangement R. When the studs A and C are in contact either with each other or with some other object placed in between, the ratchet slips over the screw cap without moving the screw forward. This avoids undue pressure between the studs or on the object.

To find the least count

To find the least count of a screw gauge, we first determine the pitch. Pitch of a screw is defined as the distance between two consecutive threads taken parallel to its axis and is measured by the distance through which the screw moves forward or backward when one complete rotation is given to the circular cap.

To find the pitch

(*i*) Rotate the circular scale H and coincide the zero mark with the reference line. Note the reading on the main scale.

(*ii*) Give four complete rotations to the circular scale and again note the reading on the main scale.

(*iii*) Calculate the pitch by the formula

$$\text{Pitch} = \left(\frac{\text{Distance travelled on the main scale}}{\text{Number of rotations}}\right)$$

The **least count** of the screw gauge is defined as the distance through which the screw moves backward or forward when the cap is rotated through one division on the circular scale.

To find the least count note the number of divisions on the circular scale H, then

$$\text{Least count} = \left(\frac{\text{Pitch of the screw}}{\text{Number of divisions on the circular scale}}\right)$$

Zero correction

If we bring the jaws A and C in contact without applying any undue pressure, the zero of the circular scale may not coincide with the reference line. In such a case there is a zero error. In some instruments the zero mark goes beyond the reference line, while in others it is left behind.

To find zero correction count the number of divisions on the circular scale by which the zero mark has advanced beyond or is left behind the reference line. Multiply this number with the least count. This is the zero correction. If the zero of the circular scale has advanced beyond the reference line the zero correction is positive but if it is left behind the reference line it is negative.

Diameter of the wire

To find the diameter of a wire insert the wire between the jaws A and C and turn the cap till the wire is just held between them without undue pressure. The ratchet stop R begins to turn freely when a contact has been established. Note the reading on the circular scale against the reference line. Multiply it by the least count and add it to the number of complete divisions visible on the main scale. Therefore total reading is T. R. = M. S. R. + (C. S. R. × L.C.)

If the pitch is half millimetre and there are 50 divisions on the circular scale, care should be taken to note whether the last uncovered divisions is less than half or more than half. If it is more than half, add 50 to the circular scale reading before multiplying it by the least count. Turn the wire through a right angle and note the reading again at the same point.

0.5.3 Spherometer

A spherometer is an instrument used for measuring very small distance. It also works on the principle of a micrometer screw. It is generally used for measuring the thickness of a thin plate and for determining the radius of curvature of spherical surfaces such as lenses or mirrors.

Construction: It consists of a metal framework supported on three fixed legs of equal length. The ends of the three legs lie at the corners of an equilateral triangle. An accurately cut screw works through a threaded hole at the centre of the framework. The screw terminates at the top into a milled head and carries a large graduated disc as shown in Fig. 0.3.

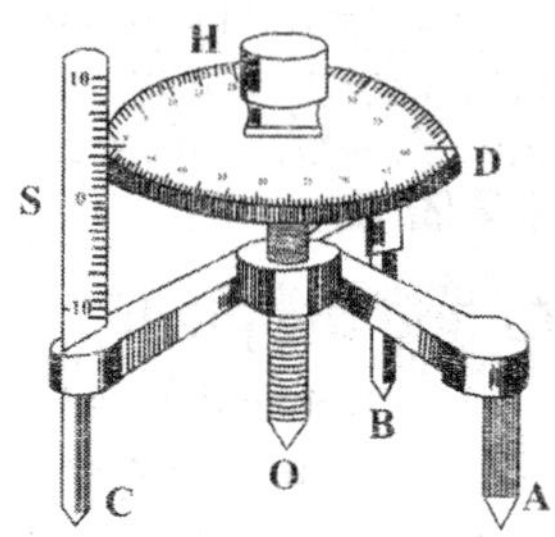

Fig. 0.3

The lower end of this screw forms the central leg of the instrument.

A small vertical scale marked in millimetres or half millimetres is fixed at one end of the frame with its graduations close to those on the circular disc. The edge of the circular disc is divided into a large number of equal divisions, generally 50 or 100.

Least count

The least count of a spherometer is the smallest distance that can be measured with it. It is the distance moved by the screw when it is turned through one division on the circular scale.

To find the least count of the spherometer

(*i*) Bring the zero of the circular scale against a division mark on the main scale.

(*ii*) Give four complete rotations to the screw and find the distance moved by it on the millimetre scale.

(*iii*) Divide this distance by 4 to get the 'pitch' of the screw.

(*iv*) Divide the pitch by the total number of divisions on the circular scale to get the 'least count' of the spherometer.

(*v*) Determine the radius of curvature (R) of the lens with the help of a spherometer using the formula

$$R = \frac{a^2}{6h} + \frac{h}{2}$$

where a is the distance between two legs of the spherometer and h is the height of the most bulged point of the lens from plane surface.

0.6 BACKLASH ERROR

In the screw gauge or a spherometer or any other instrument in which the principle of a screw is used to measure small distances, the accurately cut screw moves in a nut. There is always a misfit or play between the two. This misfit increases with constant use. Due to this play when the direction of motion of the screw is reversed axial motion of the screw does not take place until the screw has been rotated through a certain angle.

This lag between the linear motion of the screw and the circular motion given to its head is called the backlash error. This error can be avoided by turning the screw in the same direction while taking an observation. If the screw is to be rotated back it is first given a few rotations in the backward direction and then it is moved forward to the desired value. This eliminates the possibility of an error.

0.7 SOURCES OF LIGHT

Various sources of monochromatic and white light are employed in different optical experiments.

0.7.1. Bunsen's Flame

It is the simplest method of producing spectrum of various inorganic substances, *e.g.*, sodium, potassium, lithium, calcium, barium, strontium etc. A brush of asbestos is impregnated with a solution of the salt and then held in the flame. Owing to the high temperature the salt volatilizes and gives the characteristic spectrum of the metal.

0.7.2. Discharge Tube

It consists of a glass tube having two aluminums electrodes sealed at the two ends as shown in Fig. 0.4. The tube is filled with a gas at a pressure ranging between 5 mm and 10 mm of mercury.

Fig. 0.4 Discharge tube

When connected to the secondary of an induction coil or a high voltage transformer, a discharge passes and the gas gives its characteristic spectrum. The colour of discharge depends upon the nature of the gas, being red for neon, pink for helium and so on.

0.7.3. Sodium Vapour Lamp

It consists of a U-shaped glass tube L (Fig. 0.5) having two electrodes of tungsten in the form of a spiral coated with barium oxide enclosed inside a vacuum jacket V. It is filled with neon at a pressure of about 10 mm of mercury and some specks of metallic sodium are deposited on the sides of the tube. To start the discharge a potential of about 400 volts is required which is obtained from a step up transformer.

The lamps are available in 45 W, 80 W and 140 W powers. The discharge first passes through the neon gas which gives its characteristic red colour. The heat generated causes the metallic sodium to evaporate and gradually the tube is filled with the persistent golden yellow light of sodium. The sodium lamp is connected to the mains through a special transformer called the auto-leak transformer.

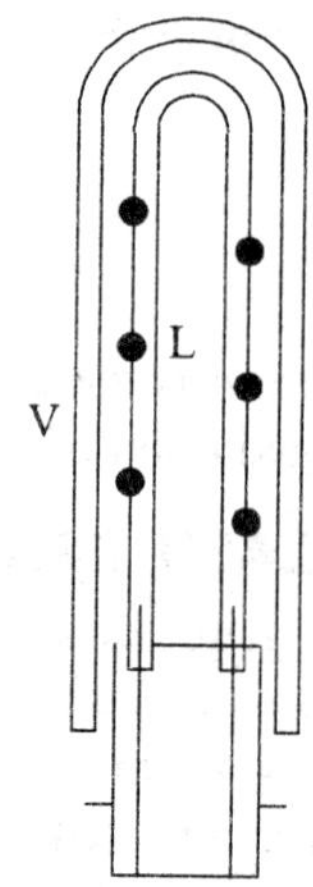

Fig. 0.5 Sodium vapour lamp

0.7.4. Mercury Vapour Lamp

The basic principle of mercury vapour lamp is the same as that of sodium vapour lamp. A high pressure lamp of the modern form essentially consists of an outer bulb A and an inner bulb B, the space in between being highly evacuated (Fig. 0.6). The inner bulb contains a small quantity of argon and mercury, two main electrodes E_1 and E_2 and a starting electrode S placed very close to E_1. The auxiliary electrode S is connected to the other electrode E_2 through a high resistance R of about 500 Ω. Each of the main electrodes E_1 and E_2 consists of a tungsten wire spiral containing a rod of barium compound which when heated gives out a copious supply of electrons.

When the lamp is switched on, full mains voltage is applied across the electrodes E_1 and S so that a discharge passes between the two, thereby starting the ionization of the gas. As the lamp warms up, the mercury gradually vaporizes and the pressure increases. The mercury discharge between E_1 and E_2 fills the whole tube. The argon discharge becomes effective because the resistance of the starting circuit is very high as compared to that of the parallel circuit. Such lamps are available in 250-500 watts capacity.

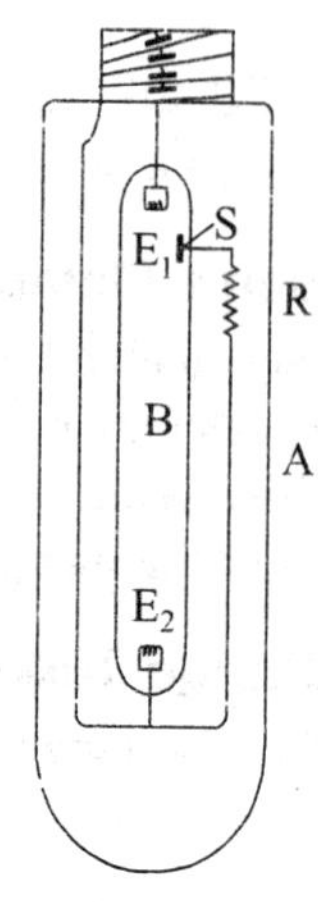

Fig. 0.6 Mercury vapour lamp

The lamp is worked on an A.C. mains with the help of a choke which serves as stabilizer. A condenser is also placed across the mains to improve the efficiency of the lamp. If the lamp is switched off it cannot be started again at once because mercury must condense before the starting electrode can function. This requires about 10 minutes after which the lamp can be started as usual.

0.8 CATHODE RAY OSCILLOSCOPE

The cathode ray oscilloscope (Commonly abbreviated as CRO) is an electron device which is capable of giving a visual indication of a signal wave form. With the help of an oscilloscope, the wave shape of a signal can be studied with respect to amplitude distortion and deviation from the normal. In addition, the oscilloscope can also be used for measuring voltage, frequency and phase shift.

In an oscilloscope, the electrons are emitted from a cathode accelerated to a high velocity, and brought to focus on a fluorescent screen. The screen produces a visible spot where an electron beam strikes. The electron beam can be deflected over the screen in response to the applied electrical signal. As electrons have negligible mass, therefore, they respond almost instantaneously when acted upon by an electrical signal and can trace almost any electrical variation no matter how rapid.

0.8.1 Block Diagram

The block diagram of a CRO is shown in (Fig. 0.7)

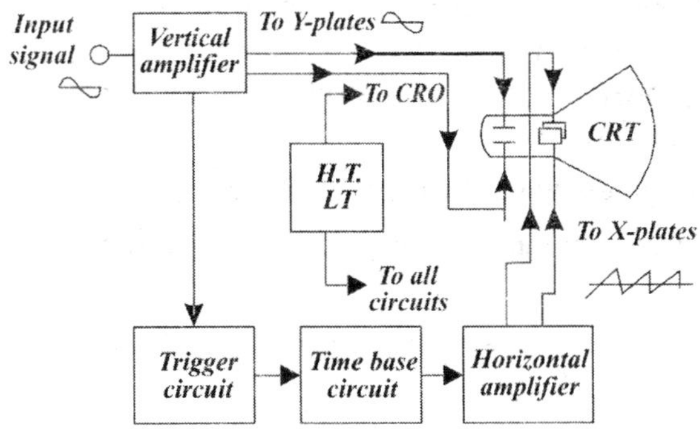

Fig. 0.7 Block diagram of CRO

As seen from the block diagram, an oscilloscope consists of the following scale system:

1. Cathode ray tube (CRT)
2. Vertical amplifier
3. Horizontal amplifier
4. Sweep generator
5. Trigger circuit
6. High voltage power supply
7. Low voltage power supply

0.8.2 Cathode Ray Tube

A cathode ray tube (commonly abbreviated as CRT) is a high vacuum tube of special geometrical shape which converts an electrical signal into a visual signal. A CRT provides plenty of electrons which are accelerated to high velocity and are brought to focus on a fluorescent screen. The electron beam produces a spot of light wherever it strikes the fluorescent screen.

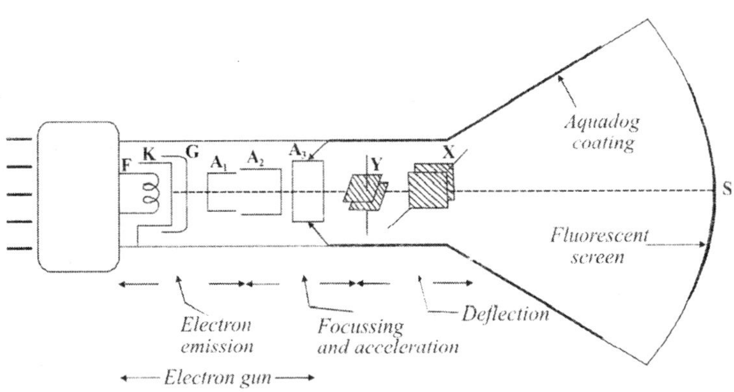

Fig. 0.8 Schematic diagram of a cathode ray tube

Construction

The schematic diagram of a cathode ray tube, showing its various parts, is shown in Fig. 0.8. A CRT is a self contained unit like any electron tube with a base through which leads are brought out for different pins. It consists of following four major components.

1. An evacuated glass envelop with a fluorescent screen which produces bright spot when struck by a high velocity electron beam.
2. An electron gun for producing a stream of electrons.
3. Focussing and accelerating anodes for producing a narrow and sharply-focussed beam of electrons.
4. Horizontal and vertical deflecting plates for controlling the path of the beam.

1

SEXTANT

OBJECT: To determine the height of an inaccessible object using a sextant.

APPARATUS: A sextant, a measuring tape (to mark the position of telescope from the wall or tower).

1.1 DESCRIPTION OF THE APPARATUS

1.1.1 Sextant

This is an instrument which enables us to determine the angle subtended by the two distant point objects at the place of observation. If the distance of one of the points from the sextant is known, the distance between two points can be calculated. It consists of a graduated arc AB connected to two fixed radial arms BC and AC which meet at a point C which is the centre of the circle of which the arc forms a part (Fig.1.1). The arc subtends an angle of 60° at the centre. But on the scale 120° are marked. Each degree on the scale is marked twice to make the instrument direct reading. If the mirror rotates through 5°, the scale reads 10°.

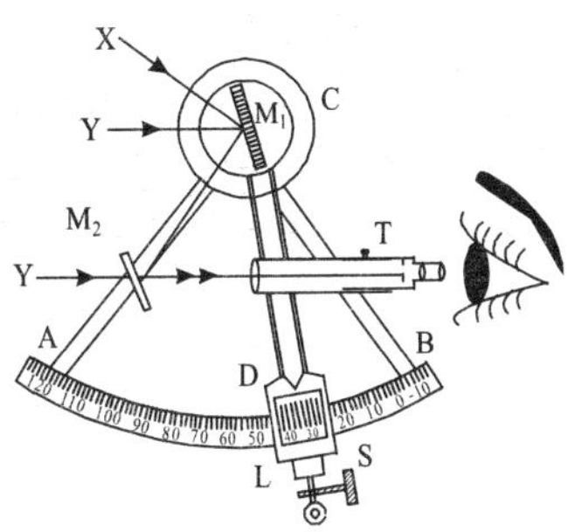

Fig. 1.1 Sextant

An *index arm* CD which is free to rotate round an axis passing through C carries a plane mirror M_1 called the *index glass* at the end C fixed in a plane perpendicular to the plane of the scale. At the other end of the arm at D there is vernier scale V which can slide along the main scale. A tangent screw S is provided for fine adjustments of the vernier.

A glass mirror M_2 called the horizon glass is fixed on the arm AC with its plane perpendicular to the plane of the scale in such a manner that when the index arm is at zero this is parallel to the mirror M_1. Lower half of this glass plate is silvered whereas the upper half is transparent. A telescope T fixed to the arm BC with its axis parallel to the plane of the scale and passing through the centre of the horizon glass.

A screw is provided at the side of the telescope (not shown) to move it in a plane perpendicular to the plane of the scale. There are screws fixed at the back of the mirrors M_1 and M_2 to set their planes perpendicular to the plane of the arc.

1.1.2 Working of Sextant

When the arm CD occupies the position CB, the planes of M_1 and M_2 are parallel to each other. Therefore, BC marks the zero position of the circular scale AB. A wooden handle is provided with the sextant to hold it in hand.

A stand is also provided on which the sextant can be fixed. In this position, if the telescope is pointed towards a distant object, two images will be seen of that object coinciding with each other. One image is seen through the transparent portion while the other is due to the rays which have been doubly reflected, once at M_1 and then at M_2. If the mirrors are parallel, the rays will be parallel and consequently the two images will coincide as shown in Fig. 1.2.

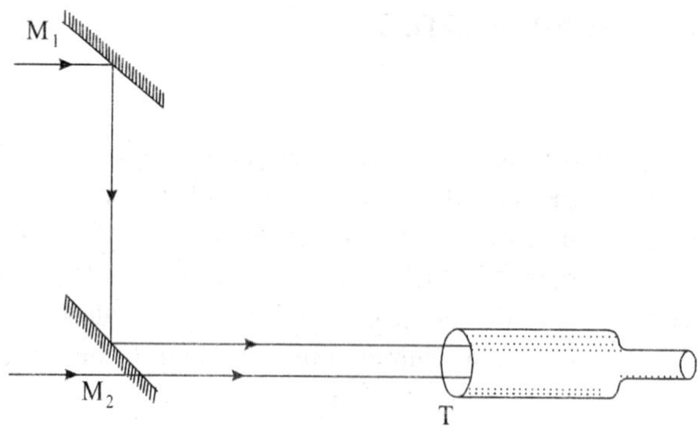

Fig. 1.2

Sextant is employed for determing the angular separation between two points or the angular elevation oı an inaccessible object. Angular separation then gives the height between the two points under consideration. If height of an inaccessible object is to be determined, a reference mark is made on the ground level below the object or at the eye level.

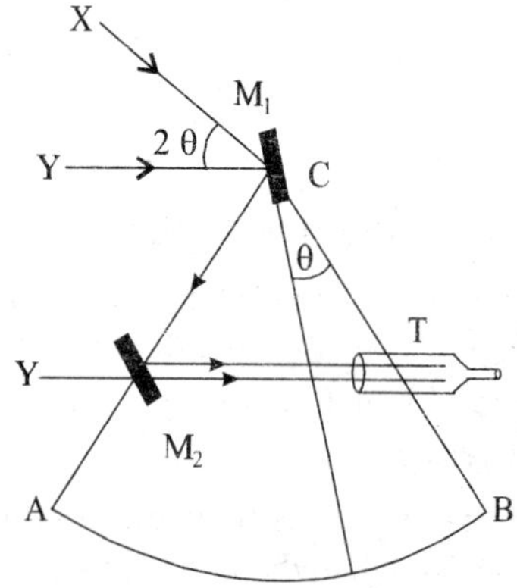

Fig. 1.3

If the telescope T is directed towards this mark, its image will be seen through the transparent portion of the plate M_2. Now the arm CD is moved so that M_1 moves and receives the rays coming from the high object. These rays will be first reflected at M_1 and then after getting reflected at M_2 enter the telescope and forms the image of the object. The arm CD is so adjusted that the image of the mark as seen directly through M_2 and the reflected image of the high object coincide.

Then the angular separation of the reference mark from the top will be the angle $\angle XCY$ (Fig.1.3). The angle θ through which the mirror is turned is measured on the scale. Naturally 2θ will be the angular separation.

1.2 THEORY

Let h be the height of the object at X. Let α and β be the angles subtended by the height of the object at two places R and S respectively. Let the distance between R and S be d, then by simple geometry we have

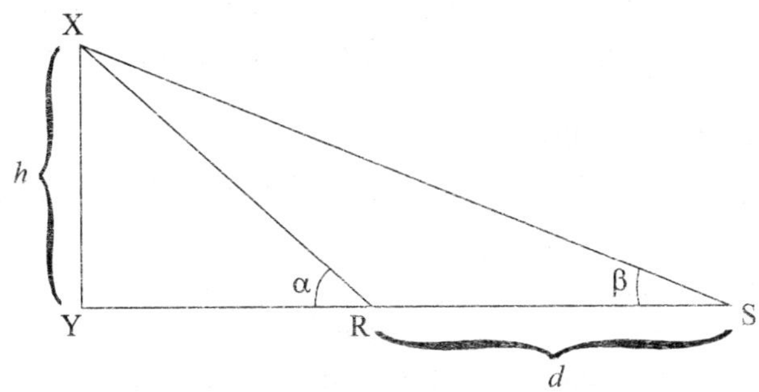

Fig. 1.4

$$h \cot \beta = YR + d \qquad (1.1)$$

$$h \cot \alpha = YR \qquad (1.2)$$

Subtracting Eq. (1.2) from Eq. (1.1.) and solving for h, we get

$$h = \frac{d}{\cot\beta - \cot\alpha} \qquad (1.3)$$

α and β are determined with the help of a sextant.

1.3 PROCEDURE

1.3.1 To Adjust the Sextant

(*i*) Bring the arm CD near the middle of the scale, and keeping your eye near the index glass, look in it for the front part of the scale. If this image is in the same plane as the scale, the plane of the index glass becomes perpendicular to the plane of the arc. If it is not so, move the screw provided at the back of the mirror, and make this adjustment.

(*ii*) Observe any object through the telescope and adjust it in such a way that its two images coincide at the centre of the field of view. Now tilt the sextant so that these images lie near the edge of the field of view and note if they still coincide. Tilt it in the opposite direction so that the images lie near the opposite edge. If the coincidence persists, the axis of the telescope is parallel to the scale. If it is not so, adjust the telescope by means of the screws provided.

(*iii*) Point the sextant towards a distant object. You will see two images one direct through M_2 and the other reflected one after reflection at M_1 and then at M_2. Adjust the movable arm so that the two images coincide. The image which moves by moving the movable arm is the reflected image. If the coincidence is perfect, the horizon glass is parallel to the index glass and perpendicular to the plane of the circular scale. If these two are not parallel, one image will lie sideways with respect to each other. To remove this defect move the screws provided at the back of M_2 so that M_1 and M_2 are set parallel to each other, and the two images coincide.

(*iv*) Determine the number of divisions marked on the circular scale. Usually they are sixty, and therefore, vernier constant of this scale will be one minute.

(*v*) Now determine the number of divisions marked on the vernier scale. They are usually four in number, therefore vernier constant of the instrument will be 15 seconds.

1.3.3 To Determine Zero Reading

(*i*) As described in theory, make a reference mark on the building or the ground level. Move away from the building by a considerable distance and select some place on the ground. Mark the position of this place. Let it be *S* (Fig. 1.4). Now point the telescope towards the reference mark. Direct image of the mark will be visible through the transparent portion of M_2. Move the arm *CD* and get the reflected image of the same reference mark. Coincide these two images. Read the vernier and the main scale. It gives zero reading at that place.

(*ii*) Next point the telescope towards the reference mark so that its image is clearly visible through the transparent portion. Now rotate the arm *CD* so that the rays from the high object fall on the mirror M_1 and in this way obtain the image of the object in the telescope. In this portion, you will see two images, (*a*) the direct image of the reference mark, and (*b*) the doubly reflected image of the object. Clamp the arm *CD* and by tangent screw, adjust *CD* so that these two images completely coincide on the cross wires. Read the vernier and the main scales. This gives the angular elevation of the high object with respect to the reference mark. Add or subtract from it the zero reading (as the case may be). It gives the angle α.

(*iii*) Now move away or towards the mark say through a distance of 5 to 10 m. Again find the zero reading at this place. Repeat the procedure described in step (*ii*) and determine the angular elevation β at this place.

(*iv*) Find out the distance between the two places with the help of a measuring tape. It gives *d*.

(*v*) Knowing α, β and *d*, determine *h* using Eq. (1.3). It will be height of the object from the reference mark. To get the actual height find out the height of the mark from the bottom and add it to *h*. This is one set.

(*vi*) To get another set of readings change *d* and take more sets and determine height *h* of the object.

(*vii*) Determine the mean height.

1.4 OBSERVATIONS

1. For least count of the instrument

   The smallest division on the main scale = 1°

   No. of divisions on the circular scale of the micrometer $x = 60$

   Least count of circular scale of the micrometer $= \left(\frac{1}{60}\right)^{\circ} = 1'$

   No. of divisions on the vernier scale of the micrometer $(y) = 5$

   Least count of the instrument $= 1^{\circ}/xy = \left(\frac{1}{300}\right)^{\circ}$

   $= \left(\frac{1}{5}\right)$ min = 12 sec

   Total reading $= \text{M.S.R.} + \left(\frac{1}{60}\right)^{\circ} \text{C.S.R.} + \left(\frac{1}{300}\right)^{\circ} \text{V.S.R.}$

2. Observation table for α and β

S. No.	Distance (*m*)	Zero Error (*a*)			Angular Elevation (*b*)			Angle of Elevation (*b-a*)	(β & α) for different distances
		M.S.R.	V.S.R.	Total (*a*)	M.S.R.	V.S.R.	Total (*b*)		
1	5							α =...	
2	10							β =...	for 5 m
3	15							β =...	for 10 m
4	20							β =...	for 15 m

3. Distance between two places (d) = ...m
4. Height of the reference mark from the ground (e) = ...m

1.5 CALCULATIONS

Elevation $\alpha = b - a =$ ....

Elevation $\beta = b - a =$ ....

Height of the object

$$h = \frac{d}{\cot\beta - \cot\alpha} = \text{............ m.}$$

∴ Height of the object from the reference mark (h) = ...m.

∴ Height of the object from the ground level = $h + e$ = ...m.

1.6 RESULT

The height of the given object = ...m

% Error = ...

1.7 SOURCES OF ERROR

1. The centre of the graduated arc and of the moving arm may not coincide.
2. The index mirror and the horizon mirror may not be parallel in zero reading.
3. The plane of the index mirror may not be perpendicular to the plane of the circular arc.
4. The axis of the telescope may not pass through the centre of the horizon glass.

1.8 PRECAUTIONS

1. The zero reading change from place to place. Therefore, it should be freshly determined at the place of observation.
2. The two mirrors M_1 and M_2 must be made parallel to each other and perpendicular to the plane of the arc with the help of levelling screws.
3. The axis of rotation of the mirror M_1 should lie at the centre of the graduated arc.
4. The reference mark and the two places of observation should lie in the same straight line.
5. The axis of the telescope should be parallel to the plane of the circular scale and must pass through the centre of the horizon glass.

QUESTIONS WITH ANSWERS FOR VIVA-VOCE

Q.1. Why is this instrument called a sextant?

Ans. It is called a sextant because its scale is only one-sixth of the circle, *i.e.*, its scale forms an arc of 60°.

Q.2. On what principle is the action of the sextant based?

Ans. The action of the sextant is based on the principle of rotation of a plane mirror. This principle states that when a plane mirror is rotated through an angle θ about an axis perpendicular to the plane of incidence, the reflected ray rotates through an angle 2θ, the direction of the incident ray remaining unchanged.

Q.3. Is the incident ray fixed here as mirror is rotated?

Ans. No, the incident ray is not fixed, but the reflected ray is fixed in this experiement. On account of the reversibility of light-path, the incident and reflected rays are interchangeable. Sextant incorporates this principle.

Q.4. What mechanical adjustments should the two mirrors satisfy?

Ans. These mirrors should satisfy two conditions. Firstly, their planes should be perpendicular to the plane of the arc. Secondly these glasses should be parallel to each other when there is no error in the instrument.

Q.5. What is the purpose of the screws attached to the back of the frames carrying mirrors M_1 and M_2?

Ans. With the help of these screws, these mirrors are made normal to the bed of the instrument.

Q.6. What is the function of light filters in this instrument?

Ans. They are used to cut down the intensity of light in case the object happens to be too dazzling.

Q.7. What type of telescope is fitted in your sextant? Does it give an erect or inverted image ?

Ans. This is a Galilean type of telescope. Its eye-piece has a concave lens and hence it gives an erect image.

Q.8. What is the total angle which can be read by this instrument?

Ans. The divisions are marked up to 120°. Hence an angle up to this magnitude can be measured.

Q.9. But the actual angle subtended by the arc appears to be an acute angle. Is it not so?

Ans. Yes, the actual angle is acute.

Q.10. How is this anomaly explained?

Ans. The values marked on the scale are doubled, so that 10° is written as 20°. The angle subtended by the object is twice the angle of rotation of the index glass. To record the angular elevation, therefore, the observed value should be doubled. To avoid this calculation, the manufacturers have written the values after doubling them and hence the instrument becomes direct reading.

Q.11. Why are two images seen when the sextant is pointed towards an object?

Ans. One image is formed by the rays directly entering the telescope through the transparent portion of the horizon glass and the second by those rays which enter the telescope after reflections from the index glass and the horizon glass.

Q.12. What do you mean by zero error in a sextant?

Ans. The instrument is manufactured in such a way that when the glasses are exactly parallel the zero of the vernier should coincide with the zero of main scale. If it does not, the instrument is said to have a zero error.

Q.13. How this zero error determined?

Ans. This can be determined by pointing the sextant towards a distant object, and by coinciding the direct image with the image formed by double reflection at the two glasses. If the reading of the instrument is zero, there is no zero error. If not, the reading is noted which is the required zero error.

Q.14. Why coloured glasses are provided with the sextant?

Ans. The sun or any other brilliant object gives intense glare in the telescope. Hence to diminish the intensity of light entering the telescope, these coloured glasses have been worn in the path of light.

Q.15. Does the zero error depend on the distance of the object?

Ans. Yes, it depends upon the distance of the object. Nearer the object, greater is the divergen ce of the beam received from it and thus greater is the value of zero error. In other words, zero error is inversely proportional to the distance of the object.

Q.16. How are the two points of observation chosen?

Ans. The two points of observation are chosen in such a way that they are collinear with the reference mark.

Q.17. To what practical uses the sextant is put?

Ans. Sextant is used by navigators for measuring the angular altitudes of the sun and stars and thus they locate their exact position on the earth. It is also used to determine the vertical and horizontal distances between terrestrial objects.

2

DISPERSIVE POWER

OBJECT: To determine the dispersive power of the material of a prism by spectrometer.

APPARATUS: A spectrometer, a spirit level, a glass prism, an eye-piece and source of light (mercury lamp).

2.1 DESCRIPTION OF APPARATUS

2.1.1 Spectrometer

It is a compact apparatus for obtaining a pure spectrum. It is used for spectral studies and for finding the refractive index of a material in the form of a prism. It consists of the following main parts (Fig. 2.1.)

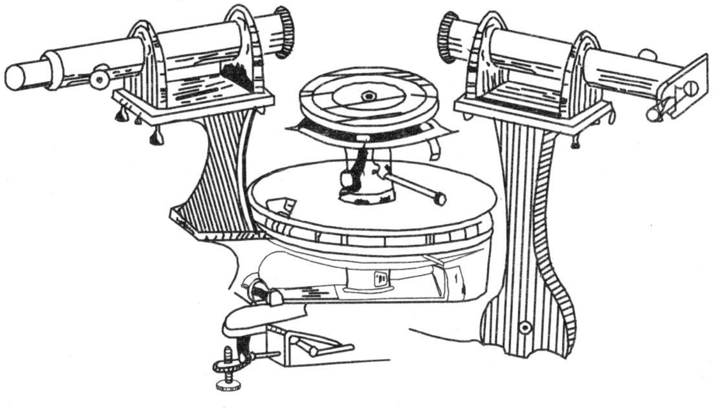

Fig. 2.1 Spectrometer

2.1.2 The Collimator

The purpose of collimator (C) is to produce a parallel beam of light. It consists of a tube mounted horizontally on the arm of the spectrometer. The tube has a converging achromatic lens at one end and a sliding tube having an adjustable vertical slit at the other end. The focal length of the lens is almost equal to the length of the collimator tube. The distance between the slit and the lens can be altered by a rack and pinion arrangement or by sliding the inner tube to obtain parallel rays. The tube rests on two screws C_1 and C_2 by which it can be slightly tilted up or down if necessary.

The vertical slit consists of two sharp edges. One of the edges is fixed while the other can be moved parallel to it with the help of a screw S_1 provided at its side.

2.1.3 The Telescope

It is an astronomical telescope (T) having an achromatic objective and a Ramsden eye-piece. It is mounted on another arm fixed rigidly to the circular scale graduated in half degrees. The telescope along with the scale can be turned round a vertical axis passing through the centre of the spectrometer. It can be fixed in any position by the screw T_1 and then can be given a slow rotation or fine adjustment by a tangential screw T_2. The position of the telescope can be read by the two verniers V_1 and V_2, 180° apart fixed to the prism table. The telescope rests on the screws T_3 and T_4 by which it can be slightly tilted up or down for adjustment of its axis, if necessary. A rack and pinion arrangement is provided on the side of the telescope for focusing.

2.1.4. The Prism Table

It consists of an upper plate A and a lower plate B separated by three springs through which pass the levelling screws P, Q and R. A set of parallel equidistant lines are engraved on the upper plate as shown if Fig. 2.1. The lines are parallel to the lines joining any two of the screws, say P and Q. The prism is always placed with one of its reflecting faces perpendicular to these lines. A series of circles concentric with the axis of rotation of the prism are also ruled on this plate. The height of the prism table can be adjusted by a clamping screw A_1 which fixes the prism table to the verniers V_1 and V_2. The table can rotate about a vertical axis and may be fixed in any desired position by means of the screw A_2. When fixed by the screw A_2 it can be turned very slowly by a tangential screw A_3 placed at the base of the spectrometer. The position of the table can be accurately read by the verniers moving on the circular scales.

2.2 ADJUSTMENTS OF THE SPECTROMETER

Before starting work with the spectrometer, it requires some preliminary adjustments without which readings cannot be taken with it. The necessary adjustments are given below and these should be performed in the given order.

2.2.1 Levelling of Spectrometer

(*i*) A spirit level is placed on the graduated circular scale parallel to the line joining any two levelling screws provided at the base say L_1 and L_2. Adjust by turning the screws in opposite direction till the air bubble is in the centre. Now place the spirit level perpendicular to this position and adjust the third screw L_3 if necessary.

(*ii*) Place the spirit level on the tube of telescope and adjust the screws T_3 and T_4 to make it horizontal.

(*iii*) Now place the spirit level on the collimator tube and adjust its levelling with the help of the screws C_1 and C_2.

(*iv*) **Levelling of Prism Table**

The prism table has three separate levelling screws attached to its base. The ruled lines on its surface are parallel to a line joining two of these screws. The spirit level is placed parallel to these lines and the air bubble is brought to the centre by turning the screws lying at the ends of the parallel lines, in opposite directions. Then the spirit level is placed perpendicular to the lines again and the bubble is brought to the centre by turning the third screw. This makes the prism table top horizontal.

2.2.2. Adjustment of the Telescope

(*i*) **Focussing of the Cross-wire:** The telescope is turned towards an illuminated source and when viewed through the eye-piece, a blurred image of the cross-wire will be seen. The eye-piece is adjusted till the cross-wire can be seen distinctly without any strain on the eyes. The cross-wires should be kept horizontal-vertical.

(*ii*) **Focussing for Parallel Rays**

(*a*) **Direct Method:** In order to focus the telescope and the collimator for parallel rays by direct method, the telescope is turned towards a distant object whose image is focussed on the cross-wire without any parallax. The collimator and the telescope are brought in a straight line and the image of the slit attached to the collimator is sharply focussed with the help of the screw attached to the collimator. There must not be any parallax between the image of the slit and the cross-wire. Thus in the first arrangement, the telescope is focussed for parallel rays, while in the second operation, the collimator produces a beam of parallel rays.

(*b*) **Schuster's Method:** The prism is set on prism table with its refracting edges vertical and the centre of its base lines at the centre of the table. The refracting face AC is turned towards the collimator and through the face AB, the image of the slit is viewed (Fig. 2.2).

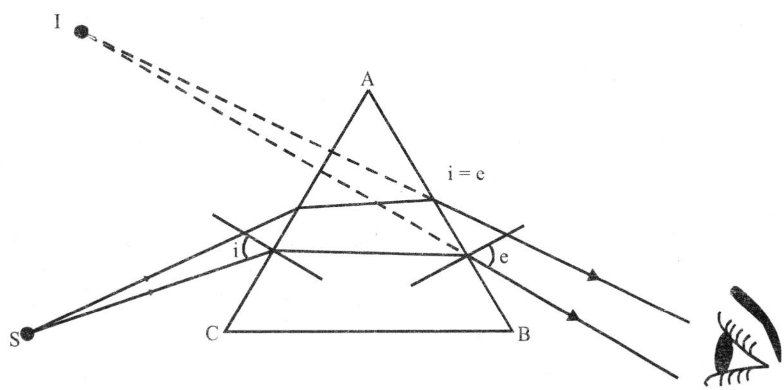

Fig. 2.2 Position of minimum deviation

With the eye fixed on the slit image, the table with the prism is slowly rotated in the appropriate direction so that the image (I) is brought as far as possible near the collimator. From this position, in whatever direction the prism is turned, the image (I) will move away from the collimator. When the image is at I (Fig. 2.2), then the prism is at the minimum deviation position and at this position, the angle of incidence ($< i$) on the prism is equal to the angle of emergence ($< e$) of the ray.

When the prism is brought to the position A_1 B_1 C_1 by rotation, the angle of incidence ($< i$) of the ray on it is greater than the angle of emergence ($< e$). This position of the prism is known as the '*slant position*' (Fig. 2.3). In this position, the distance of the image I_1, formed is further away from the position of the image formed at the minimum deviation position. The image I_1 also appears blurred. It is focussed sharply with the help of the screw attached to the telescope. The image at this position appears to be thin.

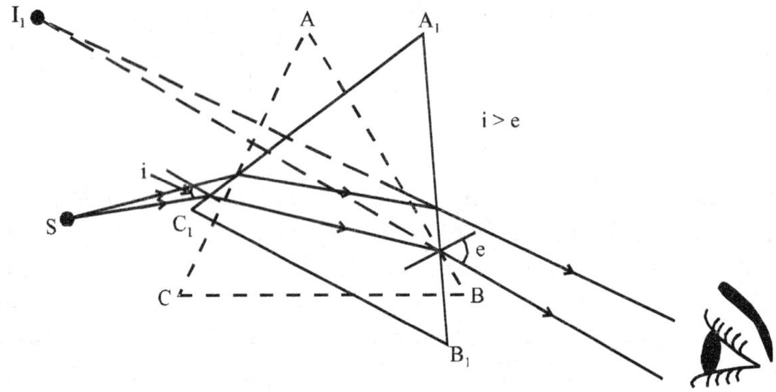

Fig. 2.3 Slant position of prism

Next the prism table is rotated on the other side and slit image is again viewed through the telescope. This time the image I_2 appears to be broad (Fig. 2.4). In this position $(A_2B_2C_2)$, the angle of incidence $(<i)$ is less than the angle of emergence $(<e)$ and the position is known as the '*normal position*'.

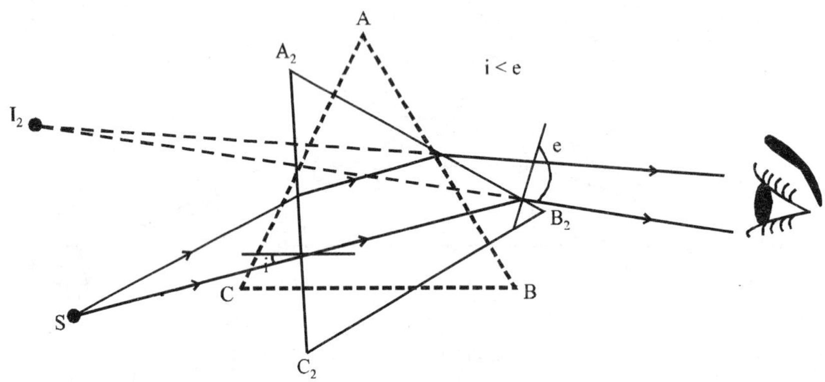

Fig. 2.4 Normal position of the prism

The distance of the image I_2 formed is less than that of the image formed at the minimum deviation position. Again this image also appears blurred and it is focussed with the help of the screw attached to the collimator. The focussed image appears to be broad in size. If the prism table is again rotated in the opposite direction to keep the prism in slant position, the image again appears thin and out of focus. As before, the screw attached to the telescope is operated to focus the image. In general, it will be necessary to repeat these operations several times in succession till the image appears in sharp focus at both positions. Finally the image is formed at infinity and the spectrometer is focussed for parallel rays. In the final focussed position, there must not be any parallax between the image and the cross-wires.

2.2.3. Adjustment of Collimator

First the spectrometer is placed with collimator slit near the window of the source so that slit image can be seen very clearly with brightness when viewed through the telescope and the collimator. Once the position of the source is adjusted, it should not be disturbed further. The opening of the slit is then adjusted by means of screws attached to it. Care should be taken that all the portions of the slit get equal illumination from the source. This can be checked by viewing the slit image through the collimator. If necessary, the slit image can be properly focussed with the help of the rack and pinion arrangement attached to the collimator tube.

2.2.4. Adjustment of the Slit

The telescope is turned and brought in line with the collimator. When viewed through the telescope, generally a blurred image of the slit is observed. The focussing screw of the telescope and of the collimator should be turned until the image of the slit appears in sharp focus. The width of the slit opening should not be very large and the boundary edges of the slit should be sharply defined when viewed through the telescope.

2.3 SOURCES OF LIGHT

Mercury vapour lamp is used as a source of white light.

2.3.1. Mercury Vapour Lamp

The basis principle of mercury vapour lamp is the same as that of sodium vapour lamp. A high pressure lamp of the modern form essentially consists of an outer bulb A and an inner bulb B, the space in between being highly evacuated (Fig. 2.5). The inner bulb contains a small quantity of argon and mercury, two main electrodes E_1 and E_2 and a starting electrode S placed very close to E_1. The auxiliary electrode S is connected to the other electrode E_2 through a high resistance R of about 500 Ω. Each of the main electrodes E_1 and E_2 consists of a tungsten wire spiral containing a rod of barium compound which when heated gives out a copious supply of electrons.

When the lamp is switched on, full mains voltage is applied across the electrodes E_1 and S so that a discharge passes between the two, thereby starting the ionization of the gas. As the lamp warms up, the mercury gradually vaporizes and the pressure increases. The mercury discharge between E_1 and E_2 fills the whole tube. The argon discharge becomes effective because the resistance of the starting circuit is very high as compared to that of the parallel circuit. Such lamps are available in 250-500 watts capacity.

The lamp is worked on an A.C. mains with the help of a choke which serves as stabilizer. A condenser is also placed across the mains to improve the efficiency of the lamp. If the lamp is switched off it can not be started again at once because mercury must condense before the starting electrodes can function. This requires about 10 minutes after which the lamp can be started as usual.

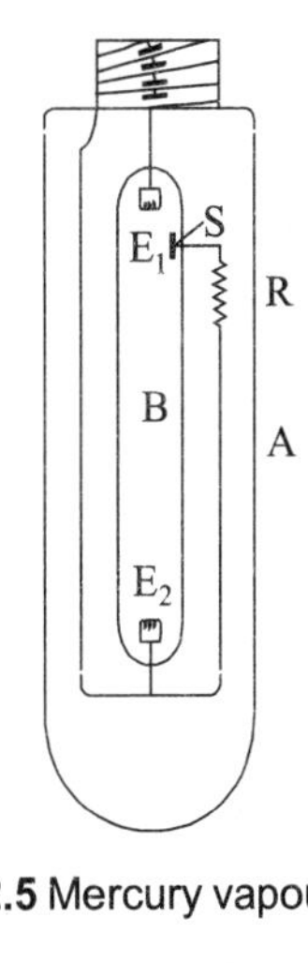

Fig. 2.5 Mercury vapour lamp

2.4 THEORY

2.4.1 Angular Dispersion

When a ray of white light passes through a prism it splits into its constituent colours.

This phenomenon of splitting up of white light into its constituent colours is called dispersion.

The refractive index of the material of prism is given by

$$\mu = \frac{\sin(A+\delta_m)/2}{\sin A/2} \tag{2.1}$$

where A is the angle of prism, δ_m is the angle of minimum deviation. For a small angled prism, the angle of deviation is also small and almost constant. Therefore, for a small angled prism

$$\mu = \frac{(A+\delta_m)/2}{A/2} \qquad (\because \sin\theta = \theta \text{ when } \theta \text{ is small})$$

or $$\delta = (\mu - 1)A \tag{2.2}$$

As refractive index for light of red colour is smaller than that for light of violet colour, the angle of deviation for red colour is the least and deviation for violet colour is maximum. For other colours of the spectrum the deviation lies in between.

The total angle through which the colours of a spectrum are spread is a measure of dispersion. *Hence angular dispersion is defined as the difference between the deviations produced in the extreme colours of the spectrum.*

If δ_v and δ_r are the deviations for violet and red colours respectively then, angular dispersion

$$\alpha = \delta_v - \delta_r$$

$$= (\mu_v - 1)A - (\mu_r - 1)A$$

$$= (\mu_v - \mu_r)A \tag{2.3}$$

2.4.2 Mean Deviation

The mean deviation is defined as the deviation produced for the mean colour of the spectrum. The mean colour is generally taken as the yellow colour and the mean refractive index as the refractive index for yellow light μ_y or the average of the refractive indices of the two extreme colours μ_v and μ_r.

Mean deviation $\delta = (\mu_y - 1)A = (\mu - 1)A$

where $$\mu = \frac{\mu_v + \mu_r}{2} = \mu_y \tag{2.4}$$

2.4.3 Dispersive Power

The dispersive power of the material of a prism is defined as the ratio of angular dispersion to mean deviation.

$$\text{Dispersive power } \omega = \frac{(\mu_v - \mu_r)A}{(\mu_y - 1)A}$$

or
$$\omega = \frac{(\mu_v - \mu_r)}{(\mu_y - 1)} \tag{2.5}$$

It is clear from the above expression that dispersive power depends upon the nature of the material of the prism and is independent of the angle of the prism.

2.5 PROCEDURE

2.5.1 Adjustment of the Spectrometer

The necessary mechanical and optical adjustments of the spectrometer for obtaining pure spectrum are made as described in Article 2.2. Once these adjustments are completed, they should not be disturbed through out the experiment.

2.5.2 Determination of Angles of Minimum Deviation for Various Lines

(*i*) Illuminate the slit of the collimator by means of light from a mercury lamp. The lamp should be along the axis of the collimator.

(*ii*) Put the prism symmetrically at the centre of the prism table with its ground face away from the collimator and set the telescope at about 45° to the axis of the collimator. Now turn the prism table towards left or right such that the ground face of the prism moves towards the collimator and becomes almost parallel to the line joining the collimator and telescope lenses. On turning the telescope a little, the spectrum will be seen.

(*iii*) Slowly rotate the prism table such that the spectrum moves nearer and nearer to the collimator axis (in this case the angle of deviation will go on decreasing). Move the telescope also to keep the spectrum in the field of view. A stage comes when the spectrum just starts returning back. This is minimum deviation position. By careful movement of the prism table ascertain this position. Stop at this stage and clamp the prism table (Fig. 2.6).

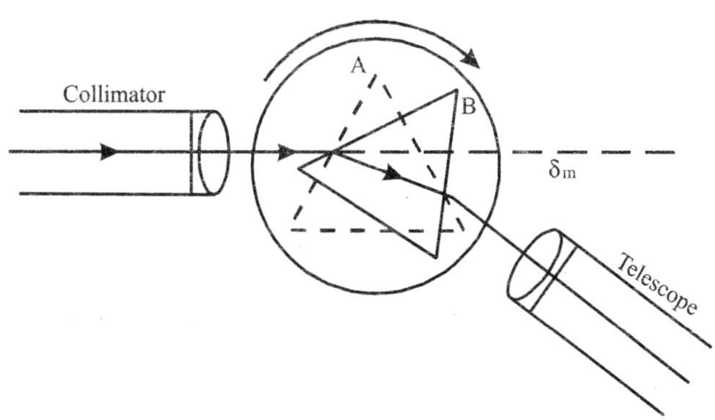

Fig. 2.6

(*iv*) Now turn the telescope and set its cross-wire on the red line (6234Å) of the spectrum. Clamp the telescope and use its tangent screw to bring the cross-wire exactly on this line. Note the readings of both the verniers.

(*v*) In a similar manner set the cross-wire of the telescope on the yellow (either of the doublet 5790Å, 5770Å) and violet (4047Å) lines respectively of the spectrum. You can test for the minimum deviation position for each line. (It will almost be the same). Thus for each line note the position of the telescope on both the verniers.

(*vi*) Now, without clamping the prism table remove the prism, release the telescope and turn it to obtain the direct image of the slit on the cross-wire. Note the readings of both the verniers.

(*vii*) By taking the difference of the direct image reading and the minimum deviation position readings of the various spectral lines for repective verniers, find the value of angle of minimum deviation δ_m for each line.

(*viii*) Turn the telescope to the other side, put the prism on the table to get spectrum on this side. As in previous case obtain the minimum deviation position on this side also. As before, determine the mean value of δ_m for various lines.

(*ix*) Using the values of δ_m obtained in steps (*vii*) and (*viii*) find the mean value of δ_m for each line.

2.5.3. Determination of the Angle of the Prism

(*i*) Put the prism on the prism table with its refracting edge at the centre and ground face nearly perpendicular to the collimator axis as shown in Fig. 2.7. Clamp the prism table firmly.

(*ii*) Reflected light from each face of the prism gives rise to an image of the slit. These are to be received on the cross-wire of the telescope in turn. For this see the reflected light from the face on the right side. Turn the telescope to receive this light. An image of the slit will be seen in the field of view. Clamp the telescope and using tangent screw set the cross-wire on this image. Note the readings of both the verniers.

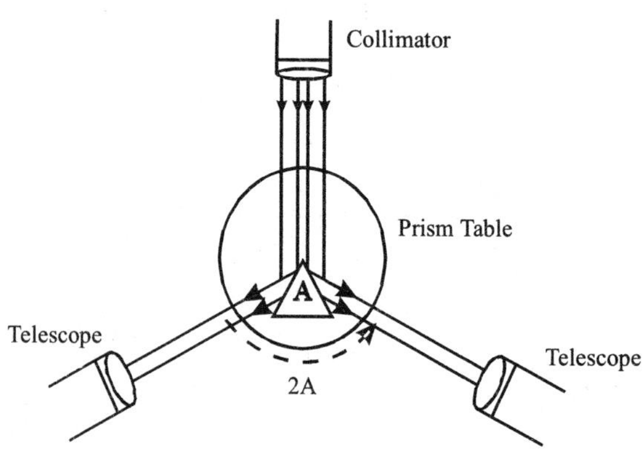

Fig. 2.7

(*iii*) Next see for a similar image on the left side formed by reflection of light from the other face of the prism. Fix the cross-wire on this image and note the readings of both the verniers.

(*iv*) The difference of the readings of the respective verniers taken in setps (*ii*) and (*iii*) will give angle 2 A, where A is the refracting angle of the prism.

(*v*) Disturb the setting of the prism, reset it and repeat the observations of steps (*ii*) to (*iv*) to find 2 A. Repeat the process several times and hence find the mean value A of the refracting angle of the prism.

2.6 OBSERVATIONS

Value of one division of main scale (x) =

Number of divisions on vernier scale (n) =

Least count of the vernier = $\frac{x}{n}$ = ...

Total reading = M.S.R. + L.S.R. × V.S.R.

[A] Determination of the angle of minimum deviation (δ_m)

S. No.	Colour	Vernier	Reading in the minimum deviation Position			Reading in the Position of direct image			δ_m = (a~b)	Mean δ_m
			M.S.R.	V.S.R.	Total Reading (a)	M.S.R	V.S.R.	Total Reading (b)		
1	Red	V_1 V_2								
2	Yellow	V_1 V_2								
3	Violet	V_1 V_2								

[B] Determination of the angle of prism (A).

S. No.	Vernier	Reading for reflection from face *AB*			Reading for reflection from face *AC*			Difference (a~b)	Mean Value 2A	Mean Value A
		M.S.R.	V.S.R.	Total Reading (a)	M.S.R.	V.S.R.	Total Reading (b)			
1	V_1 V_2									
2	V_1 V_2									

2.7 CALCULATIONS

Using the value of δ_m determined above for various lines of mercury spectrum and the value of refracting angle A, calculate the refractive indices of the material of the prism for each line with the help of the formula given in Eq. (2.1).

Using the above calculated values of the refractive indices for Red, Yellow and Violet lines, compute the dispersive power of the material of the prism with the help of formula given in Eq. (2.5).

(*i*) For red (6234 Å) line

$$\delta_{mr} = \ldots\ldots\ldots \qquad A = \ldots\ldots\ldots$$

$$\mu_r = \frac{\sin\dfrac{A+\delta_{mr}}{2}}{\sin\dfrac{A}{2}}$$

(*ii*) For yellow (5770 Å) line

$$\mu_{my} = \ldots\ldots\ldots \qquad A = \ldots\ldots\ldots$$

$$\therefore \mu_y = \frac{\sin\dfrac{A+\delta_{my}}{2}}{\sin\dfrac{A}{2}}$$

(*iii*) For violet (4047 Å) line

$$\mu_{mv} = \ldots\ldots\ldots \qquad A = \ldots\ldots\ldots$$

$$\therefore \mu_v = \frac{\sin\dfrac{A+\delta_{mv}}{2}}{\sin\dfrac{A}{2}}$$

(*iv*) Dispersive power of the material of the prism.

$$\omega = \frac{\mu_v - \mu_r}{\mu_y - 1}$$

2.8 RESULT

The refractive indices of the material of the prism for prominent lines of mercury spectrum are as follows:

$$\mu_r = \ldots\ldots\ldots\ldots\ldots\ldots$$

$$\mu_y = \ldots\ldots\ldots\ldots\ldots\ldots$$

$$\mu_v = \ldots\ldots\ldots\ldots\ldots\ldots$$

Dispersive power ω = ...

Standard Value for the crown glass $(\omega) = 0.015$

For flint glass $(\omega) = 0.024$

Error =.........%

2.9 SOURCES OF ERROR

1. The slit may not be perfectly vertical.
2. The edge of the prism may not be accurately vertical.
3. The cross-wire may not be properly focussed at the slit.
4. The axis of rotation of the telescope may not pass through the centre of the circular scale.

2.10 PRECAUTIONS

1. The axis of the telescope, the collimator and the plane of the prism table should be horizontal.
2. The position of the eyepiece should be adjusted so that the cross-wires are clearly visible without any strain.
3. The telescope should be focussed for infinity and the collimator should be adjusted to give a parallel beam of light.
4. The prism should be placed with its centre coinciding with the centre of the prism table when finding the angle by rotating the telescope.
5. The prism table should be levelled so that the reflected image from both the reflecting faces lies symmetrically with respect to the horizontal cross-wire.
6. Both the verniers should be read to avoid error due to nonconcidence of the centre of the circular scale with the axis of rotation of the telescope or the table.
7. The position of minimum deviation should be noted when the image of the slit remains stationary even when the tangent screw of the table is given one rotation either way.
8. The circular scale table should be kept fixed during the experiment. Only the prism table should be rotated to change the observation.

QUESTIONS WITH ANSWERS FOR VIVA-VOCE

Q.1. What do you mean by dispersion of light?

Ans. When a ray of white light passes through a prism, it is separated into rays of its constituent colours. This phenomenon is known as dispersion.

Q.2. Can any other device also disperse the light?

Ans. Yes, the diffraction grating also disperses white light.

Q.3. What is meant by dispersive power of a material?

Ans. The dispersive power of a material is its ability to disperse various colours constituting the incident light.

Q.4. On what factors does it depend?

Ans. It depends on the nature of material and on the wavelength for which it is determined.

Q.5. What is meant by angular dispersion?

Ans. Angular dispersion for any two colours (wavelengths) is the difference between the deviations of those wavelengths (colours).

Q.6. What is the best source of light for determining the dispersive power?

Ans. Neon lamp is the best source for this purpose because it gives spectral lines in all the regions of the visible spectrum.

Q.7. How does a ray pass through a prism at minimum deviation?

Ans. Inside the prism the ray passes parallel to the base of the prism.

Q.8. How does the deviation of a ray vary with its angle of incidence?

Ans. The deviation is minimum at a particular angle of incidence but it increases both for greater and smaller angles of incidence than that particular value.

Q.9. Does the deviation depend on the colour of light?

Ans. Yes, the deviation is less for red than for violet light.

Q.10. What is the condition for obtaining minimum deviation?

Ans. The deviation is minimum when the angles of incidence and the emergence are equal.

Q.11. Does the deviation depend on the angle of the prism?

Ans. Yes, greater the angle of the prism, more is the deviation.

Q.12. Why are two verniers provided with the telescope?

Ans. To avoid the error which may arise due to non-coincidence of the axis of rotation of the telescope with the centre of the circular scale.

Q.13. What kind of image is produced by the telescope?

Ans. It is a virtual image formed at infinity.

Q.14. Is there any limit for the value of the angle of the prism?

Ans. Yes, if the refracting angle of the prism exceeds twice the critical angle for the material of the prism and air, no ray will emerge out of it.

Q.15. Do the light rays of different colours travel with the same velocity while passing through the prism?

Ans. No, since different colours have different values of refractive index for the same prism, they travel with different velocities through the prism. Violet rays travel with less velocity than that of red.

Q.16. Does the deviation depend on the length of the base of the prism?

Ans. No, it is independent of the length of the base.

Q.17. Then, why, some prisms are made of larger base?

Ans. Only to increase their resolving power.

Q.18. On what factors does the value of μ depend?

Ans. It depends on the nature of the pair of media and also on the colour (wavelength) of light used.

Q.19. When a beam of light falls on one face of the prism, is the whole of it is refracted?

Ans. No, although the major portion of incident light is refracted some fraction is reflected and some is absorbed at the face.

Q.20. Will the angle of minimum deviation change, if the prism is immersed in water?

Ans. Yes. The refractive index of glass in water is less than that in air. Hence the angle of minimum deviation is less.

Q.21. What do you mean by pure spectrum?

Ans. A spectrum in which there is no overlapping of colours is known as pure spectrum. Each colour occupies a separate and distinct position.

Q.22. How does refractive index vary with wavelength?

Ans. Higher is the wavelength, smaller is the refractive index.

Q.23. What is the working temperature of a mercury lamp?

Ans. The working temperature of mercury lamp is about 600°.

Q.24. What is an eye-piece?

Ans. Eye-piece is a magnifier designed to give more perfect image than obtained by a single lens.

Q.25. Which eye-piece is used in the telescope of a spectrometer?

Ans. Ramsden's eye-piece.

Q.26. What is the construction of Ramsden's eye-piece?

Ans. It consists of two plano-convex lenses each of focal length f separated by a distance equal to $2f/3$.

Q.27. What is the construction of Huygen's eye-piece?

Ans. It consists of two plano-convex lenses, one having focal length $3f$ and the other with focal length f and separated by a distance $2f$.

Q.28. What is the reason for which Ramsden's eye-piece is used with a spectrometer?

Ans. In this eye-piece, the cross-wire is outside the eye-piece and hence mechanical adjustment and measurements are possible.

Q.29. Out of the prisms of flint and crown glasses, which one will you prefer to use?

Ans. We shall prefer a prism of flint glass because it gives greater dispersion.

Q.30. Without touching can you differentiate between microscope and telescope?

Ans. The objective of microscope has small aperture while that of telescope has a large aperture.

Q.31. What is normal dispersion?

Ans. If the refractive index μ increases with the decrease of the wavelength the dispersion is said to be normal.

Q.32. What is anomalous dispersion?

Ans. If the refractive index is higher for longer wavelengths, the dispersion is anomalous.

Q.33. What is normal spectrum?

Ans. When the rate of change of deviation with wavelength of light $(d\delta/d\lambda)$ is the same for all parts of the spectrum, the spectrum is said to be normal.

Q.34. What kind of image is produced by the telescope in this experiment?

Ans. It is the virtual image formed at infinity.

Q.35. Do the light rays of different colours travel with the same velocity in air?

Ans. In air, or more strictly in vacuum, light rays of different colours travel with the same velocity.

3

NUMERICAL APERTURE

OBJECT: To measure the numerical aperture of an optical fibre.

APPARATUS: Laser source, optical fibre, holders, optical bench, screen with concentric circles drawn on it or ruler, power meter, mounting platform *etc.*

3.1 THEORY AND FORMULA

3.1.1 Optical Fibre

In today's information age, the need for fast, accurate and high volume communication is more important. Since optical frequencies are extremely large ($\approx 10^{15}$ Hz) as compared to the conventional radio waves ($\approx 10^{6}$ Hz) and microwaves ($\approx 10^{10}$ Hz) so a light beam acting as a carrier wave is capable of carrying for more information than radio waves and microwaves. Therefore in order to have an efficient and dependable communication system, one would require a guiding medium in which the information carried by light waves could be transmitted. This guiding medium is known as optical fibre, which is a glass or plastic pipe as thin as human hair, that guides light waves from one place to another place in the same way that copper wire conducts electricity or pipes conduct water.

3.1.2 Structure of Optical Fibre

Physically an optical fibre is a very thin and flexible medium, having a cylindrical shape consisting of three sections: (*i*) the core (*ii*) the cladding and (*iii*) the jacket.

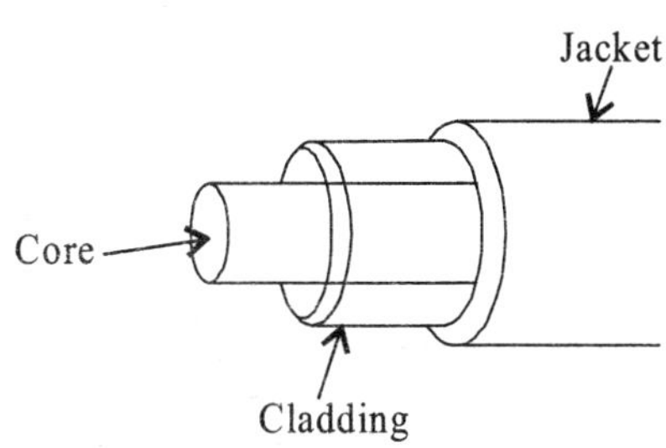

Fig. 3.1 Structure of optical fibre

Out of these, the core is the innermost section and is made of glass or plastic. This is the actual fibre and has the remarkable proeprty of conducting an optical beam. It is surrounded by its own cladding- a glass or plastic coating, that has optical properties which are very different from those of the core. The outer seciton is called the jacket made of plastic or polymer and other materials and is provided for protection against moisture, absorption, crushing and other environmental dangers.

The core acts like a continuous layer of two parallel mirrors. A signal is first enclosed into a light beam, which is then passed in between the two boundaries and propagated as a result of multiple internal reflections.

3.1.3 Propagation of Light Waves through Fibre

(*a*) **Mechanism:** If light wave enters at one end of a fibre in proper conditions most of it is propagated down the length of the fibre and comes out from the other end of the fibre. There may be some loss due to a small fraction leakage through the side walls of the fibre. The reason of confining the light beam inside the fibre, is the total internal reflection of light waves by the inside surface of the fibre. Fibre obeys law of reflection and refraction of light waves. The light which enters at one end of a fibre at a slight angle to the axis of the fibre, follows a zig-zag path due to series of reflections down the length of the fibre.

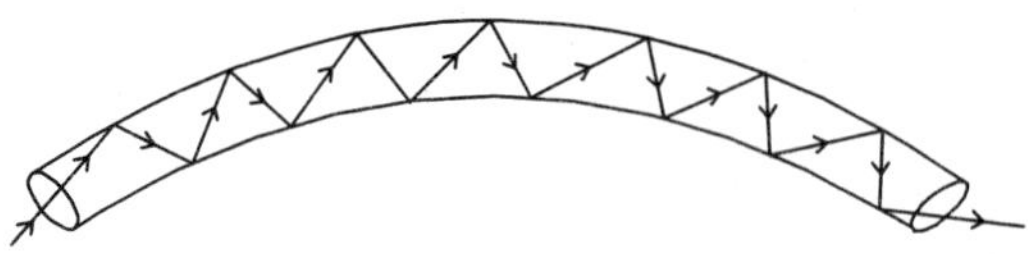

Fig. 3.2 Propagation of light in optical fibre

The angle of incidence at which total reflection first occurs is called the critical angle θ_c for the two media. Light waves incident at angles greater than θ_c will be totally reflected.

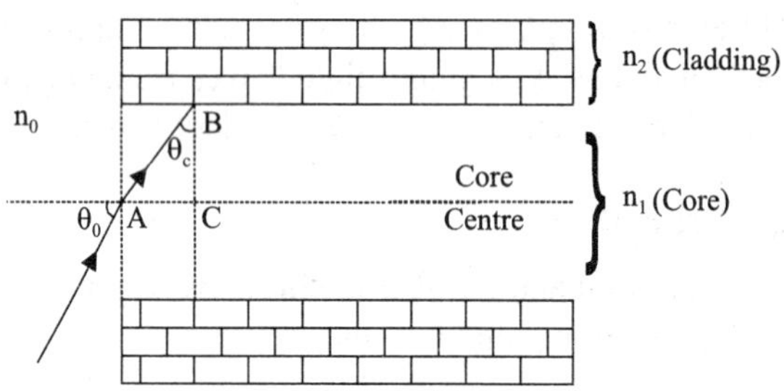

Fig. 3.3

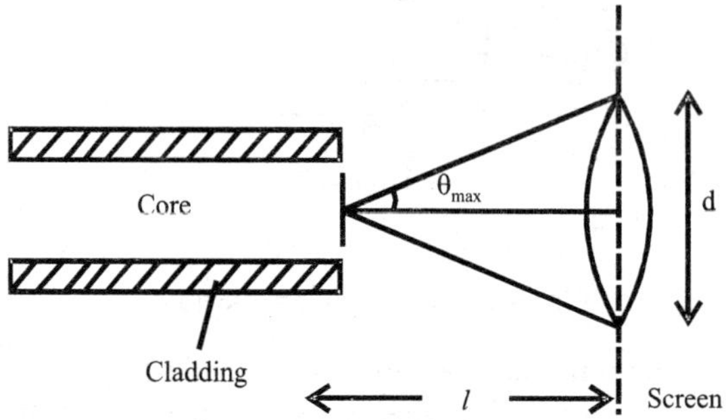

Fig. 3.4

3.1.4 Acceptance Angle and Numerical Aperture

The numerical aperture (NA) is a measure of how much light can be collected by an optical system like optical fibre. It is related to acceptance angle, which indicates the size of a cone of light that can be accepted by the fibre. The numerical aperture is defined as

$$NA = n_o \sin \theta_{max} \tag{3.1}$$

where n_o = Refractive index of incidence medium (1 for air) and θ_{max} = Maximum entrance angle (acceptance angle) at which optical ray can enter and propagate in the fibre.

Thus $$NA = \sin \theta_{max} \tag{3.2}$$

For step- index fibre the NA is given by

$$NA = \sqrt{n_1^2 - n_2^2} \tag{3.3}$$

where n_1 = refractive index of core

and n_2 = refractive index of cladding

From Fig. (3.4)

$$\therefore \quad \sin \theta_{max} = \frac{d/2}{\sqrt{l^2 + \frac{d^2}{4}}} = \frac{d}{\sqrt{4l^2 + d^2}} \tag{3.4}$$

Therefore from Eqs. (3.2) and (3.4), we get

$$NA = \frac{d}{\sqrt{4l^2 + d^2}} = \sin \theta_{max} \tag{3.5}$$

where d = Diameter of the circle on screen and

l = Separation between optical fibre end and screen.

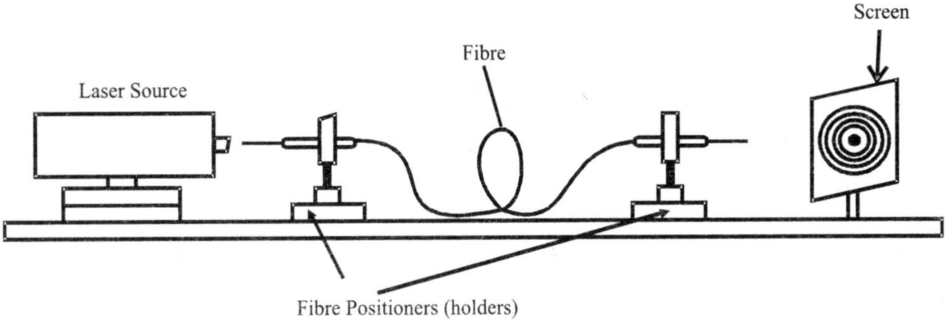

Fig. 3.5

3.2 PROCEDURE

(*i*) Make arrangement as shown in Fig. (3.5)

(*ii*) A pointed laser source is kept fixed and screen is placed at some distance a may from the source. This screen can be prepared on a plastic/metal sheet with concentric circles of different diameters (5, 10, 15, 20 mm) drawn on it.

(*iii*) Mount the fibre in the holder such that it remains horizontal.

(*iv*) Switch on laser and see the position of the light spot coincides with one of the circles.

(*v*) Record distance l and diameter (d) of the spot.

(*vi*) Repeat this procedure for other circles on the screen and take readings of L and D.

3.3 OBSERVATIONS

S. No.	l (mm)	d (mm)	NA	θ_{max}
1				
2				
3				
4				

3.4 CALCULATIONS

$$NA = \frac{d}{\sqrt{4l^2 + d^2}} = \frac{d}{\left(4l^2 + d^2\right)^{1/2}} = \sin\theta_{max}$$

3.5 RESULT

(*a*) The numerical aperture of the given fibre = ...

(*b*) The maximum angle of acceptance = ...

3.6 SOURCES OF ERROR AND PRECAUTIONS

(*i*) Do not look directly at laser beam because it may damage your eyes.

(*ii*) Attach the optical fibre cord property.

(*iii*) The laser source is to be switched off after taking observations.

(*iv*) Distinguish the outer and inner pink light spots and thus, make measurement of D.

QUESTIONS WITH ANSWERS FOR VIVA-VOCE

Q.1. Define laser.

Ans. Laser is a device which generates very intense, coherent, highly directional and monochromatic source of light.

Q.2. What is meant by laser?

Ans. Laser stands for light amplification by stimulated emission of radiation.

Q.3. What is the basic principle of laser?

Ans. It is based on the stimulated emission of light.

Q.4. What is the wavelength of red light emitted by He-Ne laser?

Ans. 6328 Å.

Q.5. What are basic requirements of a laser?

Ans. There are three main requirements (*a*) population inversion (*b*) optical resonator and (*c*) active medium of desired energy levels.

Q.6. What is function of optical resonator?

Ans. It provides wgh photon density to stimulate emission.

Q.7. Why ordinary light is incoherent?

Ans. Here spontaneous emission dominates stimulated emission.

Q.8. What is optical fibre?

Ans. It is very thin (like hair) fibre made of glass, quartz or transparent plastic materials.

Q.9. What is use of the optical fibre?

Ans. It is used for transmission of optical frequency to a very long distance without much attenuation.

Q.10. What is the principle of optical fibre?

Ans. The propagation of light through the fibre is based on the phenomenon of total internal reflection.

Q.11. What are the various parts of optical fibre?

Ans. There are three main parts: (*a*) core (*b*) cladding and (*c*) jacket.

Q.12. Why cladding is used in optical fibre?

Ans. It is used to confired the light in case, so the refractive index of cladding is always lower than that of the core.

Q.13. What is physical meaning of numerical aperture?

Ans. It means light gathering capacity.

Q.14. What is acceptance angle?

Ans. It is maximum entrance angle at which optical ray can enter and propagate in the fibre.

Q.15. What is relation between NA and fractional refraction index?

Ans. $\text{NA} = n_{core}\sqrt{2\Delta}$

where $\Delta = \dfrac{n_{core} - n_{clad}}{n_{core}}$ = Fractional refractive index.

Q.16. What are advantages of optical fibre communication over traditional copper cable communication?

Ans. (1) Large bandwidth. (2) Free from electromagnetic interference. (3) Smaller and lighter for the same information carrying capacity. (4) Improved security.

4

MAGNETIC HYSTERESIS

OBJECT: To obtain hysteresis curve for a given ferromagnetic material on a C.R.O. using solenoid and to find magnetic parameters from the B-H loop.

APPARATUS: A C.R.O., step down transformer (6-12 *V*, 2-3 *A*), A.C. ammeter (0-3 *A*), a rheostat two resistances (1Ω each), carbon resistance (47 kΩ), two capacitors ($1\mu F$ and $2\mu F$), a solenoid and a specimen rod.

4.1 DESCRIPTION OF THE APPARATUS

4.1.1 Solenoid

The solenoid consists of a primary coil and a secondary coil on the central part of the primary. The primary coil consists of a non-metallic, non-magnetic frame of approximate diameter 2-3 cm and length 30 cm, with two or three layers of winding of enamelled copper wire S.W.G. 22 on it. The secondary is wound on a spool of about 10 cm length and 4.5 cm diameter and consists of 10,000 truns of enamelled copper wire S.W.G. 48.

4.1.2 Specimen Rod

The specimen rod consists of 40-50 thin iron wires grouped to form a rod. The cycle spokes can serve the purpose. The rod is inserted completely in the solenoid.

4.1.3 Integrating Circuit

The 47 kΩ resistance in series with a $0.2\ \mu F$ capacitor where the output is drawn from the capacitor forms an integrating circuit.

4.2 THEORY

The specimen is placed inside the primary of the solenoid which produces a varying magnetic field due to the alternating current passing through it. The value of the varying magnetic field is given by

$$H = n\,I = n\,I_0 \cos \omega t$$

$$\Rightarrow \qquad H = \sqrt{2} n I_{RMS} \cos \omega t \qquad \textbf{(4.1)}$$

where n is the number of turns per unit length of the primary of the solenoid. I_{RMS} is the R.M.S. value of current as measured by A.C. ammeter. This current also develops on A.C. voltage across the resistance $R_l = 1\Omega$, which is applied to the X-X plates of the C.R.O. This A.C. voltage produces an X-deflection on the screen of the C.R.O. As the voltage across R_1 as well as the magnetic field produced by the solenoid are both proportional to I_{RMS}, the X-deflection is proportional to the varying magnetic field H.

This magnetising field H in the primary produces a magnetic induction B in the specimen. As H is a varying magnetic field, B also changes at the rate dB/dt and induces an e.m.f. in the secondary proportional to dB/dt.

The circuit consisting of 47 kΩ resistance connected in series with $2\mu F$ capacitor acts as an integrating circuit, which therefore produces a potential difference across $2\mu F$ capacitor proportional to magnetic induciton B. This potential difference is applied to the Y-Y plates of the C.R.O. due to which there is a Y-deflection proportional to B.

Thus we find that the X-deflection in the C.R.O. is proportional to the magnetising field H and Y-deflection to the magnetic induction B, which results in producing the hystersis (B-H) loop on the C.R.O. screen.

4.3 PROCEDURE

4.3.1 Obtaining the Hysteresis Curve for a Given Sample on a C.R.O. Screen

1. Make the circuit connections as shown in Fig. 4.1. Turn the switch on. Some pattern will appear on the C.R.O. screen.

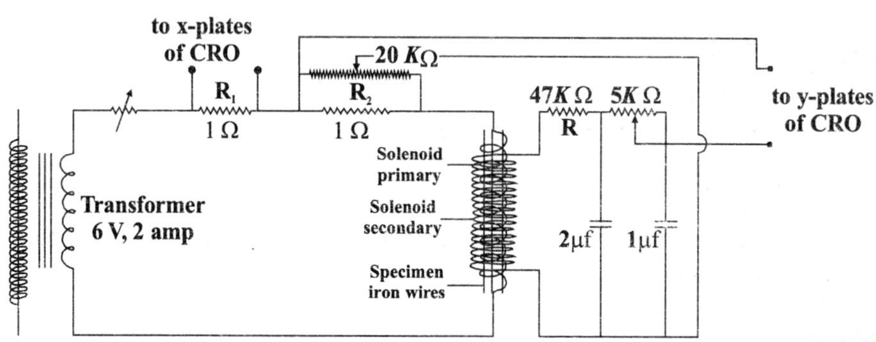

Fig. 4.1 Circuit diagram to study magnetic hysteresis

2. Adjust the X and Y amplifiers of the C.R.O. to get a pattern within the screen.
3. Shapes of *B-H* curve like those shown in Fig. 4.2 (*a*) to (*d*) may appear, (The secondary leads to the C.R.O. may have to be interchanged to bring the curve in the proper quadrants).
4. Ajust the 5 $k\,\Omega$ rheostat (Fig. 4.1) in the phase correcting circuit in such a way that the loops formed at the tips of the pattern disappear and flatness is restored. This would give a curve

like that shown in Fig. 4.2 (*c*) with sharp tips. But these ends may not be parallel to the H-axis (horizontal axis). This is due to the contribution of the flux in the air core.

5. Then, adjust the 20 $k\,\Omega$ potential divider in the primary circuit to make the ends of the curve horizontal. (The connections at R_2 may have to be reversed).

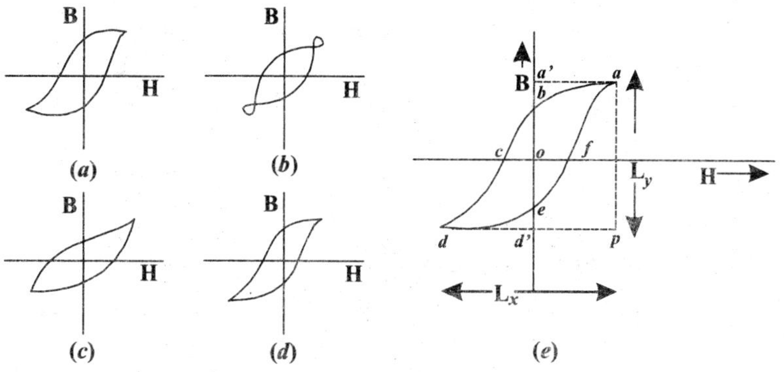

Fig. 4.2 Possible figures seen on C.R.O. (*a*) to (*d*) stages of improvement (*e*) the appropriate B-H curve

6. If the tips of the curve still do not show a sharp overlapping region [see Fig. 4.2(d)], this means that magnetic saturation has not been achieved. Increase the current by using the rheostat in the primary.

4.3.2 Marking the Axes

After getting the proper hysteresis pattern on the screen mark the *X*-axis (*H*) and *Y*-axis (*B*) as explained below.

1. **To Mark *H*-Axis:** Set *Y*-amplifier gain to zero without disturbing the *X*-amplifier gain. You will get a horizontal straight line. This is *H*-axis. Mark this line on the screen. Bring the *Y*-amplifier gain to its original value.
2. **To Mark *B*-Axis:** Now set *X*-ampllfier gain to zero without disturbing the *Y*-amplifier gain. You will now get a vertical straight line. This is *B*-axis. Mark this line also on the screen.
3. After marking the *H*-axis and *B*-axis, adjust the *X* and *Y*-amplifier again to get proper pattern on the screen. Place a tracing paper on the screen and trace the *B*-*H* hystersis loop (pattern) on it along with *H*-axis and *B*-axis.

   Note the r.m.s. value of current in the solenoid as indicated by the ammeter A.

4.3.3 Calibration of *X*-axis in Terms of *H*

Measure the maximum *X*-deflection on trace paper L_X. Magnetising field produced by the solenoid primary is given by

$$H = nI \quad \frac{I_0}{\sqrt{2}} = I_{RMS}$$

$$\Rightarrow \qquad H = nI_0 \cos\omega t = \sqrt{2}\, n\, I_{RMS} \cos\omega t$$

Its maximum value is

$$H_{\max} = \sqrt{2}\, n\, I_{RMS}$$

Trace length L_x corresponds to $2H_{\max}$.

Hence unit trace length corresponds to the conversion factor (F) given by

$$F_{\max} = \frac{2H_{\max}}{L_X} = \frac{2\sqrt{2}\, n\, I_{RMS}}{L_X}\ \text{Amp/m}^2 \qquad \textbf{(4.2)}$$

where n is the number of turns per unit length in the primary, I_{RMS} is the current in ampere as indicated by A.C. ammeter A and L_X is the length in meter.

4.3.4 Calibration of *Y*-axis in Terms of *B*

Reduce X-gain to zero and adjust 20 K potentiometer to zero. A line along Y-trace will be left. Meausre this trace as L_y. In Fig. 4.2 (*e*) it is *b-e*. Slowly move the specimen out of the tube until L_y on the screen reduces to $L_y/2$. Increase Y-gain to restore back to L_y. This implies Y-gain is now double of its previous value. Repeat the process once or twice more so that Y-gain may be 4-times, 8-times *etc.* of its initial value. Let this factor be $F' = 2, 4, 8, 16, \ldots\ldots\ldots.$

Now take out the specimen completely. Measure the trace h on the screen in metre. The length of this trace with original amplification will be h/F'. The deflection h/F' corresponds to double the maximum magnetic flux due to air core, *i.e.*,

$$2\left[\phi_{air}\right]_{\max} = 2H_{\max}\pi r^2 \qquad \textbf{(4.3)}$$

where r is the radius of the primary coil in metre, the flux for unit length gives the conversion factor for flux measurement.

$$F'' = \frac{2H_{\max}\pi r^2}{h/F'}\ \text{Weber/metre}$$

If B is magnetic inducion in the iron wire and S their area of cross-section, then $B = \dfrac{\phi_m}{S}$.

Hence calibration constant for B is

$$F_y = \frac{F''}{S} = \frac{2H_{\max}\pi r^2 F'}{hS}\ \text{Weber/metre}^3 \qquad \textbf{(4.4)}$$

where $\quad H_{\max} = \sqrt{2}\, n\, I_{RMS}\ \text{Amp/m}$

4.4 OBSERVATIONS

Diameter of the primary of solenoid $\qquad D = \ldots\ldots\ldots$ m

Radius of the primary of the solenoid $r = D/2 = \ldots\ldots\ldots$ m

Diameter of one specimen wire = ... m

Radius of one specimen wire = ... m

Total number of wires (spokes) used = n_1 = ...

Total area of cross-section of the specimen rod

$$S = \pi r^2 n_1 = \text{..........m}^2$$

Number of turns per metre of the primary of the solenoid n = ...

Y-amplification factor F' = ...

Length of the vertical trace h = ... m

Current in the primary circuit I_{RMS} = ... Amp.

4.5 CALCULATIONS

(*i*) $H_{max} = \sqrt{2} n I_{RMS}$ = ...Amp/m

(*ii*) From B-H loop :

(a) Breadth of B-H loop, $dp = L_X$ = ...m

∴ Calibration constant for H-axis

$$F = \frac{2H_{max}}{L_x} \text{..........Amp/m}^2$$

(b) Height of B-H loop, $ap = L_y$ = ...m

∴ Calibration constant for B-axis

$$F_y = \frac{2H_{max}\pi r^2 F'}{hS} = \text{..........Tesla/m}$$

(*iii*) Saturation value of magnetic induction

$$B_{max} = \frac{1}{2} F_y L_y = \text{..........Tesla}$$

(*iv*) Remanence or remanent induction or retentivity
(from B-H loop) = ob = ...m

$= ob \times F_x$ = ...Tesla

(*v*) Coercivity or coercive force
(from B-H loop) = oc = ...m

$= ob \times F_x$ = ...Amp/m

(*vi*) **Loss of energy per unit volume per cycle due to hysteresis**

To find out loss of energy per unit volume per cycle of hystersis calculate the area on B-H loop from the graph paper. Find the area of one small sqaure on the graph. Count the total number of squares on the graph within the B-H loop taking each square more than half as one, ignoring each square less than half and taking two half squares as one. Then hysteresis loss per unit volume per cycle

= Area of hysteresis loop

= ....Joule/cycle/m^3

4.6 RESULT

Remanence or remanent induction or retentivity = ... Tesla

Coercivity or coercive force = ... Amp/m

Loss of energy per unit volume per cycle due to hysteresis

$$= \text{............} \left(\frac{\text{Joule}}{\text{m}^3}\right)\frac{1}{\text{cycle}}$$

4.7 PRECAUTIONS

1. All magnetic materials should be kept away from the experiment.
2. The magnetising current in the primary coil of the solenoid should be large enough to properly magnetise the given material.
3. The B-H loop should be stable.
4. The A.C. supply should be switched on only while taking observation and switched off immediately thereafter.
5. The C.R.O. controls should be operated carefully and wihtin specified limits.

QUESTIONS WITH ANSWERS FOR VIVA-VOCE

Q.1. Define magnetic field.

Ans. The magnetic field (H) is simply ni, where n is the number of turns per unit length of a solenoid and i is the current passed through it.

Intensity of magnetic field at any point in a magnetic field is equal the force acting on a hypothetical unit north pole placed at that point.

Q.2. Define magnetic flux density.

Ans. Magnetic flux density or magnetic induction (B) at a point inside a magnetic material is the total number of lines of force, due to both the magnetic field (H) and the magnetisation (I) of the material, passing through unit area positioned around the point and perpendicular to the lines of force.

Q.3. What is intensity of magentisation?

Ans. Intensity of magnetisation (I) of a magnetised material is the magnetic moment per unit volume of the material.

Q.4. Define permeability.

Ans. Permeability (μ) is defined as the ratio of magnetic induction to magnetising field $\mu = B/H$.

Q.5. What is susceptibility?

Ans. Susceptibility of a magnetic material is the ratio of the intensity of magnetisation (I) to the magnetising field H.

$$X = I/H$$

Q.6. What is the relation that connects B, H and I ?

Ans. The relation connecting the quantities B, H and I is $B = \mu_0(I + H)$.

Q.7. What is the relation that connects magnetic permeability to susceptionability ?

Ans. The relation connecting the quantities μ and X is $\mu = \mu_0(1 + X)$.

Q.8. What is relative permeability?

Ans. The ratio of magnetic permeability of the material to that of free space is defined as relative permeability.

Q.9. What is differential permeability?

Ans. The differential permeability at a given magnestising field is the slope of *B-H* curve at that field. It is different at different points of the *B-H* curve.

Q.10. What are the SI units of magnetic flux density and magnetising field?

Ans. The SI units of B is *Tesla* and that of H is Ampere/metre.

Q.11. What are C.G.S. units of B and H and how are they related to their SI counterparts?

Ans. The C.G.S. units of B and H are *Gauss* and *oersted* respectively.

$$1 \text{ Tesla} = 10^4 \text{ Gauss}$$

$$1 \text{ Amp/m} = 4\pi \times 10^{-3} \text{ oersted.}$$

Q.12. What do you mean by hysteresis?

Ans. Hysteresis of a magnetic material is the property due to which the magnetic induction B lags behind the magnetising force (H) when the material undergoes a cycle of magnetisation.

Q.13. What is hysteresis loop?

Ans. B is a double valued function of H. When H changes from $(-H_1)$ to $(+H_1)$, B does not follow the same curve that it does when H changes from $(+H_1)$ to $(-H_1)$. As a result, a loop is formed in the *B-H* plane [as in Fig. 4.2. (*e*)]. This loop is known as the hysteresis loop.

Q.14. What do you mean by hysteresis loss?

Ans. The hysteresis loss is the loss of energy per cycle when a magnetic material is subjected to an alternating magnetic field. This energy loss is proportional to the area of the hysteresis loop and is dissipated in the form of heat.

Q.15. What is retentivity?

Ans. Retentivity of a magnetic material is the magnetisation retained by it when the magnetising field is withdrawn and the material is left undisturbed.

Q.16. What is coercivity?

Ans. Coercivity of a magnetic material is measured by the reversed magnetic field required to demagnetise the material.

Q.17. Which has a smaller hysteresis loop area-soft iron of steel?

Ans. Soft iron.

Q.18. When the differential permeability nearly zero?

Ans. The differential permeability is nearly zero at magnetic saturation.

Q.19. What type of material should be used in the construction of a permanent magnet?

Ans. A material with high retentivity and high coercivity, such as alnico, should be used.

Q.20. What are the comparative values of relative permeability for a diamagnetic and a ferromagnetic substances?

Ans. The relative permeability (μ_r) of a diamagnetic substance is slightly less than unity while for a paramagnetic substance is slightly greater than unity. For a ferromagnetic substance μ_r is much larger than unity.

Q.21. What is Curie temperature?

Ans. The temperature, above which a ferromagnetic substance behaves as a paramagnetic material, is referred to as Curie temperature. For iron its value is 770°C.

Q.22. Is the phenomenon of hysteresis shown by diamagnets and paramagnets?

Ans. No, it is exhibited by ferromagnets only.

Q.23. Do you spend any energy in the magnetisation of the rod?

Ans. Yes, the process of magnetisation requires some energy. Actuallly the energy is spent in alligning the molecular magnets in the direction of magnetising field.

Q.24. What is the utility of finding the hysteresis loop?

Ans. In the construction of armatures of motors and dynamos, the materials are so selected that their hysteresis loops are very small. So that the heat generated is very small.

5

CHARGING & DISCHARGING OF A CONDENSER

OBJECT: To study the charging and discharging of a condenser and hence determine time constant (Both current and voltage graph are to be plotted).

APPARATUS: Battery, voltmeter, milliammeter (25 – 0 – 25), resistances of various values (kΩ), condensers of different capacities (Ω*f*), switches (S_1, S_2, S_3), stop watch etc.

5.1 THEORY AND FORMULAE

5.1.1 For Charging

Take a resistor and a capacitor and complete the circuit as shown in Fig. 5.1. When the switch S_1 and S_3 are kept on and switch S_2 is kept off, the condenser C charges, at time t, the charging of a condenser can be represented by the relation

$$Q = Q_0\left(1 - e^{-t/RC}\right) \tag{5.1}$$

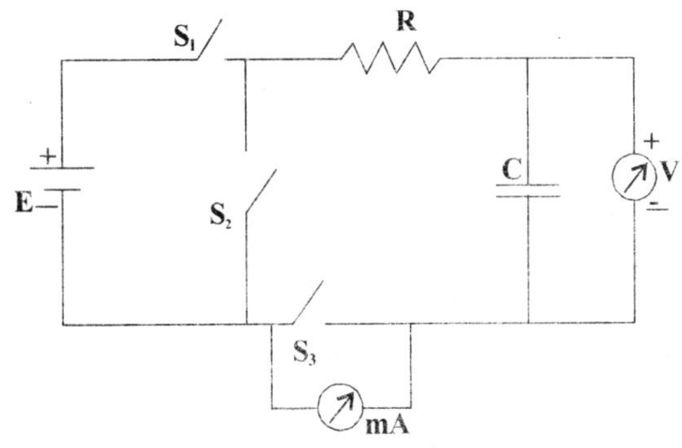

Fig. 5.1 Circuit diagram for charging and discharging of a condenser

where RC is time constant and Q_0 is maximum charge acquired by the condenser.

The charging current is given by

$$I = \frac{dQ}{dt} = \frac{Q_0}{RC}\, e^{-t/RC}$$

$$\Rightarrow \qquad I = I_0\, e^{-t/RC} \qquad (5.2)$$

5.1.2 For Discharging

When S_1 is kept off and switches S_2 and S_3 are kept switched on, the condenser will start discharging across R. The discharging of a capacitor can be given by

$$Q = Q_0\, e^{-t/RC} \qquad (5.3)$$

and the discharging current is given by

$$I = \frac{dQ}{dt} = -\frac{Q_0}{RC}\, e^{-t/RC}$$

$$\Rightarrow \qquad I = -I_0\, e^{-t/RC} \qquad (5.4)$$

From equations (5.1), (5.2), (5.3) and (5.4) it is clear that the charging and discharging processes of a condenser are exponential in nature as shown in Fig. 5.2 and 5.3 and rate at which the charge or discharge accurs depends on the RC-value of the circuit.

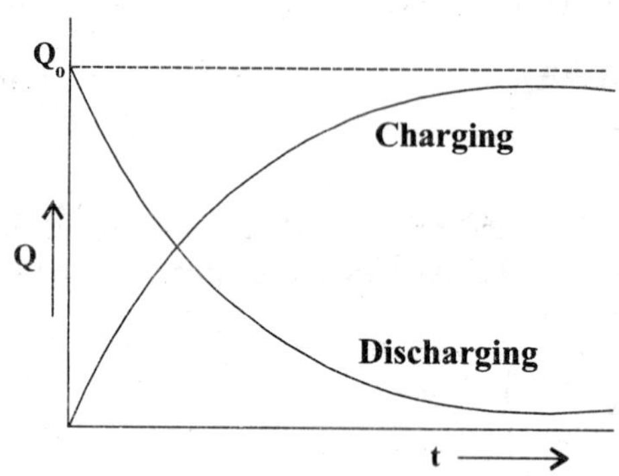

Fig. 5.2

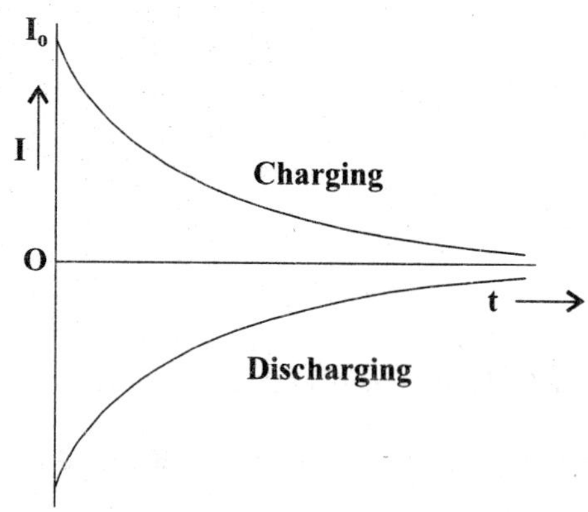

Fig. 5.3

where $t = RC = \tau$ then from equation (1) and (2), we get

$$Q = 0.63\, Q_0$$

and $I = 0.37\, I_0$.

Thus the time constant is that time during which the charge rises on the condenser plates to 63% of its maximum value during charging process or time during which current in the circuit falls to 37% of its maximum value during charging process as shown in Fig. 5.3.

5.2 PROCEDURE

5.2.1 For Charging of Condenser

(*i*) Connet the circuit as shown in Fig. 5.1. Keep S_1 and S_2 switches off while S_3 switched on.
(*ii*) Now turn on switch S_1 and simultaneously start a stop watch.
(*iii*) Reading in voltmeter will increase while current in milliammeter will decrease. Record reading of voltmeter and milliammeter after every 10 sec, till voltage reaches a maximum and current a minimum.
(*iv*) Repeat the experiment by changing the values of R and C.
(*v*) Plot the graphs between voltage (V) and time (t) as well as current (I) and time (t).

5.2.2 For Discharging of Condenser

(*i*) Connect the circuit as shown in Fig. 5.1. keep S_1 and S_2 switches off while S_3 switched on.
(*ii*) Now turn on switch S_1 till reading in voltmeter reaches to maximum value.
(*iii*) Now turn off switch S_1 and turn on switches S_2 and S_3. Simultaneously stop watch is started. The condenser will start dischargeing accross R. Reading in voltmeter will decease while in milliammeter will increase (in - ve side).
(*iv*) Record readings of voltmeter and milliammeter after every 10 sec. till voltage reaches a minimum and current a maximum.
(*v*) Repeat the experiment by changing the values of R and C.
(*vi*) Plot the graph between voltage (V) and time (t) as well as the graph between current (I) and time (t).

5.3 OBSERVATIONS

(*a*) For charging of condenser

S. No.	Time (t) sec.	R_1= - -kΩ, C_1= - -μf		R_2= - -kΩ, C_2= - -μf		R_3= - -kΩ, C_3= - -μf	
		V (volts)	I (mA)	V (volts)	I (mA)	V (volts)	I (mA)
1.	0						
2.	10						
3.	20						
4.	30						
5.	40						
6.	50						
7.	60						
-	-						
-	-						
-	-						

***(b)* For discharging of condenser**

S. No.	Time (t) sec.	R_1= - -$k\Omega$, C_1= - -μf		R_2= - -$k\Omega$, C_2= - -μf		R_3= - -$k\Omega$, C_3= - -μf	
		V (volts)	I (mA)	V (volts)	I (mA)	V (volts)	I (mA)

5.4 CALCULATIONS

5.4.1 For Charging of Condenser

Plot a graph in voltage and time, and other in current and time. Find time constant which is the time corresponding to the point where the voltage is 0.63 V_0, compare it with theoretical value $\approx RC$.

5.4.2 For Discharging of Condenser

Plot one graph in voltage and time, and other in current and time. Find the time constant which is the time corresponding to the point where voltage is $0.37\,V_0$. Compare it with theoretical value $\approx RC$.

5.5 RESULT

The values of time constant from graphs

τ_1 = ... sec.

τ_2 = ... sec.

τ_3 = ... sec.

and the theoretical values

τ_1 = ... sec.

τ_2 = ... sec.

τ_3 = ... sec.

5.6 SOURCES OF ERROR AND PRECAUTIONS

(*i*) Condensers should not be leaky.

(*ii*) It is better to use voltmeter and milliammeter having small least counts.

(*iii*) The electrical connections should be made carefully.

(*iv*) The product of *R* and *C* should be large.

(*v*) It is not possible to take voltmeter reading and ammeter reading simultaneously. Hence experiment must be performed in two parts. First take voltmeter readings with respect to time and then ammeter readings.

(*vi*) The reading of current should be taken quickly and accurately, because current decreases quite quickly in the process of discharge of condenser.

QUESTIONS WITH ANSWERS FOR VIVA-VOCE

Q.1 What is condenser?

Ans. It is a pair of conductors of opposite charges, on which sufficient quantity of charge may be accommodated.

Q.2 Define the capacity of a condenser.

Ans. The capacity of a condenser is numerically equal to that electric charge which raises its potential by unity.

Q.3 What is the unit of capacity?

Ans. The unit of capacity is *farad.*

Q.4 What is time constant of *R-C* circuit?

Ans. The product *R-C* is called time constant of the circuit. It is equal to the time taken by the condenser to raise its charge to 0.63 of its maximum value.

Q.5 Can you define time constant in terms of discharge of capacitor?

Ans. Yes. It is equal to the time taken by the condenser to reduce its charge to 0.37 of its maximum value.

Q.6 What value of time constant will you choose if quick discharge is desired?

Ans. Low value of *RC.*

Q.7 Mention the factors on which the capacity of a condenser depends.

Ans. It depends upon the following factors.

(*a*) Shape or area of the plates.

(*b*) Distance between the plates.

(*c*) Dielectric medium between the plates.

Q.8 What will happen if *R* is reduced to zero in an *RC* circuit?

Ans. The charging or discharging will take place instantly.

Q.9 What is the function of a dielectric?

Ans. It increases ε_r times the capacity of a condenser, where ε_r is the dielectric constant of the medium. It can be defined as

$$\varepsilon_r = \frac{\text{permittivity of medium}}{\text{permittivity of vacuum}} = \frac{\varepsilon}{\varepsilon_0}$$

It is a dimensionless quantity and is different for different media.

Q.10 Mention one important application of *RC* circuit.

Ans. It is used to quench the discharge in nuclear detectors specially in given counter.

Q.11 Does any other current correspond to charging and discharging of a capacitor?

Ans. Yes, it is called as displacement current. Due to time varying electric field between the plates, an electric current also flows across the space between the plates of condenser. This current is known as the displacement current and is defined as $I_D = \varepsilon_0 \frac{d\phi}{dt} = \varepsilon_0 A \frac{d\vec{D}}{dt}$

6

HIGH RESISTANCE BY METHOD OF LEAKAGE

OBJECT: To determine high resistance by method of leakage of a condenser with the help of ballistic galvanometer.

APPARATUS: A ballistic galvanometer, lamp and scale arrangment, a standard condenser (of the order of 1/3 to $1\mu F$), high resistance which is to be determined, lead accumulator, a stop watch, three condenser keys etc.

6.1 DESCRIPTION OF APPARATUS

6.1.1 Ballistic Galvanometer

A ballistic galvanometer is a specially designed moving coil galvanometer for measuring the quantity of electricity or charge. When such a transient current (charge) passes through a galvanometer the moving part of the galvanometer experiences an angular impulse and begins to perform angular oscillations about its axis of suspension.

It consists of rectangular coil made by winding a large number of turns of fine insulated copper wire on a light non-conducting frame (Fig. 6.1). The coil is suspended by a thin phosphor-bronze suspension fibre from a torsion-head which is connected to a terminal screw T_1. To the lower portion of the suspension fibre is attached a small light mirror which rotates along with the fibre and its deflection can be measured by a lamp and scale arrangement. The lower end of the coil is connected to a fine, loosely wound, spring which is connected to another, terminal screw T_2.

The coil hangs in the space between the pole-pieces of a powerful permanent horse shoe magnet NS. The pole-pieces of the magnet are cut concave cylindrical. In between the pole pieces, within the coil, is fixed a cylinder of some non-conducting material called the 'core'. The core does not touch the coil anywhere. The whole arrangement is enclosed in a non-magnetic box with a glass window. Three levelling screws are provided at the base of the box.

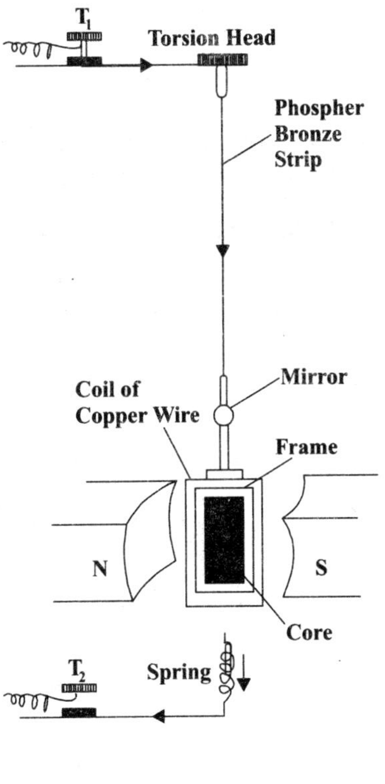

Fig. 6.1

6.2 THEORY

If a charged condenser is connected to a resistance, it begins to discharge. Further, if the charge on the condenser at any instant t be Q coulomb, the potential equation of the circuit is if there is no applied voltage in the circuit.

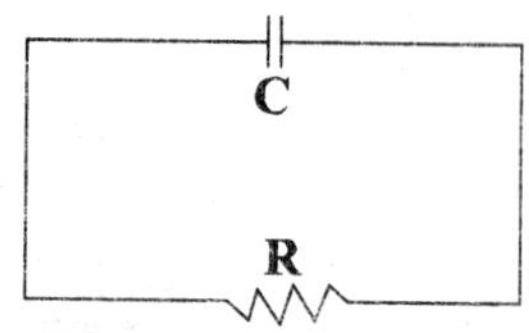

Fig. 6.2

$$\frac{Q}{C} + R\frac{dQ}{dt} = 0 \tag{6.1}$$

Integrating Eq. (6.1), we get

$$\int \frac{dQ}{Q} = -\int \frac{dt}{RC}$$

$$\Rightarrow \qquad \log_e Q = -\frac{t}{RC} + C_1$$

where C_1 is the constant of integration at time $t = 0,\ Q = Q_0$

$$\therefore \qquad C_1 = \log_e Q_0$$

Hence, the solution of Eq. (6.1) is

$$Q = Q_0 e^{-t/RC} \tag{6.2}$$

Now, let the condenser be given charge Q_0 and then immediately discharged through the ballistic galvanometer. During the discharge of the condenser, the charge passing through the galvanometer,

$$Q_0 = K\theta_1\left(1 + \frac{\lambda}{2}\right) \tag{6.3}$$

where θ_1 is the first throw of the galvanometer, K is the ballistic constant of the galvanometer and λ the logarithmic decrement of the galvanometer.

If the same condenser is again charged to the same value Q_0 and discharged through the same galvanometer after allowing the condenser to leak its charge through a high resistance R for a known time t, the charge passing through the galvanometer is given by

$$Q = K\alpha_1\left(1 + \frac{\lambda}{2}\right) \tag{6.4}$$

where α_1 is the first throw of the galvanometer. From Eqs. (6.2), (6.3) and (6.4)

$$\frac{Q}{Q_0} = \frac{\alpha_1}{\theta_1} = e^{-(t/RC)}$$

$$\Rightarrow \qquad R = \frac{t}{C\log_e\left(\frac{\alpha_1}{\theta_1}\right)}$$

$$\Rightarrow \qquad R = \frac{t}{2.3026C\log_e\left(\frac{\alpha_1}{\theta_1}\right)} \tag{6.5}$$

The dielectrics which are generally used in the condensers are not perfect insulators, they have very high resistance. Hence some charge of the condenser leaks through it. This resistance is therefore, known as self leakage resistance. It is denoted by r (say).

Due to self leakage resistance of the condenser, the calculated value of R from Eq. (6.5.) will stand now for equivalent resistance of the parallel combination of r and R. Let the equivalent resistance is denoted by R', then

$$\Rightarrow \qquad \frac{1}{R} = \frac{1}{R'} + \frac{1}{r} \tag{6.6}$$

The condenser is charged to maximum value Q_0 and then allowed to leak through the dielectric for the same time t. If after self leakage the condenser is discharged through the same ballistic galvanometer, then let the first throw in it is α_2, then

$$\frac{1}{r} = \frac{2.3026C}{t}\log_{10}\left(\frac{\theta_1}{\alpha_2}\right)$$

Similarly
$$\frac{1}{R'} = \frac{2.3026C}{t}\log_{10}\left(\frac{\theta_1}{\alpha_1}\right)$$

Putting these values in Eq. (6.6), we get

$$\frac{1}{R} = \frac{1}{R'} - \frac{1}{r}$$

$$= \frac{2.3026C}{t}\left[\log_{10}\left(\frac{\theta_1}{\alpha_1}\right) - \log_{10}\left(\frac{\theta_1}{\alpha_2}\right)\right]$$

$$\Rightarrow \qquad R = \frac{t}{2.3026C\log_{10}\left(\frac{\alpha_2}{\alpha_1}\right)} \text{ ohm} \tag{6.7}$$

This equation is used for the calculation of given high resistance.

6.3 PROCEDURE

1. The ballistic galvanometer is levelled and set free. The lamp and scale arrangement is also adjusted so that a fine spot is formed on the scale. The initial reading of the spot, when the coil is in the rest position, is noted and adjusted at the zero of the scale.
2. The tapping key K, known as damping key (connected across the galvanometer) may be used by pressing and releasing it a number of times in succession, to bring the galvanometer coil to its rest position.

3. The high resistance, standard condenser, ballistic galvanometer, key are connected according to the circuit diagram shown in Fig. 6.3.

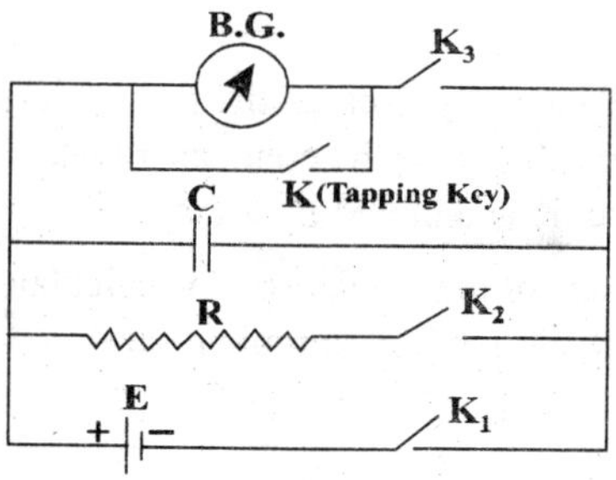

Fig. 6.3

4. Now the key K_1 is pressed for few seconds to charge the condenser fully. While charging the condenser, the key K_2 and K_3 are kept open.
5. Immediately after charging the condenser, the key K_3 is pressed after releasing key K_1. In this way, the charge of the condenser will pass through the galvanometer without self leakage. The first throw (θ_1) of the galvanometer is recorded.
6. The galvanometer coil is brought to its rest position by the damping key K. The above procedure is repeated at least for three or four times.
7. Release key K_3 and press key K_1 for sometime (say 20 sec.) to charge the condenser fully. Thereafter the key K_1 is released and simultaneously a stopwatch is started. The charge of the condenser is allowed to leak for sometime t (say 15 or 20 or 30 sec). At the end of time t the key K_3 is pressed and the corresponding first throw (α_2) is found out. During this key K_2 should remain open.
8. The key K_3 is disconnected, the key K_1 is again pressed for say 20 sec. for charging the condenser again, then K_1 is released. The tapping key K_2 is pressed and simultaneously the stop watch is started. The charge is allowed to leak through the high resistance R for the same time t and then the tapping key K_2 is released. Immediately after releasing K_2, the key K_3 is connected. The remaining charge on the condenser will pass through the galvanometer. The throw (α_1) of the galvanometer is recorded.
9. The above process is repeated at least for three times for different values of t.

6.4 OBSERVATIONS

[A] Capacity of the condenser C =.........................farad

[B] Measurement of θ_1, α_1 and α_2

S. No.	Rest Position of spot of light in *mm* (*a*)	First throw of the galvanometer			Leakage time (t) in *sec.*	θ_1 in *mm* $(a \sim b)$	α_2 in *mm* $(a \sim c)$	α_1 in *mm* $(a \sim d)$	$\frac{\alpha_2}{\alpha_1}$	$\log_{10}(\alpha_2/\alpha_1)$
		Corres-ponding to full charge (*b*)	After self leakage (*c*)	After leakage through resistance (*d*)						
1										
2										
3										
4										

6.5 CALCULATIONS

$$R = \frac{t}{2.3026C\log_{10}\left(\frac{\alpha_2}{\alpha_1}\right)} = \text{..........ohm}$$

6.6 RESULT

The value of given high resistance R = ... ohm.

6.7 SOURCES OF ERROR AND PRECAUTIONS

1. The e.m.f. of the cell used for charging the condenser must be constant.
2. The condenser used in this experiment must be a standard one.
3. There should be no leakage of charge through the keys and connections, hence the connections must be tight.
4. The leakage time, t, should be accurately measured by the stop watch.
5. The condenser should be of such a capacity that it may ensure satisfactory rate of leakage through the high resistance.
6. The leakage resistance of the condenser must be high, otherwise its resistance will have to be accounted for.
7. Care should be taken in noting down the first throw, as discharge is very quick. A slight mistake may appreciably affect the result.

QUESTIONS WITH ANSWERS FOR VIVA-VOCE

Q.1. What is condenser?

Ans. It is a device in which a large amount of charge can be stored. It consists of two metal plates placed parallel to each other at a short distance. The intermediate space is filled with some dielectric.

Q.2. What is function of a dielectric?

Ans. It increases the capacity of a condenser.

Q.3. Why the condenser is charged to some length of time say 20 or 30 sec each time?

Ans. The condenser does not take up maximum charge q_0 immediately after connecting it to a cell. At the beginning it takes some charges. Thereafter, this tends to oppose the introduction of further charge to the condenser from the cell. In other words, a back e.m.f. is set up in the circuit. The main current takes sometime to overcome the back e.m.f.

Q.4. On what factors does the capacity of a condenser depend?

Ans. It is directly proportional to the area of the plates and the dielectric constant of the medium, and inversely proportional to the plate separation.

Q.5. Does the condenser get totally discharged immediately after connecting it with resistance?

Ans. The discharge of a condenser takes place according to the formula $Q = Q_0 e^{-t/RC}$. The discharge of the condenser is not instantaneous. Theoretically, it will take infinite time to get totally discharged.

Q.6. Define time constant for discharge of a condenser.

Ans. The time constant for discharge of a condenser is defined as that time taken by a conderser to reduce its charge to 37% of the initial maximum value.

Q.7. What is the order of the high resistance which can be measured by this method?

Ans. It is of the order of megaohms ,*i.e.*, $10^6\ \Omega$.

Q.8. Why observations are taken in two parts?

Ans. To account for the self leakage of the condenser through its dielectric.

Q.9. Why the condenser does not achieve its maximum charge at once when the circuit is completed?

Ans. The charge on the condenser increases exponentially and achieves maximum value at infinite time. The charge Q on the condenser at any time t is given by

$$Q = Q_0\left(1 - e^{-t/RC}\right)$$

where Q_0 is the maximum charge.

Q.10. On what factors does the growth of charge on the condenser depend?

Ans. It depends on the resistance R and the capacitance C.

Q.11. Can this high resistance be measured by a P.O. Box or metre-bridge or by similar method?

Ans. No, because these methods are based on Wheatstone Bridge principle where the bridge is sensitive only when the resistances of four arms are of the same order.

Q.12. What is meant by time-constant of the circuit?

Ans. It is the product of the capacitance and the resistance of the circuit.

Q.13. Should the possible leakage of charge through the dielectric of the condenser be accounted for?

Ans. Yes, because the dielectric may not be perfect insulator and some charge may leak out through the condenser even without connecting to R. This is known as self leakage resistance of the condenser.

Q.14. What is the suspension fibre made of?

Ans. The suspension fibre usually used is made of phosphor bronze, an alloy of copper (92.5%), tin (7%) and phosphorus (0.5%).

Q.15. Why is it taken in the form of a strip and not a wire?

Ans. The twisting couple per unit twist in case of a strip of rectangular cross-section is much smaller than in a wire of circular cross-section. Hence for a given couple, the deflection obtained by a strip will be much greater. It also provides a larger surface area for getting cooled if heated by a current.

7
RESOLVING POWER

OBJECT: To verify the expression for the resolving power of a telescope.

APPARATUS: A telescope, a rectangular aperture with width variable with the help of a micrometer screw, a black cardboard with narrow white strips drawn on it and a micrometer microscope.

7.1 DESCRIPTION OF APPARATUS

7.1.1 Variable Rectangular Aperture

The variable rectangular aperture is an adjustable micrometer slit in which the width of the slit can be controlled by means of a micrometer screw.

This slit can be mounted either at the collimator lens or the telescope objective to check the width of parallel beam emerging from the collimator or the parallel beam received by the telescope. An alternative to this arrangement is to use an ordinary adjustable slit and to measure its width with the help of a travelling microscope.

7.1.2 Strip Board

This is a cardboard sheet of size $8'' \times 10''$ on which are drawn four or five sets of parallel narrow strips with white ink in black background. Each set contains parallel strips (Fig. 7.1) of equal width and equal separation. The separation of lines is different in different sets and is either given or measured with a travelling microscope or vernier calliper.

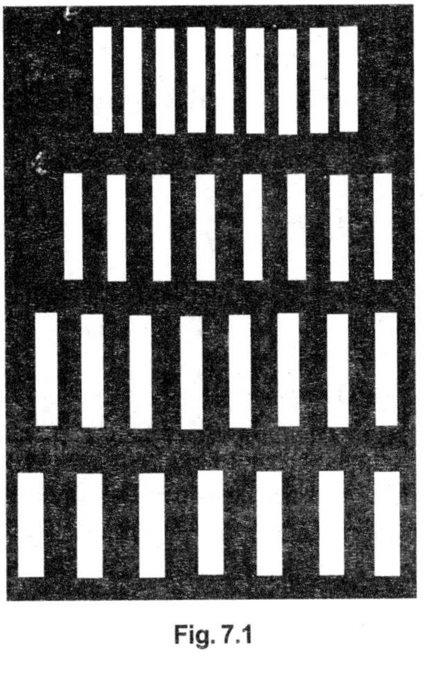

Fig. 7.1

7.2 THEORY

When a telescope is employed to view a distant point object, its objective lens behaves like a circular aperture, and consequently the image is formed by diffraction. It is for this reason that the image of a point object is actually never a point but it is a circle, surrounded by very faint concentric dark and bright rings.

To calculate the radius r of the image, let us refer to the accompanying (Fig. 7.2) where P is a point where all the rays parallel to the axis coming from a distant point object are focussed. The point which is the centre of the image, will obviously have maximum brightness since all the rays arrive here in the same phase.

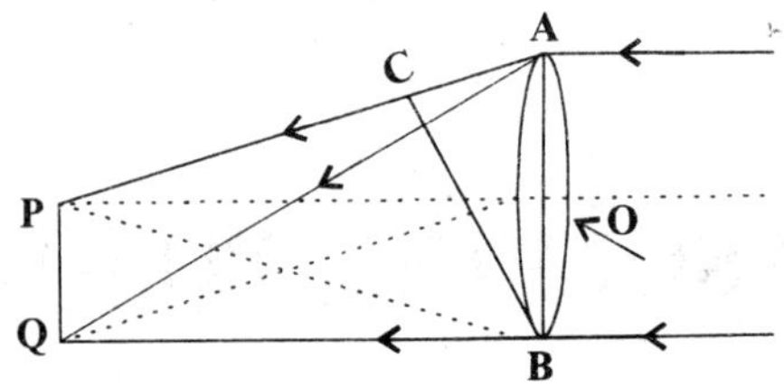

Fig. 7.2

The point Q is situated at the edge of the image and will be a point with minimum intensity. This will be so when AQ-BQ = λ because under this circumstance the sets of rays coming from the two halves, OA and OB of the lens will differ in phase by $\lambda/2$.

Now, with Q as centre and QB as radius, draw an arc BC. This will be sensibly a straight line, and consequently, the triangle ABC may be regarded as a right angled triangle. Hence, the angle

$$\angle ABC = \frac{\lambda}{a}$$

where $AC = \lambda$

and AB = a (the aperture of the lens).

But $\angle ABC = \angle POQ$

Thus $\frac{\lambda}{a} = \frac{r}{f}$

where r = PQ (the radius of the image)

f = OP (the focal length of the lens)

$$\text{Thus } r = f\frac{\lambda}{a} \quad \textbf{(7.1)}$$

The intensity distribution of the image is shown in Fig. 7.3 if, instead of a single point object, we have two distant point objects, their images will now be two circles each of radius r, and the distance x between their centres is given by

$$x = f \tan\theta \quad \textbf{(7.2)}$$

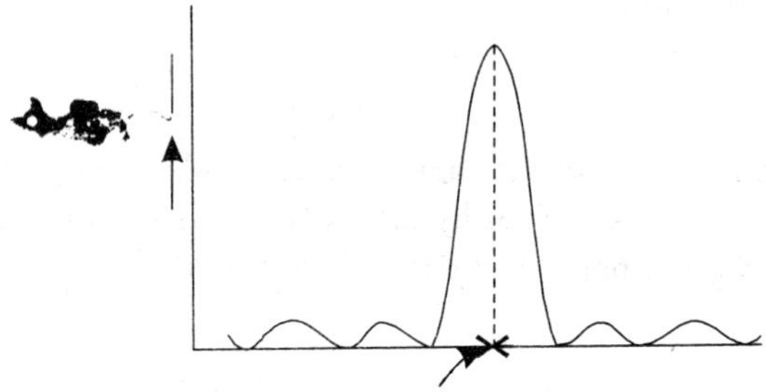

Fig. 7.3 Intensity distribution of an image

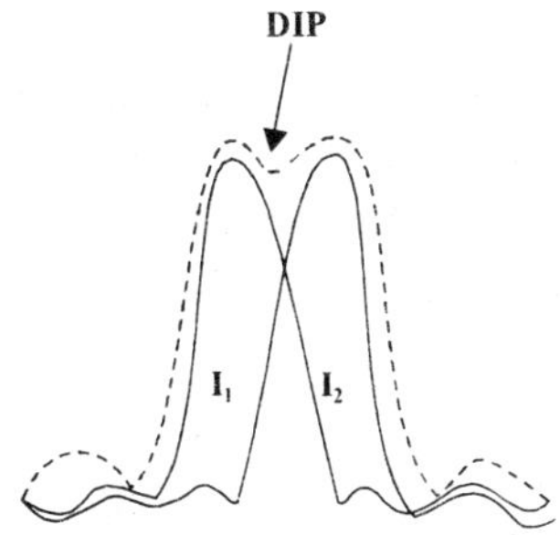

Fig. 7.4 Intensity distribution of two images

where θ is the angle subtended by the two objects at the objective of the telescope.

Now, it is clear that the two images will appear distinctly separate when the edge of one image touches that of the other, *i.e,.* when $d = 2r$. If the images come closer than this, overlapping of the two shall begin and the two images shall cease to be distinguished as separate. The intensity distribution of the two might then be represented as shown in Fig. 7.4. The two continuous lines represent the intensity distribution of the separate images, while the upper dotted curve shows the two compounded.

The resultant has two pronounced maxima with an appreciable dip in between. It will thus be possible, with the aid of this falling off in intensity between two bright regions, to distinguish the two objects, or the two objects will just be resolved. Lord Rayleigh has shown that the resolution of the images is possible only when *'d'* is not less than *'r'*,*i.e.,* the centre of one image may, at the most, lie on the edge of the other. If d is less than r, the dip seen in the figure shall disappear and the curve shall resemble one for a single point object. Hence, under this circumstance, the image shall not be resolved. Thus, in order to secure resolution of the two images we must have $x = r$; *i.e.,*

$$r = f \tan\theta \cong f\theta \qquad \text{(Since } \theta \text{ is small)}$$

$$\therefore\ \theta = r/f \tag{7.3}$$

This is taken as the limiting case, and the limit of resolution is measured by this ratio, But from Eq. 7.1

$$\frac{r}{f} = \frac{\lambda}{a}$$

The resolving power is defined as the reciprocal of the limit of resolution.

$$\therefore \text{ The theoretical resolving power } = \frac{a}{\lambda} \tag{7.4}$$

Now this angle θ is also equal to d/D where d is the distance between the two objects and D is their distance from the objective lens. Hence

$$\text{Practical resolving power } = \frac{D}{d} \tag{7.5}$$

The object of this experiment is to compare the theoretical and practical resolving powers.

7.3 PROCEDURE

(*i*) Mount the telescope on a stand and adjust such that its axis lies horizontally and meets normally the cardboard which provides the two sources. This cardboard is mounted on another stand and is placed at a distance of 15–20 ft. from the telescope. Adjust the cardboard in such a way that the white rectangluar objects are vertical.

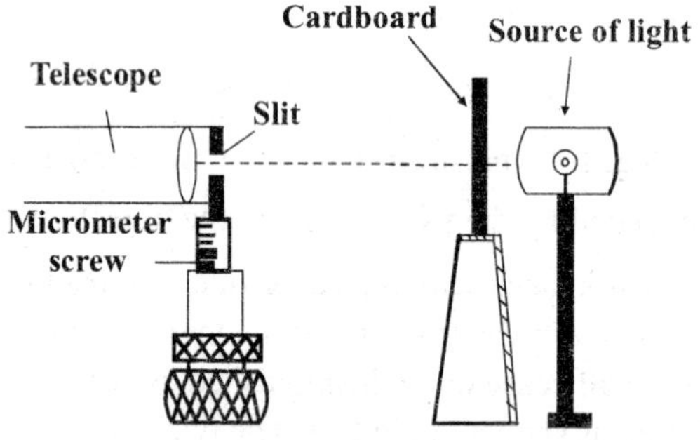

Fig. 7.5 Resolving power of a telescope

(*ii*) Keep the slit wide open and focus the telescope on two white rectangles in the first row. By a horizontal movement of the telescope at right angles to its axis adjust such that the two images lie symmetrically with respect to the intersection of the cross-wires in the field of view.

(*iii*) Now, gradually diminish the width of this aperture till the two images just cease to appear as two. Take the reading of the micrometer screw. Close that slit completely by operating the screw further and take the reading again. The difference of the two readings gives the width '*a*' of the slit which is just sufficient to resolve the white lines situated in the first row of the cardboard.

(*iv*) Measure the distance between the centres of the two consecutive lines in the first row with a travelling microscope. This gives '*d*' of the formula.

(*v*) Then measure with a tape the distance between the cardboard and the objective lens of the telescope. This gives '*D*' of the formula.

(*vi*) Repeat the experiment with objects lying in the second and the third rows and thus obtain different values of '*a*' and '*d*'.

(*vii*) The only quantity to be known now is λ, the wavelength of light employed. In the present case, this can be taken as 6×10^{-7} m, which is the mean wavelength of diffused sunlight. If sodium lamp is used then take mean wavelength as 5893 Å.

7.4 OBSERVATIONS

[A] (*i*) Pitch of the micrometer screw =.................cm

(*ii*) No. of divisions on the cap =......................

(*iii*) Least count of the micrometer screw =.................cm

[B] Mean value of $\lambda = 6 \times 10^{-7}$ m.

[A] Table for width of slit (a)

S. No.	Distance *D* (cm)	Micrometer Reading						$a = Y - X$ cm
		Just Resolved			Closed			
		M.S.R. (cm)	C.S.R.	Total X (cm)	M.S.R. (cm)	C.S.R.	Total Y (cm)	
1	100							
2	250							
3	200							
4	250							

[B] Table for distance between two objects (d)

Micrometer Reading						$d = Y - X$ cm
ONE END			OTHER END			
M.S.R. (cm)	V.S.R.	Total X (cm)	M.S.R. (cm)	V.S.R.	Total Y (cm)	

7.5 RESULT

The thoretical and experimental values of resolving power of the telescope for the two objects are given in the following table.

Distance (*D*)	Theoretical resolving power (λ/a)	Practical resolving power (*d/D*)

7.6 SOURCES OF ERROR

1. The axis of the telescope may not be horizontal.
2. The slit may not be parallel to the object lines on the cardboard.
3. The chief source of error in this experiment lies in the fact that the exact position, when the two images just cease to be distinguished as two, is difficult to obtain in practice.

7.7 PRECAUTION

1. The axis of the telescope should be horizontal and it should meet the black cardboard normally.
2. The two objects drawn on the cardboard should be vertical and the slit in front of the objective should be parallel to these lines.
3. Avoid back-lash error while measuring the width of the aperture and the distance between two linear sources.

4. The slit should not be closed too tightly, otherwise it will be damaged.
5. To get two readings for the same distance, it is better to measure the width of the slit also by opening the slit for just resolution of the two objects.
6. The minimum width of slit for resolution should be adjusted very carefully.

QUESTIONS WITH ANSWERS FOR VIVA-VOCE

Q.1. Define the resolving power of a telescope.

Ans. It is defined as the reciprocal of the angular separation that two distant object points must have, so that their images will appear just resolved according to Rayleigh's criterion.

Q.2. What is limit of resolution of a telescope?

Ans. It is the minimum angle subtended at the objective of the telescope by two distant objects which can be just resolved by it. This angle is the measure of the resolution by the telescope. Smaller is this value, higher will be the resolving power.

Q.3. On what factors does it depend?

Ans. It is directly proportional to the wavelength and inversely proportional to the diameter of the aperture.

Q.4. Why the telescope is fitted with objective of large diameter?

Ans. The telescope is fitted with objective of large diameter to increase the resolving power.

Q.5. What is Rayleigh's cirterion of resolution?

Ans. According to this criterion the resolution of two images is possible when the centre of one may, at the most, lie on the edge of the other. If the images come closer than this, they cannot be resolved and they will give the sensation of one image only.

Q.6. How can the resloving power of a telescope be increased?

Ans. By increasing the aperture of the lens.

Q.7. If the focal length of the objective is doubled, what will be the effect on the resolving power of the telescope?

Ans. No effect, because the expression for resolving power is independent of the focal length.

Q.8. Should the telescope not have objective of large focal length?

Ans. It must have an objective of large focal length, but it will not increase its resolving power rather it increases its magnifying power.

Q.9. What is magnifying power of a telescope?

Ans. It is the ratio of the angle subtended by the image at the eye to the angle subtended by the object at the eye when it is seen directly without the telescope.

Q.10. What is the formula for magnifying power of a telescope?

Ans. It is given by F/f, where F is the focal tength of the objective and f is the focal length of the eye-piece.

Q.11. What is the limit of resolution of normal human eye?

Ans. For a healthy eye it is nearly one minute.

Q.12. Suppose two close lines are just resolved by a telescope and then it is receded away further from the lines. Will the two lines be still resolved?

Ans. No, the lines will not remain resolved when the distance of the objects from the telescope is increased. In the first case the angle subtended by the two lines at the objective lens was equal to the angle of minimum resolution. When the telescope is further taken away, the angle subtended by the two lines becomes less than the angle for minimum resolution. Hence, the telescope shall now fail to resolve the two lines.

Q.13. Then what can be done to resolve the two lines again?

Ans. Now the lines can again be resolved by increasing the width of the slit.

Q.14. What do you mean by 40 inch telescope?

Ans. By this we mean that the diameter of the objective of the telescope is 40 inch.

Q.15. What will be the limit of resolution of such a telescope?

Ans. Limit of resolution

$$d\theta = \frac{1.22\lambda}{a}$$

$$= \frac{1.22 \times 6 \times 10^{-5} \times 180 \times 60 \times 40}{40 \times 2.54 \times 3.14}$$

or $$d\theta = 0.15 \text{ sec.}$$

Q.16. Why do you get theoretical value of resolving power different from the practical one?

Ans. This is because the resolution limit is different for defferent eyes. There is enough arbitrariness about resolution limit.

8

CAREY-FOSTER'S BRIDGE

OBJECT: To determine the specific resistance of the material of a given wire with the help of Carey-Foster's bridge.

APPARATUS: Carey-Foster's bridge, Laclanche cell, galvanometer, decimal resistance box, two approximately equal resistances, copper strips, keys, given resistance wire etc.

8.1 DESCRIPTION OF APPARATUS

Carey-Foster's bridge is based on the principle of Wheatstone's bridge. It is the modified form of meter bridge. The bridge wire is divided in two segments by a jockey and extra resistances are connected with each segment and thus it is converted into a Carey-Foster bridge. Thus in such a bridge the effective length of the wire is increased without actually increasing the length of the wire. The sensitivity of the bridge is proportional to the length of the bridge wire, therefore the sensitivity of Carey-Foster's bridge is higher than that of the metre-bridge. The effect of end correction due to end resistances is also eliminated in this method.

In a Carey-Fosters bridge one metre long resistance wire of uniform cross-section is fixed on a wooden board. The resistivity of the wire is high and its temperature coefficient of resistance is low. A metre scale is also fixed parallel to the wire. The ends of the wire are so fixed with the brass or copper strips *S* and *T* that its one end is at zero mark and another end is at 100 cm mark of the scale. Three brass or copper strips A, B and C are also fixed on the board parallel to the wire and these form four gaps together with the end strips as shown in Fig. 8.1.

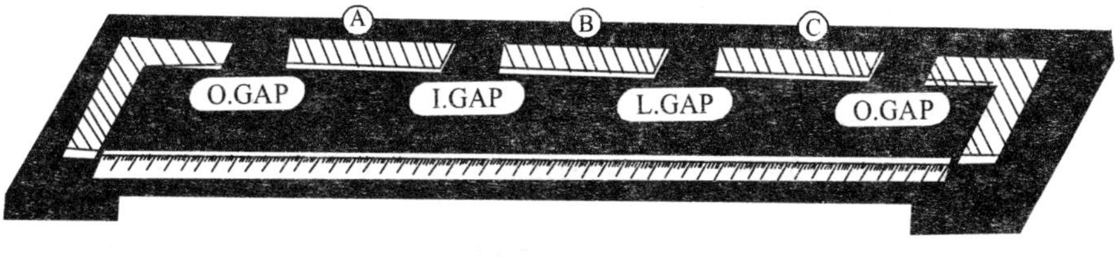

Fig. 8.1 Carey-Foster's bridge

Terminals are attached at the ends of these strips. Resistance coils of nearly equal values are connected in two inner gaps (generally one ohm coils are connected in these gaps). Known and unknown resistances X and Y are connected in the outer gaps. Jockey is connected over the bridge wire. Laclanche cell with a plug key K is connected between the strips A and C. Galvanometer G is connected between the strip B and jockey.

8.2 THEORY

The circuit diagram of Carey-Foster's bridge is shown in Fig. 8.2.

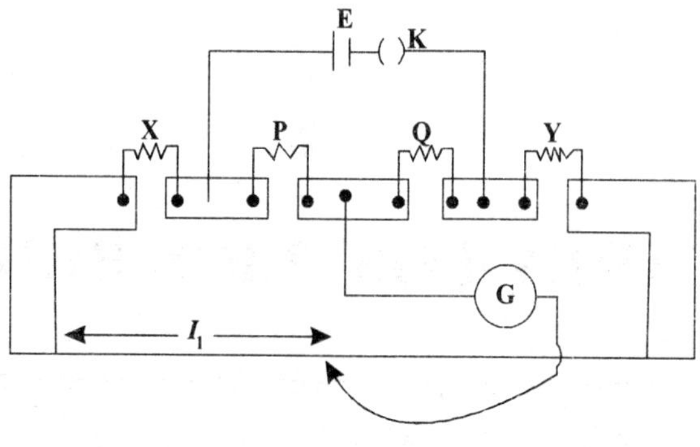

Fig. 8.2 Circuit diagram of Carey-Foster's bridge

Let α and β be the end resistances and let the null point be obtained at a distance l_1 measured from left end. From Wheatstone's bridge formula, we get

$$\frac{P}{Q} = \frac{X + \rho l_1 + \alpha}{Y + \rho(100 - l_1) + \beta} \tag{8.1}$$

where ρ is the resistance per unit length of the bridge wire.

Suppose, on interchanging the resistances X and Y, the null point is obtained at a distance l_2, measured from the left end, we get

$$\frac{P}{Q} = \frac{Y + \rho l_2 + \alpha}{X + \rho(100 - l_2) + \beta} \tag{8.2}$$

Comparings Eqs. (8.1) and (8.2), we get

$$\frac{X + \rho l_1 + \alpha}{Y + \rho(100 - l_1) + \beta} = \frac{Y + \rho l_2 + \alpha}{X + \rho(100 - l_2) + \beta}$$

Adding 1 to both sides and simplifying, we get

$$\frac{X + Y + 100\rho + \alpha + \beta}{Y + \rho(100 - l_1) + \beta} = \frac{X + Y + 100\rho + \alpha + \beta}{X + \rho(100 - l_2) + \beta}$$

$$\Rightarrow Y + \rho(100 - l_1) + \beta = X + \rho(100 - l_2) + \beta$$

$$\Rightarrow X - Y = \rho(l_2 - l_1) \tag{8.3}$$

Thus, the difference between resistances X and Y is equal to the resistance of the bridge wire between two balance points.

Let Y be replaced by a thick copper strip and X by a fractional resistance R of a resistance box. If the balancing lengths are denoted by l'_1 and l'_2, then

$$\rho = \frac{R}{l'_2 - l'_1} \qquad \textbf{(8.4)}$$

since $Y = 0$ for thick copper strip. From Eqs. (8.3) and (8.4) the unknown resistance Y is given by

$$Y = X - \frac{R(l_2 - l_1)}{(l_2' - l_1')} \text{ ohm.}$$

$$Y = X - \rho(l_2 - l_1) = X + \rho(l_1 - l_2) \text{ ohm.} \qquad \textbf{(8.5)}$$

8.3 SPECIFIC RESISTANCE

At constant temperature the resistance of a wire is

(*i*) directly proportional to the length of the wire, *i.e.*,

$$Y \propto l$$

(*ii*) inversely proportional to the area of cross section of the wire, *i.e.*,

$$Y \propto l/A$$

Combiningly

$$Y \propto l/A$$

$$\Rightarrow \qquad Y = K\,l/A$$

where K is a constant which depends upon the material of the wire and temperature. This constant is called the specific resistance or the resistivity of the material of the wire.

If the radius of the unknown resistance wire is r, then

$$A = \pi r^2$$

$$\therefore \qquad Y = K\frac{l}{\pi r^2}$$

$$\Rightarrow \qquad K = \frac{YA}{l} = \frac{Y\pi r^2}{l} \text{ ohm-meter} \qquad \textbf{(8.6)}$$

8.4 PROCEDURE

8.4.1 Measurement of the Resistance Per Unit Length (ρ) of the Bridge Wire

(*i*) Approximately equal resistances P and Q (preferably 1 Ω coils) are introduced in the inner left and right gaps of the bridge respectively. Other connections are made according to Fig. 8.2.

(*ii*) For determining the resistance per unit length of the bridge wire, a resistance box is introduced in one of the outer gaps and the remaining gap is short-circuited by a thick copper strip. The balance point is obtained for a suitable value of R of the resistance box by shifting the jockey on the bridge wire and the corresponding balancing length l_1' is noted. The positions of the copper strip and the resistance box (having the same resistance R) are interchanged and the balancing length l_2' is determined again.

(*iii*) The observations are repeated for different values of R of the resistance box.

8.4.2 Measurement of the Unknown Resistance of the Exprimental Wire

1. The wire, whose resistance is to be determined is connected at the right outer gap and the decimal resistance box in the left outer gap.
2. The balancing length, l_1, is determined for a suitable value of X of the resistance box.
3. The positions of the resistance box and the unknown resistance wire are interchanged and the balancing length, l_2 is determined.
4. The observations are repeated for different values of X of resistance box.
5. For each set calculate the unknown resistance and find its mean value.
6. Measure the length of resistance wire between the points the wire leaves the terminals.
7. Find the radius of the wire using a screw gauge by measuring the diameter at different places and mutually perpendicular directions.

8.5 OBSERVATIONS

[A] *Table for the determination of* ρ.

S. No.	R in ohm introduced in the resistance box	Distance of the balance point from the left end when R is in		$l_2' - l_1'$ (cm)	$\rho = \frac{R}{l_2' - l_1'}$	Mean ρ
		Left gap l_1' (cm)	Right gap l_2' (cm)			
1.						
2.						
3.						
4.						

[B] *Table for the determination of unknown resistance Y.*

S. No.	Resistance introduced in the resistance box X ohm.	Distance of the balance point from the left end when X is in the		$l_1 - l_2$ in (cm)	$Y = X + \rho(l_1 - l_2)$ ohm	Mean Y ohm
		Left gap in cm (l_1)	Right gap in cm (l_2)			
1.						
2.						
3.						
4.						
5.						

[C] Length of the resistance wire l = ... cm

[D] Determination of radius of the wire
Pitch of the screw = ... cm
Least count of the screw gauge = ... cm

S.No.	Diameter of the wire		Mean diameter
	In one direction	In perpendicular direction	

Radius of the wire $r = \dfrac{diameter}{2} = \ldots$ cm

8.6 CALCULATIONS

$$\rho = \frac{R}{l_2' - l_1'} = \ldots \text{ ohm/cm.}$$

Calculate ρ for each set and find the mean value of ρ.

$$Y = X - \rho(l_2 - l_1) = X + \rho(l_1 - l_2)$$

Calculate Y for each set using the mean value of ρ. Find mean value of Y.

$$\text{Specific resistance } K = \frac{Y\pi r^2}{l}$$

$$= \ldots \text{ ohm-cm}$$

8.7 RESULT

The specific resistance of the material of the wire = ... ohm-cm.

Standard value of specific resistance of wire = ... ohm-cm.

% Error = ... %

8.8 SOURCES OF ERROR AND PRECAUTIONS

1. While making connections see that undue resistance due to loose contact or some coating on the ends of the connecting wire, do not get involved in the experiment.
2. The resistance of the four arms of the bridge must be of the same order to achieve high sensitiveness.
3. The cell key must be closed only when observations are to be taken. Unnecessary flow of current will cause heating of the wire and will change the resistance per unit length of the wire.
4. The wire is assumed to be uniform with a constant value of ρ through out the length of the wire. The value of R while determining ρ should therefore be changed for different sets so that $(l_2' - l_1')$ covers practically the whole length of the wire. The value of ρ thus determined will be the average value for the whole wire and the error due to uniformity of wire will be minimised.
5. The value of R while determining ρ and the value of $(X - Y)$ while determing the unknown resistance should not exceed the resistance of the whole length of the wire.
6. When the bridge is out of balance a shunt should always be used with the galvanometer, but for finding the exact balance point the shunt must be removed.
7. The jockey should be pressed gently and it should not be used keeping it pressed. This will introduce non-uniformity in the wire.
8. To increase the accuracy the balance point must be near the middle of the wire.
9. The unknown resistance wire should not be directly connected between the terminals as every time when the connections are interchanged a different length of wire will be coming in between the terminals. The ends of the wire should be soldered to two copper strips and connections should be made by these strips.

QUESTIONS WITH ANSWERS FOR VIVA-VOCE

Q.1. What is the difference between a metre bridge and a Carey-Foster's bridge?

Ans. Both bridges are based on the principle of Wheatstone's bridge. The accuracy and sensitivity of the metre bridge is increased by modifying it to Carey-Foster's bridge.

Q.2. What type of wire is used in this bridge?

Ans. The bridge wire is made of manganin since it has uniform area of cross-section.

Q.3. Why it is necessary to measure the average resistance per unit length?

Ans. Due to non-uniformity in the cross-section of the bridge wire, the value of ρ is different at different portions of the wire. Therefore the average value of ρ is considered.

Q.4. Why the null points are taken at the extrimities of the bridge wire?

Ans. When the null points lie at the extrimities, the difference $(l_2 - l_1)$ is fairly large and the measurment of resistance is more accurate.

Q.5. Does the presence of end errors affect the resul t?

Ans. While deriving the formula the end error cancelled and therefore the measurement is independent of end errors.

Q.6. What should be the values of the resistance used in the inner gap?

Ans. Their values must be nearly equal for maximum sensitivity of the bridge wire.

Q.7. What will happen if current is passed for a long time in the bridge wire?

Ans. The resistance of the bridge wire will change due to continuous flow of current and therefore the null point will shift.

Q.8. What do you mean by the resistance of a conductor?

Ans. The ratio of the potential difference between the ends of a conductor to the current flowing in it is called the resistance of the conductor.

Q.9. On what factors does it depend?

Ans. Resistance of a conductor is directly proportional to its length (l), inversely proportional to the area of cross-section (A). It also depends upon the nature of material and temperature of the conductor.

Q.10. What is specific resistance?

Ans. Specific resistance of a substance is defined as the resistance of a piece of that substance having unit length and unit area of cross-section.

$$K = \frac{RA}{l}$$

If $A = 1$ and $l = 1$, then $K = R$

Q.11. What is the unit of specific resistance?

Ans.

$$K = \frac{RA}{l} \rightarrow \frac{\text{ohm-metre}^2}{\text{metre}} \rightarrow \text{ohm-metre}$$

Thus the unit of specific resistance is ohm-metre.

Q.12. Is specific resistance same for all materials?

Ans. No. It depends on the nature of the material. Hence, it is different for different materials.

Q.13. What is the effect of temperature on resistance?

Ans. It increases with increase in teperature.

Q.14. What is the minimum difference in resistance that can be measured with this bridge?

Ans. It is equal to the resistance of the one metre length of the bridge wire.

Q.15. What is the maximum difference in resistance that can be measured with this bridge?

Ans. It is equal to the resistance of the total length of the bridge wire.

Q.16. What is the working principle of Carey-Foster's bridge?

Ans. When the resistance in the outer gaps of the bridge are interchanged, the position of null point shifts. This difference in the value of two resistances is equal to the resistance of that much length of the bridge wire by which the null point has shifted.

Q.17. What will happen if the two resistances in the inner gaps are not equal?

Ans. If the difference between them is large, the sensitivity of the bridge will be reduced and the null point will be pushed towards one end. The percentage error in the measurement of length will increase.

Q.18. Why should you use decimal ohm box?

Ans. The resistance R introduced in resistance box should be such that when the resistances in outer gaps are interchanged, the null point should be covered on the length of the bridge wire. Thus the maximum resistance is equal to the resistance of the bridge wire. In order to have several readings for this part of the experiment, a deci-ohm box is more suitable.

Q.19. Instead of using two coils in the inner gaps, can we use a rheostat?

Ans. It is more preferable to use a rheostat for the following two reasons:

(*i*) The resistances obtained with the rheostat are equally affected by external conditions.

(*ii*) The null point can be easily obtained on any portion of the bridge wire simply by altering the position of the variable contact. This is obviously not possible with two separate coils having fixed resistances.

Q.20. But by using a rheostat we do not know the values of two resistances in the inner gap. Is it not?

Ans. This is not necessary. In the formula for the bridge the values of inner resistances do not occur at all.

Q.21. Is it possible to compare two resistances without determining the value ofρ for the bridge wire?

Ans. Yes. It can be determined. Some resistance, say R_1 is introduced in one of the outer gaps and the other gap is short circuited. The shift of the balance point is found in the usual manner. Then

$$R_1 - 0 = \rho l_1 \Rightarrow R_1 = \rho l_1$$

where l_1 is the shift of the balance point.

The experiment is then repeated with the other resistance R_2.

Let the shift of null point in this case is l_2, then $R_2 = \rho l_2$.

Hence $\dfrac{R_1}{R_2} = \dfrac{l_1}{l_2}$

APPENDICES

Appendix-A

1. Physical constants

Speed of light in vacuum	$c = 3 \times 10^{8}$ m/s
Planck's constant	$h = 6.062 \times 10^{-34}$ J–s
Avogadro's constant	$N = 6.023 \times 10^{23}\,\text{mol}^{-1}$
Boltzmann's constant	$K = 1.380 \times 10^{-23}\,\text{J K}^{-1}$
Gas constant	$R = 8.314\,\text{J mol}^{-1}\text{K}^{-1}$
Bohr magneton	$\mu_B = 9.273 \times 10^{-24}\,\text{Am}^2$
Coulomb force constant	$1/4\pi\varepsilon_0 = 9 \times 10^{9}\,\text{Nm}^2\text{C}^{-2}$
Acceleration due to gravity	$g = 9.81\,\text{ms}^{-2}$
Ice point	$0°\text{C} = 273.15\,\text{K}$
Bohr radius	$r = 0.529$ Å
Triple point of water	$T_{tr} = 273.16\,\text{K}$
Electric permittivity of free space	$\varepsilon_0 = 8.85 \times 10^{-12}$ F/m
Electric charge to mass ratio	$e/m = 1.76 \times 10^{11}$ C/kg
Elementary charge	$e = 1.6 \times 10^{-19}\,\text{C}$
Gravitational constant	$G = 6.67 \times 10^{-11}\,\text{N m}^2/\text{kg}^2$
Permeability of free space	$\mu_0 = 4\pi \times 10^{-7}$ H/m
Atmospheric pressure	$p = 1.01 \times 10^{5}\,\text{N/m}^2$
One atomic mass unit	$1\,\text{a.m.u.} = 1.66 \times 10^{-27}\,\text{kg} = 931\,\text{MeV}$
Rest mass of electron	$m_e = 9.10 \times 10^{-31}$ kg
Rest mass of neutron	$m_n = 1.67 \times 10^{-27}$ kg
Rest mass of proton	$m_p = 1.67 \times 10^{-27}$ kg
Ratio of mass of proton & electron	$\dfrac{m_p}{m_e} = 1840$
Solar constant	$S = 1370\,\text{W/m}^2$
Rydberg's constant	$R_a = 10{,}973{,}371\,\text{m}^{-1}$

Stepan's Boltzmann constant $\sigma = 5.67 \times 10^{-8}\,\mathrm{Wm^{-2}K^{-4}}$

Volume of one mole of gas at NTP $V = 22.4 \times 10^{-3}\,\mathrm{m^3/mole}$

Wein's constant $0.0028\,\text{meter-K}$

Appendix-B

1. Conversion factors

$1\,\text{micron} = 10^{-6}\,\text{m}$

$1\,\text{nm} = 10^{-9}\,\text{m}$

$1\,\text{Å} = 10^{-10}\,\text{m}$

$1° = \dfrac{1}{57.3}\,\text{rad}$

$1\,\text{eV} = 1.602 \times 10^{-19}\,\text{J}$

$1\,\text{erg} = 10^{-7}\,\text{J}$

$1\,\text{eV / entity} = 96.49\,\text{K J mol}^{-1}$

$1\,\text{Calorie} = 4.18\,\text{J}$

$1\,\text{Watt} = 1\,\text{Joule/sec}$

$1\,\text{hp} = 746\,\text{Watt}$

$1\,\text{atmosphere} = 1.01325 \times 10^5\,\text{Pa}$ or $\left(\dfrac{N}{m^2}\right)$

$1\,\text{torr}\,(= 1\,\text{mm of Hg}) = 133.3\,\text{Nm}^{-2}$

$1\,\text{bar} = 10^5\,\text{Nm}^{-2} = 10^{-1}\,\text{MPa}$

$1\,\text{psi} = 6.90\,\text{K N m}^{-2}$

$1\,\text{dyne/cm}^2 = 0.1\,\text{N m}^{-2}$

$1\,\text{Poise} = 0.1\,\text{Pas}$

$1\,\text{mol/cm}^2\text{/sec} = 10^4\,\text{mol m}^2\text{s}^{-1}$

$1\,\text{mol/cm}^3 = 10^6\,\text{mol m}^{-3}$

$T°\text{C} = (T + 237.15)\,\text{K}$

$1\,\text{erg/cation} = 1.43933 \times 10^{13}\,\text{K cal/mol}$

$1\,\text{electronic charge} = 1.6 \times 10^{-19}\,\text{C}$

$1\ \text{Wb} = 10^8\ \text{Maxwell}$

$1\ \text{Tesla} = 1\ \text{Wb/m}^2 = 10^4\ \text{Mx/cm}^2 = 10^4\ \text{Gauss}$

$1\ \text{A/m} = 4\pi \times 10^{-3}\ \text{Oe} = 1.26 \times 10^{-2}\ \text{Oe}$

2. Prefixes for power of 10

Power of 10	Name	Symbol
10^{24}	*yotta*	*Y*
10^{21}	*zetta*	*Z*
10^{18}	*exa*	*E*
10^{15}	*peta*	*P*
10^{12}	*tera*	*T*
10^{9}	*giga*	*G*
10^{6}	*mega*	*M*
10^{3}	*Kilo*	*k*
10^{2}	*hecto*	*h*
10^{1}	*deka*	*da*
10^{-1}	*deci*	*d*
10^{-2}	*centi*	*c*
10^{-3}	*milli*	*m*
10^{-6}	*micro*	μ
10^{-9}	*nano*	*n*
10^{-12}	*pico*	*p*
10^{-15}	*femto*	*f*
10^{-18}	*atto*	*a*
10^{-21}	*zepto*	*z*
10^{-24}	*yoeto*	*y*

Appendix-C

1. Specific Resistances and Temperature Coefficients of Resistances

Substance	Sp. Resistance ($\times 10^{-6}$ ohm $\times$ cm)	Temp. Coeff. of resistance (10^{-4} /°C)
Aluminium	3.21	38.0
Copper	1.78	42.8
Brass (70% Cu + 30% Zn)	6.3	10.0
Constantan (58.80% Cu + 40% Ni + 1.20% Mn)	49	– 0.4 to 0.1
German Silver (62% Cu + 15% Ni + 22% Zn)	33	2.5 to 6.0
Iron	13	62
Steel	19	16 to 22
Silver	1.63	40
Platinum	11	37
Gold	2.4	40
Nichrome (67.5% Ni + 15% Cr +16% Fe + 1.5% Mn)	110	1.7
Manganin (84% Cu + 4% Ni + 12 Mn)	44.5	– 0.1 to 1.0
Nickel	11.8	27
Tungsten	5.5	51
Mercury	95.8	9.0
Graphite	3000	-7.0
Carbon filament	4000	negative
Phosphorbronze	5 to 10	35
Quartz	1.2×10^{21}	-
Mica	9×10^{21}	-
Zinc	6.1	-
Tin	11.3	37

2. Dispersive Power

Type of Glass	Dispersive Power ω
Crown (Borosilicate)	0.015
Crown	0.017
Crown (Dense Barium)	0.018
Flint (Light)	0.024
Flint (Dense)	0.027
Flint (Extra dense)	0.033
Flint (Double extra dense)	0.036

3. Wavelength of prominent spectral lines (A.U.)

Hydrogen	Mercury	Neon
6562.8 (H_α)	5790.7	7245.2
4861.4 (H_β)	5769.6	6506.6
4340.4 (H_γ)	5460.7	6402.3
4101.7 (H_δ)	4959.7	5881.9
Sodium	4916.4	5852.4
5896 D_1	4358.3	5341.1
5890 D_2	4347.5	**Cadmium**
Helium	4339.2	6438.5
7065.2	4078.1	5085.8
6678.2	4046.8	4799.2
5875.6		4678.2
4471.5		
4026.2		
3888.6		

4. Wire resistances

	Diameter (*m.m*)	Resistance (ohm/meter)		
S.W.G. No.		**Copper**	**Constantan**	**Manganin**
10	3.25	0.0021	0.057	0.051
12	2.64	0.0032	0.086	0.077
14	2.03	0.0054	0.146	0.0131
16	1.63	0.0083	0.228	0.205
18	1.22	0.0148	0.495	0.361
20	0.914	0.0260	0.722	0.645
22	0.711	0.0235	1.20	1.07
24	0.559	0.070	1.93	1.73
26	0.457	0.105	0.89	2.58
28	0.374	0.155	4.27	3.82
30	0.315	0.222	6.08	5.45
32	0.274	0.293	8.02	7.18
34	0.234	0.404	11.1	9.9
36	0.193	0.590	16.2	14.5
38	0.152	0.950	26.2	23.2
40	0.122	1.48	40.6	36.3

5. Refractive index of substances for ($\lambda = 5893$ Å) D-line of sodium relative to air:

Substance	Ref. Index	Substance	Ref. Index
Glass (Crown)	1.48–1.61	Alcohol (Ethyl)	1.362
Glass (Flint)	1.53–1.96	Paraffin Oil	1.44
Glass (Soda)	1.50	Glycerine	1.473
Diamond	2.42	Kerosene	1.390
Ice	1.31	Castor Oil	1.472
Mica	1.56–1.69	Terpentine	1.47
Quartz	1.544	Olive Oil	1.46
Benzene	1.504	Linseed Oil	1.474
Water	1.33	Carbon disulfide	1.632
Canada balsam	1.55	Chloroform	1.449
Potash Alum	1.456	Plastic (Polystyrene)	1.589
Coconut Oil	1.45	Liq. Paraffin	1.48
Rock Salt	1.544	Sod. Chloride	1.546
Ether	1.354		

6. Optical rotatory power

Specific rotation S = 10 q/*lc*, where q is rotation in degree by liquid column of length *l dm* and concentration *c* (in gm/cc). All values at 20° C, + for dextro and laevo rotation.

Substance	Solvent	Specific Rotation (deg.) Right handed (+ve) & Left handed (–ve)
Cane Sugar, ($C_{12}H_{22}O_{11}$)	Water	+ 66.5
d-Camphor	Methyl alcohol	+ 43.7
d-Glucose	Water	+ 52.5
Invert Sugar	Water	– 19.7
Lactose	Water	+ 52.5
Nicotine	Water	– 79.4
Nicotine	Pure	– 122.0
Quinine	Ethyl alcohol	– 169.3
Rochelle Salt	Water	+ 29.7
Turpentine Oil	Pure	– 37.0
Fructose	Water	– 91.0

7. Variations of thermo-e.m.f. with temperature

Temperature of cold junction	0°C										
Temp. of hot junction (°C) 0°C	0	10	20	30	40	50	60	70	80	90	100
Thermo-emf (in mV) for Copper-Constantan	0	0.39	0.79	1.19	1.61	2.03	2.47	2.91	3.36	3.81	4.28
Thermo-emf (in mV) for Iron-Constantan	0	0.52	1.05	1.58	2.12	2.66	3.20	3.75	4.30	4.85	5.40

8. Electrochemical equivalent

Element	Atomic weight	Valency	E.C.E. gm./ coulomb 0°C
Aluminium	26.97	3	0.0000935
Oxygen	16.00	2	0.0000829
Silver	107.88	1	0.0011180
Copper	63.57	2	0.0003294
Zinc	65.38	2	0.0003383
Gold	197.20	3	0.0006809
Nickle	58.69	2	0.0000304
Hydrogen	1.007	1	0.0000105

9. Magnetic susceptibilities of para and diamagnetic substances

Paramagnetic substance	$\chi = 1 \times 10^{-6}$	Diamagnetic substance	$\chi = 1 \times 10^{-6}$
Nitrogen	0.013	Hydrogen	–0.063
Air	0.35	Benzene	–7.5
Oxygen	1.9	Water	–9.0
Ebonite	14	Copper	–10.3
Aluminium	23	Glass	–12.6
Tungsten	76	Rock-salt	–12.6
Platinum	360	Quartz	–15.1
Liquid oxygen	3400	Bismuth	–176

10. Bandgap in semiconductors at 300 *K* (i = indirect gap and d = direct gap)

Substance	E_g(eV)	Substance	E_g(eV)
Diamond (i)	7.2	PbS (d)	0.37
Silicon (i)	1.116	PbTe (i)	0.29
Germanium (i)	0.66	GaP (i)	2.25
GaAs (d)	1.43	AlSb (i)	1.6
GaSb (d)	0.68	CdS (d)	2.42
InAs (d)	0.36	SiC (i)	2.86
PbSe (i)	0.27	CdSe (d)	1.74
InSb (d)	0.17	CdTe (d)	1.44
Te (d)	0.33		

11. Thermo E.M.F.

Thermocouple	Thermo-e.m.f.
Copper-Constantan	41.8 Microvolt/°C
Copper-Iron	8.6 Microvolt/°C
Antimony-Bismuth	113 Microvolt/°C

12. Internal resistance and E.M.F. of cells

Cell	Internal resistance (ohm)	E.M.F. (Volt)
1. Cadmium	900 Ω (very high)	1.0183 at 20°C
2. Daniel	3–4 Ω (fairly constant)	1.08
3. Leclanche	High, increases with usage	1.46
4. Alkali accumulator	Low	1.35
5. Lead accumulator	Very Low	2.10

13. Relative permittivity of some selected substances

Dielectric	ε_r	Dielectric	ε_r
Water	81	Mica	6
Air	1.00058	Alcohol	2.6
Paraffin	2.2	Glass	5–10
Polyethylene	2.3	Porcelain	5.6
Flint glass	6.6	Ebonite	2.8
Crown glass	4.8	Polystyrene	2.6
PVC	3.0	Toluene	2.39
Poly carbonate	3.1	Transformer oil	2.2
Calcite	6.1	Quartz (silica)	3.8

14. Magnetic elements

Station	Dip	Hoersted
Agra	40°41'	0.348
Aligarh	41°50'	0.346
Allahabad	37°10'	0.353
Delhi	42°52'	0.345
Meerut	43°30'	0.339
Varanasi	37°10'	0.364
Dehra Dun	45°50'	0.332
Gwalior	39°00'	0.353
Gorakhpur	39°40'	0.358
Jaipur	40°30'	0.347
Kanpur	38°39'	0.363
Khurja	42°10'	0.343
Lucknow	40°00'	0.354
Bareily	42°20'	0.344
Bombay	25°30'	0.376
Calcutta	31°30'	0.382

15. Wavelength and refractive index of different colors

Wavelength λ nm	Colour	Iceland spar		Quartz	
		n_e	n_0	n_e	n_0
687	Red	1.484	1.653	1.550	1.541
656	Orange	1.485	1.655	1.551	1.542
589	Yellow	1.486	1.658	1.553	1.544
527	Green	1.489	1.664	1.556	1.547
486	Blue	1.491	1.668	1.559	1.550
431	Indigo	1.495	1.676	1.564	1.554
400	Violet	1.498	1.683	1.568	1.558

INDEX

A

Absorption of radiation 4.1

Acceptance angle and acceptance cone 3.17

Alpha decay 2.36

B

Black body radiation 1.1

C

Cavity resonator 4.7

Coherence 3.1

Compton effect 1.18

D

Dead time 5.13

Density of energy states 2.23

E

Energy of electron gas 2.27

Experimental demonstration of compton effect 1.24

F

Fermi distribution function 2.25

Fermi energy 2.23

Free electron gas 2.22

G

Gas filled detectors 5.1

Gas multiplication factor 5.8

Geiger-müller counter 5.11

Graded index (GRIN) optical fibre 3.16

H

Helium-Neon laser 4.8

Holography 4.1

I

Induced emission of radiation 4.2

Ionization chamber 5.1

L

Laser 4.1

M

Matter waves 1.9

Metastable states 4.5

Mode locking 4.20

N

Numerical aperture 3.1

O

Optical fibre 3.1

P

Photomultiplier tube 5.18

Photoelectric effect 1.3

Planck's quantum theory 1.2

Population inversion 4.6

Potential barrier 2.7

Q

Q-switching 4.16

Quenching 5.13

R

Recovery time 5.13

S

Scintillation detector 5.18

Semiconductor laser 4.10

Spatial coherence 3.2

Step index (SI) optical fibre 3.15

T

Temporal coherence 3.1, 3.2

Tunnel effect 2.7

U

Uncertainty principle 1.14

W

Wave function 1.33

Wave mechanics 1.35